William Hunt, Thomas George Morton

Surgery in the Pennsylvania Hospital

Being an Epitome of the Practice of the Hospital since 1756 ...

William Hunt, Thomas George Morton

Surgery in the Pennsylvania Hospital
Being an Epitome of the Practice of the Hospital since 1756 ...

ISBN/EAN: 9783337021528

Printed in Europe, USA, Canada, Australia, Japan

Cover: Foto ©berggeist007 / pixelio.de

More available books at **www.hansebooks.com**

PENNSYLVANIA HOSPITAL.
Philadelphia

SURGERY

IN THE

PENNSYLVANIA HOSPITAL

BEING AN EPITOME OF THE PRACTICE OF THE HOSPITAL SINCE 1756; INCLUDING COLLATIONS FROM THE SURGICAL NOTES, AND AN ACCOUNT OF THE MORE INTERESTING CASES FROM 1873 TO 1878; WITH SOME STATISTICAL TABLES.

BY

THOMAS G. MORTON, M.D., AND WILLIAM HUNT, M.D.

SURGEONS TO THE HOSPITAL.

WITH PAPERS BY

JOHN B. ROBERTS, M.D., AND FRANK WOODBURY, M.D.,

Late Resident Physicians in the Hospital.

PREPARED BY DIRECTION OF THE MANAGERS OF THE HOSPITAL

PHILADELPHIA:

J. B. LIPPINCOTT & CO.

LONDON: 16 SOUTHAMPTON STREET, COVENT GARDEN

1880.

PENNSYLVANIA HOSPITAL.

MANAGERS.

WILLIAM BIDDLE,
SAMUEL MASON,
SAMUEL WELSH,
WISTAR MORRIS,
JACOB P. JONES,
BENJ. H. SHOEMAKER,
ALEXANDER BIDDLE,
JOSEPH B. TOWNSEND,
JOSEPH C. TURNPENNY,
T. WISTAR BROWN,
HENRY HAINES,
JOHN B. GARRETT.

TREASURER.

JOHN T. LEWIS.

SURGEONS.

WILLIAM HUNT, M.D.,
THOMAS G. MORTON, M.D.,
RICHARD J. LEVIS, M.D.,
D. HAYES AGNEW, M.D.

PHYSICIANS.

JOHN FORSYTH MEIGS, M.D.,
JACOB M. DA COSTA, M.D.,
JAMES H. HUTCHINSON, M.D.,
MORRIS LONGSTRETH, M.D.

RESIDENT PHYSICIANS.

CHARLES BAUM, M.D.,
HENRY M. WETHERILL, M.D.,
MERCER B. CROLL, M.D.,
RICHARD H. HARTE, M.D.

PATHOLOGIST AND CURATOR.

MORRIS LONGSTRETH, M.D.

MICROSCOPIST.

JOSEPH G. RICHARDSON, M.D.

OUT-PATIENT DEPARTMENT.

SURGEONS.

CHARLES T. HUNTER, M.D.,
WILLIAM ASHBRIDGE, M.D.,
ELLIOTT RICHARDSON, M.D.,
WILLIAM C. COX, M.D.

PHYSICIANS.

JOSEPH J. KIRKBRIDE, M.D.,
ARTHUR V. MEIGS, M.D.,
MORRIS J. LEWIS, M.D.,
ROBERT M. SMITH, M.D.

STEWARD.

WILLIAM G. MALIN.

CLERK, LIBRARIAN, AND ASSISTANT STEWARD.

JONATHAN RICHARDS.

APOTHECARY.

JACOB K. HECKER.

PREFACE.

FROM the earliest foundation of the Pennsylvania Hospital, in 1752, its medical officers have endeavored to preserve more or less complete notes of the more interesting cases received, and the operations performed. It has only been within the last few years, however, that full clinical notes of all cases have been kept.

In the year 1873 a systematic method of recording cases, both medical and surgical, was begun. Paper with appropriate headings was provided, and it became the duty of the residents to note all that was valuable in the history of the cases. From time to time, as the material accumulated, it was gathered and arranged, according to date, by the librarian, and then bound in volumes. About sixteen of these are already on hand.

It has been thought by the managers and surgeons that a useful and interesting book might be made from the surgical notes thus collected, and that these might be supplemented by using the older records of the hospital, as to the more important surgical matters, such as amputations, excisions, ligations, etc., and thus bring the history of the institution in reference to them down to the present time. Some of the tables, therefore, as to this class of cases, are repetitions and continuations of what has already been published in the *American Journal of Medical Sciences* and in the two volumes of the *Pennsylvania Hospital Reports* printed in 1868 and 1869.

The managers of the hospital, always awake to the best interests of the institution, with their accustomed liberality, made the necessary contribution to defray the expenses of this volume. We hope it may be useful to the profession. To those of the laity who may be interested in such matters (and who is not interested in them?), it will show what a vast amount of work is and has been done in the old hospital for the relief of suffering humanity, and that it is, as ever, worthy of their countenance and material support.

Some of the illustrative cases have already been published, in the form of clinical reports, in various journals, and are reproduced here to place them in a more permanent form. We think not the least interesting part of the book is that which records some of the surgery which was done in the early history of the hospital.

We have to express our obligations to Mr. Wm. G. Malin, the steward of the hospital, who has been connected with the institution for fifty-six years, for his aid in collecting statistical matter. Our acknowledgments are also particularly due to Dr. Morris Longstreth for his assistance in furnishing pathological reports, to Drs. Frank Woodbury, Wm. C. Cox, Chas. H. McIlwaine, Stacy B. Collins, and to Jonathan Richards, the assistant steward and librarian. Our thanks are also due to Mr. J. W. Lauderbach for his skill, as shown in many of the engravings.

TABLE OF CONTENTS.

LIST OF ILLUSTRATIONS.

SURGERY IN THE PENNSYLVANIA HOSPITAL.

AMPUTATIONS.

Among the records of the major surgical operations performed in any general hospital, where recent injuries from gunshot wounds, railroad crushes, and machinery and other accidents, are daily received, and cases of malignant or other chronic disorders constantly demand operative interference, amputations deservedly occupy a prominent and important position. Using the term in its restricted sense, as applied to the operation of removal of one of the upper or lower extremities, in whole or in part, at the articulations or between them, a glance at the tabulated reports reveals the fact that more than a thousand amputations have been performed in this institution within the last fifty years. By far the larger proportion of the cases were those resulting from accidents, with members mangled by machinery, crushed on the railway, or lacerated by falls, and similar severe injuries; and the large number of such patients, constantly being received, has established the reputation of the Pennsylvania Hospital as the leading "recent accident" hospital for this section of the country; it is, therefore, one whose records have more than a mere statistical value.

Of the 1011 amputations recorded since 1830, 801 were for accidental injuries. Of this number, 679 were immediate operations, performed within twenty-four hours of the receipt of the injury; 122 were secondary; and the remaining number, 210, were for chronic bone disorders or other disease. 58 double amputations were performed upon 29 patients. Of the 1011 amputations, there were 8 of the hip-joint, 139 of the thigh, 42 of the knee, 330 of the leg, 80 of the foot, 30 of the shoulder-joint, 157 of the arm, 166 of the forearm, and 59 of the wrist and hand. Of this number, 764 recovered, and 247 died. The rate of mortality was 24.43 per cent.

Reception of Cases.—Of the immediate amputations, the majority were for injuries of a very serious nature, the bones, as well as the soft parts of the extremity, usually being gravely implicated. When such an accident is brought into the hospital, the patient is placed upon a bed, the mattress being covered with oiled silk or cloth. All soiled and tattered clothes are at once cut off, only one portion of the body at a time being exposed, on account of the condition of partial collapse, and in order not to delay reaction. The injured parts are carefully examined, and all sources of hemorrhage are sought for, and carefully secured by ligature, pin, or bandage and compress. The bladder, if distended, is relieved by the catheter. Where much suffering exists, morphia, by the skin, is generally administered. As more or less shock from loss of blood or other cause is commonly present, hot coffee, milk, or beef-tea is given in small doses, frequently repeated. We impress upon the resident physicians, who immediately attend to our patients upon admission, that the treatment of shock is of primary importance, *rest* being the great factor in bringing about reaction. The common practice of giving large amounts of stimulants indiscriminately in all cases, immediately upon the receipt of an accident, is wrong in principle, and should be condemned as not only inefficient but positively injurious. We avoid the use of alcohol, as a rule, and resort to it only in very exceptional cases. Shock, as the result of an injury without great loss of blood, is soon overcome by rest; but in many cases warm liquid food may be judiciously administered, if reaction be slow. In shock from loss of blood, with moderate amount of nerve injury, the remedies just indicated are sufficient, if the stomach be retentive; if vomiting be persistent, neither food nor medicine is given by mouth, but nutritious enemata are thrown high up into the bowel. In cases of threatened dissolution, we have practised transfusion of blood as a last resort.

Shock, the result of serious nerve injury, with considerable loss of blood, such as is seen in many railway crushes, requires absolute rest and the most careful administration of the remedies referred to, and stimulants, *in very small and guarded doses*, with occasional hypodermic injections of morphia, brandy, or of ether. In such cases the stomach often rejects everything, either at once, or if an accumulation take place, and vomiting occurs, it will be found that there has not been the slightest effort at digestion. The patient is, moreover, weakened by the vomiting, and greatly depressed before this occurs. There is no absorption by the gastric mucous membrane, and it is useless to pour in stimulants or food. We therefore mainly rely upon the skin, the great intestine, rest, and external heat.

If the stock of vitality has been used up from the nerve injury and hemorrhage, the chances of life are very small; and if an effort at reaction take place, it is often only temporary, and secondary shock carries off the patient.

Thus, in many cases, after grave injuries, an indefinite form of reaction may appear to have been established, and if amputation be performed in this state of suspended shock, even where no additional loss of blood has occurred, profound depression gradually recurs and the patient dies. In these cases the partial reaction is due to stimulation; the nerve-centres temporarily respond, but they have been from the first so nearly destroyed that they are unable to keep up their influence. Thus it becomes a question, in the more serious of these cases, whether it is not better to avoid a primary amputation, and thereby avert the additional shock incident to it, and wait until a secondary amputation can be performed, after the patient has become, in a measure, habituated to his condition.

In regard to hypodermic injections, we are decidedly opposed to the free use of spirits or of water of ammonia. Inflammation of the cellular tissue and abscess generally follow their introduction, their good effects are therefore questionable. The morphia salts, however, with or without atropia, are in common use for hypodermic injections, with excellent results.

The intravenous injection of a dilute solution of ammonia may be useful in cases of prostration with a flagging heart, as it is a powerful stimulant to the circulation, and as such can be held in reserve. It has not been practised at this hospital, but has been found very efficient elsewhere.

We caution our residents not to disturb a patient suffering from shock more than is absolutely necessary, for a time at least, after the first dressing, for rest, both physical and mental, is the most important element in securing the desired result.

Patients are frequently sent to us immediately after an accident, from the mining regions and distances more or less remote, by rail or other conveyance, without affording to the sufferer a few hours' rest before his journey, and not unfrequently stimulants are freely given, while food is the last thing thought of by those having patients in charge; by the time the hospital is reached, exhaustion is so great that any primary operation is impossible.

Time for Amputation.—When reaction has fairly been established, amputation is performed, and not until then do we ever operate. Of late years the rule has been observed to give in all cases, just before the operation, a full dose of quinine.

Ether is almost invariably used as an anæsthetic; with its administration the pulse often gains volume and tone. No inhaling apparatus is used, the ether being simply instilled upon a towel.

If considerable fever or delirium come on with the reaction,—the latter not unfrequently observed after serious injuries in young subjects,—we postpone amputation, and allow these cases to take their course under a supporting treatment and good nursing until the question of a secondary operation can be entertained.

Form of Amputation.—The operation usually performed with us is that known as the skin-flap with the circular division of the muscles. The flaps are necessarily modified in shape according to the part and the injury.

Occasionally the flap operation by transfixion in the leg and thigh is made, but as a rule the skin-flap with circular division of the muscles fulfils all the requirements.

Not unfrequently we are compelled to make "patchwork" amputations in order to save a joint or to give greater length to a stump, when, if a regular amputation were made, a much greater sacrifice of tissue would be the result.

It may not be out of place, while referring to the place of election in amputations, to note the fact that, after railway or other serious crushes, the tissues about the limb, near the region of the injury, may appear sound, but careful inspection often will demonstrate a coldness of the skin; and, if the parts are cut into, more or less blood will be found in the cellular tissue, and the capillaries are paralyzed and choked. Amputation through such contused parts would undoubtedly result in sloughing of the flaps, so that, although such tissues often look well, in a majority of these cases we find it is safer to operate higher up, and secure the flaps from tissues certainly uninjured.

Teale Amputation.—When this procedure can be made without sacrificing length of stump, we occasionally adopt it, but only in the leg, lower part of the thigh, and forearm. We have seldom resorted to this form of operation in cases of recent injuries, most of our amputations by this method having been performed for chronic diseases.

Hip-joint Amputations.—In hip-joint amputations the method by making skin-flaps, oval or square, with the division of the muscles close to the pelvis, has generally been resorted to, as this course gives less surface for suppuration and subsequent drain. In several of our amputations at this part the femoral artery has been permanently secured by a ligature before the division of the muscles.

Knee-joint Amputations.—In knee-joint amputations the long skin-flap is usually adopted. The patella has been occasionally left in; however, as the patella now and then becomes hypertrophied, and thus a source of irritation, we believe it safer in most cases to remove it.

In primary amputations of the knee-joint the prominent portions of the condyles of the femur are removed, but in chronic cases they are not disturbed. Retraction of the flaps and exposure of the condyles usually follow immediate amputations, if the condyles are not removed.

Ankle Amputations.—The "Syme" has been rarely practised, but the last case, performed a few months since, proved eminently successful, and was really the only kind of amputation which could be made in that case with any prospect of success. The patient was highly syphilitic, and had serious bone disease, which had involved the metatarsal and phalangeal articulations. Several operations had been performed, but after each there was a return of the necrotic condition, apparently from opening the cancellated structure. Finally, a "Lisfranc" was performed, but abscess, bone disease, and numerous fistulous tracts appeared. The "Syme" was at last undertaken; the tibia and fibula were left untouched. The case did remarkably well. A firm, rounded, painless stump, upon which the patient could bear his entire weight in walking, was the result.

Whenever practicable the "Pirogoff" amputation is resorted to; and the success attendant upon this operation, with us, has been all that could be desired. We seldom, if ever, employ this procedure in chronic cases, as the operation is better indicated for recent injuries, and we rarely operate by this method upon patients over forty years of age.

Our former hospital messenger, now upwards of twenty-five years of age, sustained a railway crush of the foot some ten years ago. A "Pirogoff" was made, an excellent recovery followed, and he has since then been in the employ of the house, constantly walking upon the stump, which rests directly upon the shoe. The stump has never given him the least trouble, and not the slightest limp is noticed by the ordinary observer.

In many cases, we think that often too much of the os calcis and tibia and fibula are removed, and that if the operation was restricted to recent injuries and young subjects, the success attendant upon it would be far greater.

Shoulder Amputations.—In shoulder-joint amputations we employ either the Larrey method, or make simply lateral skin-flaps, and divide the muscles in a more or less oval manner; the axillary structures, which are grasped by an assistant, are finally cut across.

In arm amputations we perform either the circular operation, or make the cutaneous flaps.

Arrest of Hemorrhage during Operation.—Before the introduction of the elastic bandage and tube, it was the custom with us to drain the limb of blood by elevation and manipulation, and then to apply the tourniquet, and not unfrequently a common roller bandage was used upon the limb before the tourniquet was applied.

We now use Esmarch's bandage, though not by any means as a rule. We commonly apply the tourniquet in thigh amputations, and always when amputating the thigh for chronic disease. When we have an increased number of vessels, so often seen in amputation for chronic disease, we find less blood is lost in using the tourniquet, which can be more readily loosened and tightened while securing the vessels. We have by no means discarded the tourniquet, but always have it ready upon our operating-table and in our wards. We believe the Esmarch bandage more especially useful in operations for the removal of necrosed bone, or where a careful dissection is made for the removal of tumors, etc., rather than for the arrest of hemorrhage during amputations.

In amputations of the leg and foot, we commonly apply only the elastic tube upon the thigh, low down, without the previous application of the bandage. The same applies in amputation of the arm and forearm. In the former we very often employ manual compression, by placing the finger of an assistant upon the brachial artery.

In hip-joint amputations we apply the abdominal tourniquet, and in several of our more recent operations, after making the anterior skin-flap, the femoral artery has been tied before dividing the muscles.

In connection with the subject of the prevention of hemorrhage, it is of interest to note that the abdominal tourniquet of Prof. Joseph Pancoast, now generally adopted by surgeons in this country, was devised by him, while surgeon to this hospital, for controlling the aorta during an operation of amputation at the hip-joint, which was performed in the year 1860. It is noteworthy as being the first successful case of this amputation performed in the Pennsylvania Hospital; the case was one of encephaloid disease (myeloid sarcoma) of the femur, and the patient lived for a year afterwards.

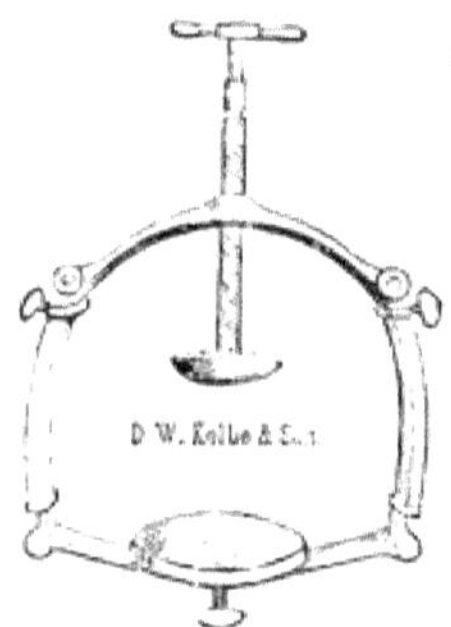

Direct digital compression of the common iliac artery, in order to shut off the circulation from the extremity during hip-joint amputation or operations upon the thigh, was recommended for the first time by Dr. Frank Woodbury while resident physician at this hospital.* With some modifications, this procedure is now in use in one of the London hospitals.

On Amputations of Fingers, etc.—It may not be out of place here to refer to the unfortunately too common practice of primarily amputating crushed fingers. Injuries of these parts, more or less severe, are brought to the hospital almost daily. Sometimes the crush involves a single phalanx, but more often an entire finger or fingers are involved; and occasionally, from knife, circular saw, or other instrument, the part is almost, if not entirely, cut through. The value of even a very small portion of a finger is so great, that it is the duty of the surgeon to save all he can, be that part ever so small.

Crushed fingers are usually attended to in hospitals by the internes; frequently they are novices, and thus the treatment of these important surgical cases (unless some definite rule of non-interference is understood) is left to the inexperienced.

Immediately upon the receipt of the injury a finger may appear hopelessly crushed, the skin may be more or less lacerated and contused, the tendon and bone exposed, while the latter is often comminuted. Such an injury, involving a limb, would doubtless demand an amputation, but repair in the bones and soft parts of a finger is so wonderful, that in a majority of such injuries a much better result can be attained than would at first be supposed. The skin should be carefully replaced, and held in position by narrow strips of plaster, and a hand-splint applied, with a dressing of carbolized oil. Indeed, we go so far as to say that it is better to avoid primary operations in all cases of crushed or lacerated fingers, and wait until time shall demonstrate how much tissue has been destroyed, and then, after suppuration has occurred, a secondary operation, if required, can be made.

* See *Amer. Jour. Med. Sciences*, Jan. 1874, p. 131; also *Transactions International Med. Congress held at Philadelphia*, 1876, p. 552.

The value of even a portion of a finger thus saved to the skilled mechanic cannot be over-estimated, while in many apparently even hopeless cases of crushes most excellent recoveries will be found.

Fingers, or portions of fingers, are occasionally saved which subsequently become such a source of annoyance that their removal is requested by the patient, but such cases are the exception, at least they are not common.

Fingers cut off, or nearly so, by sharp or cutting instruments, often do well if reapplied to the living tissues; our records show a number of such instances. We well remember, on entering the hospital some years ago, meeting one of the residents on his way to the accident ward with instruments to perform an amputation of the thumb at the metacarpo-phalangeal articulation. The patient had just been admitted. The thumb, with the exception of a half-inch bridge of skin on the dorsal aspect and the extensor tendon, was entirely cut off an hour before by a piece of sharp brass plate. Although the separated thumb was cold, we suggested that an attempt might judiciously be made to save the part; and the detached member was then secured in position by a few interrupted sutures, a hand-splint was adjusted, and a warm laudanum dressing applied. The result was an excellent recovery, with a useful finger.

Indeed, a number of cases have been reported where the finger has been completely separated for a considerable length of time, and subsequently being placed in apposition, complete union has resulted.

Ligatures.—The ordinary waxed silk ligature is usually employed in securing vessels; occasionally the animal ligature is resorted to, but rarely. Torsion was at one time quite commonly used by one of the former surgeons, but on account of the great loss of blood which so often took place before the closure of the vessel could be effected, and often secondary hemorrhage soon after the stump was closed, the method was not looked on with any favor; there would seem to be but little justification for its use, as applied to large vessels at least. The same may be said of acupressure, which also has had a fair trial. As an accessory or aid in arresting hemorrhage acupressure is very valuable, but we cannot speak well of it from our experience, for we know that much time and unfortunately, in the cases we have observed, much too great an amount of blood, has been lost before the vessels were secured; even fatal exhaustion from primary hemorrhage has, in our wards, followed the use of acupressure. Not unfrequently we have seen serious hemorrhage follow the removal of the needles from both large and small vessels, necessitating a re-introduction, with great suffering and shock to the patient, and on several occasions fatal results. We have not seen any reason for abandoning the time-honored and trusty waxed silk ligature.

Some of the surgeons use the carbolized gut ligatures, and insert a drain of some sort when closing the stump. The silk ligature, however, is still a great favorite, and is certainly more easily handled than any other. After the vessels are tied, and one end of each ligature is cut off, the combined ends, when gathered together and well greased with carbolized oil, make a most capital drain, and thus the silk efficiently serves a double purpose. The greatest care is taken to exclude all nerve-filaments; in case of doubt the tenaculum is dispensed with and the vessel is seized and drawn down with the forceps and then tied.

Dividing the Bones.—We usually employ the ordinary saw, but lately have made use of the Bonwill engine with circular saw. In sectioning the femur in a late case this method was very successful; the bone was not only rapidly divided, but it was sectioned without separating the periosteum, at the same time the bone was left unusually smooth. In this instance a primary union of almost the entire stump resulted. We think that the application of this apparatus for cutting through the bones in amputations will in time be more favorably received, for there is no doubt but that necrosis in stumps is frequently due to a clumsy saw and the consequent periosteal separation.

After-Treatment of the Stump.—Before closing the stump a stream of carbolized water is used to thoroughly cleanse the surface. If there be any capillary oozing, the best results are derived from the use of *hot water* brought in direct contact with the bleeding surfaces, or by means of a towel or cloth wrung out of hot water and then applied, which at once controls hemorrhage. The limb having been cleansed and dried, the flaps are brought together by silver wire, or strips of adhesive plaster. We have occasionally treated our cases by the open method, which has yielded excellent results.

We make one point in regard to cleanliness, and that is, to have the limb thoroughly shaved before the amputation; in fact, before all operations this is a most important but often a greatly-neglected procedure. If there are not sufficient ligatures to make an effective drain, a tent of lint thoroughly soaked in carbolized oil is inserted, the stump is then covered with charpie and lint previously immersed in carbolized oil (one part of carbolic acid to ten of olive oil), a large, coarse, but soft towel is then placed around the stump, and the sides pinned moderately firmly together. In using this simple form of dressing, we have at once a soft application and at the same time an absorbing and disinfecting substance surrounding the stump. This dressing is easily applied, and requires little or no disturbance of the stump during the dressings. Mornings and evenings the towel is unpinned, and the soiled charpie is replaced by fresh. If there be very free discharge, a sheet of waxed paper is placed upon a folded sheet and the stump allowed to rest on this, which protects the bed from being soiled.

We use but little water upon stumps for the first week or so, or until they can be handled without giving rise to pain. When, however, the stump must be elevated in order to wash the wound, the towel is first unpinned, and is firmly grasped on either side by an assistant; the stump is then lifted and a basin carried under the stump, and the carbolized water from the receiver on the dressing-carriage is allowed to flow freely over the part. Oakum is always used in place of sponges, the latter having long since been entirely discarded. A fresh towel is then laid on the bed with the carbolized charpie, and the stump is then allowed to rest, the soiled towel slipped out, and the clean one covers the part as before. Now and then we have so much muscular irritability with great pain with the least movement of the stump, that we use a small hypodermic dose of morphia before commencing the dressing. Whenever there is the least tension of the flaps, the sutures are cut.

Occasionally we use irrigation if there be much swelling or inflammation, but the difficulty experienced in keeping the patient's clothes and bed from getting soiled and wet often deters us from adopting this course.

Surgical Fever—*Septicæmia and Pyæmia.*—From the spring of 1875 to the same period in 1879, a space of four years, there were performed 108 amputations upon 100 patients. Of this number 17 died. Five of these deaths took place within the first thirty-six hours following the admission of the patient, and in each instance from recurring or continuous shock. These deaths were not, therefore, caused by any so-called "hospitalism."

Of the causes of death we note:

Tetanus	4
Exhaustion, albuminuria, etc.	2
Secondary hemorrhage	3
Delirium tremens	2
Serous cerebral effusion	1
Recurring shock	5
Total	17

We are greatly pleased to be able to report that during the past five years there has not occurred in the surgical experience of the hospital a single case of pyæmia. The most careful post-mortem examinations made by Dr. Morris Longstreth, the pathologist of the hospital, in several cases where deaths have been recorded "Exhaustion," etc., have failed to show pyæmia or any approach to this malady. We believe that this result is due to the very perfect system of forced ventilation by the fan, the scrupulous cleanliness of the wards, and the free use of carbolic acid in our dressings, etc.

In addition to this, our perfect system of dressing wounds with flowing water, dispensing with all basins and sponges from our wards, thus averting any possible contamination, has been the sure means of ridding our wards of this most terrible of hospital diseases and accomplishing the beneficial results referred to.

Table of Amputations.

No.	Age.	Name.	Admitted.	Disease or Injury.	Part amputated.	Immediate or otherwise.	Result.	Discharged.
1	58	Charles Clauser	Oct. 21, 1872	Lacerated arm, gunshot wound	Arm	Immediate	Cured	Oct. 26, 1872.
2	23	John McGonigle	Oct. 17, 1874	Railroad crush, leg	Thigh	Immediate	Cured	Dec. 19, 1874.
3	52	Abram Ruth	Nov. 28, 1874	Encephaloid, forearm	Arm	Chronic	Cured	Dec. 29, 1874.
4	14	Ellen Welsh	Nov. 30, 1874	Encephaloid, leg	Thigh	Chronic	Cured	June 6, 1875.
5	35	Gonzalo Arolo	Feb. 8, 1875	Gangrene of foot, frost-bite	Leg	Chronic	Cured	June 26, 1875.
6	20	Christopher Hicks	March 1, 1875	Gangrene of hand	Hand	Secondary	Cured	May 6, 1875.
7	33	Thomas O'Donnell	March 29, 1875	Necrosis of hand	Forearm	Chronic	Cured	May 1, 1875.
8	22	Charles Coombs	April 3, 1875	Cog-wheel crush, hand	Forearm	Immediate	Cured	May 12, 1875.
9	"	" "	" "	Necrosis and ulcer, stump	Arm	Chronic	Cured	May 12, 1875.
10	7	Frederick Anderson	April 14, 1875	Strumous synovitis	Thigh	Chronic	Cured	May 19, 1875.
11	21	John Eckert	April 22, 1875	Railroad crush	Leg	Chronic	Cured	Feb. 11, 1876.
12	15	James Devlin	June 7, 1875	Machinery crush, hand	Hand	Secondary	Cured	Aug. 17, 1875.
13	10	John Hiscock	July 6, 1875	Scrofulous synovitis	Thigh	Chronic	Cured	Oct. 25, 1875.
14	35	John Hickman	July 19, 1875	Railroad crush, leg	Leg	Immediate	Died	July 26, 1875.
15	14	Henry Cline	Aug. 20, 1875	Elevator crush, foot	Foot, Listranc	Immediate	Cured	Oct. 18, 1875.
16	7	Herman Mulock	Aug. 27, 1875	Compound com. fract., humerus	Arm	Immediate	Cured	Sept. 1, 1875.
17	13	Michael Kassan	Sept. 13, 1875	Railroad crush, hand	Forearm	Immediate	Cured	Oct. 1, 1875.
18	56	Arch. Deetz	Sept. 15, 1875	Compound com. fracture, leg	Leg	Immediate	Cured	Jan. 19, 1876.
19	34	Robert Donnelly	Sept. 27, 1875	Compound com. fracture, leg	Knee-joint	Immediate	Died	Sept. 27, 1875.
20	40	Thos. W. Humphreys	Sept. 29, 1875	Ulcer	Leg	Chronic	Died	Oct. 28, 1875.
21	50	Charles L. Diggs	Nov. 8, 1875	Necrosis	Arm	Chronic	Cured	Jan. 31, 1876.
22	16	Thos. Hoffman	Oct. 15, 1875	Compound fracture, arm	Shoulder-joint	Immediate	Cured	Jan. 8, 1876.
23	12	Charles Lloyd	Nov. 13, 1875	Elevator crush, foot	Pirogoff, foot	Immediate	Died	Nov. 24, 1875.
24	29	Christian F. Greitzinger	Nov. 30, 1875	Hand cut off	Forearm	Immediate	Cured	Dec. 22, 1875.
25	5	George Laffi	Jan. 11, 1876	Railroad crush, leg	Thigh	Immediate	Died (shock)	Jan. 13, 1876.
26	21	Ella F. Williams	Jan. 31, 1876	Frost-bite	Leg	Chronic	Cured	July 3, 1876.
27	"	" "	" "	Frost-bite	Leg	Chronic	Cured	July 3, 1876.
28	9	Henry Brown	April 6, 1876	Railroad crush, foot	Pirogoff	Immediate	Cured	July 3, 1876.
29	51	Grace Mitchell	March 14, 1876	Necrosis	Leg	Chronic	Improved	April 22, 1877.
30	"	" "	" "	Necrosis, knee-joint	Thigh	Chronic	Cured	April 22, 1877.
31	21	Daniel Welsh	April 15, 1876	Machinery crush, hand	Hand	Immediate	Cured	May 13, 1876.
32	22	John Fries	May 3, 1876	Necrosis tibia	Leg	Chronic	Died	May 16, 1876.
33	32	Joseph W. Horton	May 31, 1876	Synovitis	Thigh	Chronic	Cured	Oct. 4, 1876.
34	57	George Rhoads	June 23, 1876	Railroad crush, foot	Ankle	Secondary	Died (tetanus)	July 21, 1876.
35	27	John Howard	July 11, 1876	Crush of leg	Leg	Immediate	Cured	Oct. 9, 1876.
36	27	Joseph H. Weaver	Aug. 10, 1876	Railroad crush of leg, R.	Knee-joint	Immediate	Cured	Dec. 22, 1876.
37	"	" "	" "	Railroad crush of leg, L.	Leg	Immediate	Cured	Dec. 22, 1876.
38	15	Michael Welsh	Sept. 1, 1876	Contusion of foot and gangrene	Leg	Secondary	Cured	Dec. 8, 1876.
39	8	Henry Doyle	Sept. 28, 1876	Railroad crush, arm	Shoulder-joint	Immediate	Died	Sept. 29, 1876.
40	39	David Cannon	Nov. 1, 1876	Railroad crush, leg	Leg	Secondary	Cured	March 9, 1877.
41	30	James Williams	Nov. 3, 1876	Railroad crush, hand	Forearm	Immediate	Cured	Dec. 15, 1876.
42	18	George Mungon	Nov. 7, 1876	Gunshot wound, hand	Hand	Immediate	Cured	Dec. 22, 1876.
43	28	John C. Marshall	Nov. 11, 1876	Railroad crush, foot	Leg	Immediate	Cured	March 30, 1877.
44	49	William H. Grant	Nov. 30, 1876	Railroad crush, leg	Leg	Immediate	Cured	March 10, 1877.
45	32	John Duggan	Dec. 12, 1876	Gangrene from frost-bite	Leg	Chronic	Cured	June 4, 1877.
46	"	" "	" "	Gangrene from frost-bite	Leg	Chronic	Cured	June 4, 1877.
47	34	Michael Murphy	Dec. 6, 1876	Railroad crush, leg	Knee-joint	Immediate	Cured	Feb. 6, 1877.
48	28	Peter Muirine	Dec. 30, 1876	Necrosis	Hand	Chronic	Cured	March 1, 1877.
49	51	Gottlieb Schupher	Feb. 1, 1877	Railroad crush, foot	Pirogoff	Immediate	Cured	June 18, 1877.
50	40	Thomas H. Fenton	Feb. 5, 1877	Aneurism of leg	Thigh	Chronic	Cured (see Aneurisms)	May 18, 1877.
51	32	John Donlin	Feb. 21, 1877	Railroad crush, leg	Knee-joint	Secondary	Cured	Jan. 21, 1878.

52	21	Smith Wolverton	March 14, 1877	Lacerated hand	Hand	Immediate	Cured	April 23, 1877.
53	21	William Gove	March 20, 1877	Compound com. fracture, leg	Thigh	Immediate	Cured	May 26, 1877.
54	68	George Myers	March 21, 1877	Gangrene of foot, R	Leg	Chronic	Cured	March 19, 1878.
55	"	" "	" "	Gangrene of foot, L	Leg	Chronic	Cured	March 19, 1878.
56	15	Charles Sweeney	March 31, 1877	Railroad crush, leg	Knee-joint	Immediate	Died (transfusion)	April 8, 1877.
57	19	Charles Lehr	April 7, 1877	Railroad crush, leg	Leg	Immediate	Cured	Dec. 21, 1878.
58	19	James Noble	April 8, 1877	Railroad crush, leg	Leg	Immediate	Cured	June 16, 1877.
59	19	Ella Dougherty	April 10, 1877	Machinery crush, hand	Forearm	Immediate	Cured	June 4, 1877.
60	22	Albert Raynor	May 1, 1877	Railroad crush, foot	Leg	Secondary	Cured	June 23, 1877.
61	4	Henry Tellbe	May 30, 1877	Railroad crush, foot	Leg	Secondary	Cured	Oct. 27, 1877.
62	22	Timothy Reilly	June 5, 1877	Necrosis, ankle	Leg	Chronic	Cured	Aug. 12, 1878.
63	21	Ant. McHale	June 18, 1877	Fracture of stump	Leg	Chronic	Cured	Aug. 27, 1878.
64	24	Mich. Keeley	June 26, 1877	Railroad crush, foot	Pirogoff	Immediate	Died (delirium tremens)	July 4, 1877.
65	16	William Hales	June 29, 1877	Machinery crush, arm	Arm	Immediate	Died (delirium and exhaustion)	July 3, 1877.
66	30	John McDevitt	Aug. 7, 1877	Railroad crush, leg	Leg	Immediate	Cured	Nov. 22, 1877.
67	10	John Wolf	Aug. 10, 1877	Railroad crush, hand	Hand	Immediate	Cured	Sept. 4, 1877.
68	54	Grace Mitchell	Oct. 15, 1877	Osteomyelitis	Hip-joint	Chronic	Cured	Oct. 11, 1878.
69	29	Geo. A. Othello	Oct. 19, 1877	Railroad crush, foot	Leg	Immediate	Died (exhaustion)	Oct. 31, 1878.
70	58	James Bradley	Nov. 1, 1877	Railroad crush, arm	Arm	Immediate	Cured	Jan. 10, 1878.
71	25	Ellen Cairns	Dec. 14, 1877	Neuralgia of stump	Thigh	Chronic	Cured	April 15, 1878.
72	34	William Fricks	Dec. 26, 1877	Painful stump	Foot	Chronic	Cured	April 4, 1878.
73	42	Thomas Blackburn	Dec. 28, 1877	Necrosis	Thigh	Chronic	Cured	March 4, 1878.
74	30	Joseph Simpson	Jan. 26, 1878	Crushed hand (picker)	Hand	Immediate	Cured	Feb. 27, 1878.
75	31	Peter Marrion	Jan. 31, 1878	Compound com. fracture, leg	Leg	Immediate	Died	March 1, 1878.
76	31	Joseph Deighan	April 29, 1878	Compound com. fracture, leg	Leg	Secondary	Cured	Aug. 27, 1878.
77	35	Robert McDonald	May 17, 1878	Railroad crush	Shoulder-joint	Immediate	Died (see Transfusion, Case 5)	May 25, 1878.
78	32	Mary Shreve	June 10, 1878	Syphilitic necrosis of leg	Knee-joint	Chronic	Died	June 23, 1878.
79	45	John R. James	June 27, 1878	Necrosis of femur	Thigh	Chronic	Cured	Nov. 22, 1878.
80	11	Robt. McCullough	June 29, 1878	Railroad crush, leg	Leg	Immediate	Cured	Sept. 20, 1878.
81	52	Eliza Gales	Aug. 9, 1878	Wagon crush of leg	Leg	Immediate	Died (exhaustion and shock)	Aug. 11, 1878.
82	10	James McGee	Aug. 17, 1878	Railroad crush of leg	Thigh	Immediate	Cured	Oct. 18, 1878.
83	19	Paul Fritz	Aug. 23, 1878	Gunshot fracture, arm	Shoulder-joint	Immediate	Cured	Oct. 23, 1878.
84	20	John H. Spencer	Aug. 26, 1878	Elevator crush, foot	Leg	Immediate	Cured	Nov. 8, 1878.
85	"	" "	" "	Ulcer of stump	Leg	Chronic	Cured	Nov. 8, 1878.
86	29	James Hughes	Sept. 12, 1878	Painful stump	Foot	Chronic	Cured	Feb. 8, 1879.
87	22	William Fleming	Sept. 15, 1878	Gunshot fracture, arm	Shoulder-joint	Immediate	Cured	Oct. 16, 1878.
88	30	Julius Kling	Sept. 16, 1878	Gunshot wound of foot	Leg	Immediate	Cured	Oct. 16, 1878.
89	8	John Cummings	Sept. 17, 1878	Necrosis after excision	Thigh	Chronic	Cured	May 21, 1879.
90	7	Neil Campbell	Sept. 19, 1878	Crush of leg	Leg	Immediate	Cured	Dec. 2, 1878.
91	18	Walter H. Wright	Sept. 18, 1878	Railroad crush, arm	Arm	Immediate	Cured	Oct. 16, 1878.
92	23	John Lyons	Sept. 30, 1878	Railroad crush, foot	Leg	Immediate	Cured	Nov. 20, 1878.
93	17	John Scofield	Oct. 3, 1878	Machinery crush of hand	Forearm	Immediate	Cured	Oct. 21, 1878.
94	46	James Eveland	Oct. 5, 1878	Syphilitic caries of tarsus	Ankle (Syme)	Chronic	Cured	April 30, 1879.
95	16	Michael Goldrich	Oct. 14, 1878	Machinery crush, forearm	Arm	Immediate	Cured	Nov. 11, 1878.
96	9	Frank M. Moyemiller	Nov. 15, 1878	Railroad crush, leg and knee	Thigh	Secondary	Cured	Feb. 19, 1879.
97	24	Bassa Dosamson	Nov. 26, 1878	Osteomyelitis, right thigh	Hip-joint	Chronic	Cured	May 28, 1879.
98	18	Martin Zelley	Dec. 7, 1878	Railroad crush, arm	Arm	Immediate	Died (continued shock)	Dec. 8, 1878.
99	27	Charles Rauch	March 1, 1879	Ulcers of stump, R	Leg	Chronic	Cured	May 2, 1879.
100	"	" "	" "	Ulcers of stump, L	Leg	Chronic	Cured	May 2, 1879.
101	35	Ernest Loez	Feb. 7, 1879	Railroad crush, arm	Forearm	Immediate	Cured	March 28, 1879.
102	28	Michael Cubbein	Feb. 4, 1879	Ulcer of leg	Thigh	Chronic	Cured	April 21, 1879.
103	28	Henry Harcus	Jan. 28, 1879	Forearm cut off by circular saw	Arm	Secondary	Cured	March 21, 1879.
104	36	Thomas Murray	April 16, 1879	Railroad crush of foot	Leg	Immediate	Cured	Oct. 1, 1879.
105	22	Joseph J. Stevenson	April 21, 1879	Wound of hand and forearm	Arm	Immediate	Cured	Aug. 20, 1879.
106	17	Charles Zahn	May 12, 1879	Gunshot wound of arm	Arm	Immediate	Cured	June 2, 1879.
107	14	George Kilpatrick	May 21, 1879	Crushed foot	Foot	Immediate	Cured	July 2, 1879.
108	3	Elizabeth Minton	May 22, 1879	Wagon crush of foot	Foot	Immediate	Cured	Aug. 20, 1879.

Notes on Cases included in the Last Series of 108 Amputations.

CASE 1. *Encephaloid tumor of the leg: amputation: recovery: return of the disease in the temporal bone; death five months afterwards.*—Ellen W., aged 14, was admitted November 30, 1874. One month ago first noticed an enlargement on shin of left leg; limb increased in size. Her mother died of mammary cancer. The tumor is soft and irregular; occupies the lower and outer part of the limb. Five inches below the tuberosity of the tibia limb measures $14\frac{3}{4}$ inches; the right, $11\frac{3}{4}$; 2 inches lower the left measures 11 inches, the right $6\frac{3}{4}$.

December 15.—Temperature 102°; limb measures $15\frac{3}{4}$ inches.

January 12, 1875.—Thigh amputated.

April 1.—Stump well. A tumor has appeared on the left frontal bone, which has grown rapidly, and is very painful.

June 3.—Exhaustion excessive; pleuritis.

June 6.—Died.

Post-mortem.—The right thorax contained two quarts of serum, which had compressed the lung to the extreme upper part of the chest; the lung was crepitant, and floated. On the right side of the spinal column, between the third and seventh dorsal vertebræ, and adherent to it, was an oblong mass, irregular in outline, six inches long; on section this cut like softened bone. The tumor on the head was tightly adherent to the bone, and was somewhat softened, and encephaloid in character.

CASE 5. *Frost-bite: amputation of foot: recovery.*—Gonzalo A., aged 35, a native of Cuba, was admitted February 8, 1875. Four weeks ago was exposed to severe cold. On admission there was well-defined gangrene of the left foot; the line on dorsal surface is distinct, but it is not so clearly defined on the plantar portion; all the anterior portion of the foot is lost, and there is considerable swelling of the limbs; odor intense; general condition of health poor. Carbolized dressing was applied, and quinine, iron, milk-punch, and anodynes given, p. r. n.

February 17.—Amputation of leg was made; Esmarch and ether. Parts were vascular, and pins also were used to control the hemorrhage; about twelve ligatures applied.

February 28.—Primary union to a considerable extent.

May 26.—Stump well.

CASES 8, 9. *Crushed hand and forearm; amputation: sloughing of the flaps; necrosis: amputation of arm; recovery.*—Charles C., aged 22, was admitted April 3, 1875, suffering from a cog-wheel crush of his right hand and forearm. The crushing extended to the middle of the forearm, and amputation was made three inches below the elbow-joint. An unusual number of ligatures had to be applied, and one pin was used to control the active hemorrhage. The flaps sloughed, and necrosis followed.

April 18.—Amputation of the arm was made at the lower third by a short anterior and a long posterior skin-flap. It was not found necessary to ligate the brachial artery, which had been blocked since the former amputation. On June 2 the stump was well, and the patient was discharged.

CASE 11. *Crush of leg; attempt to save the limb: necrosis: amputation of the leg: recovery.*—John E., aged 24, was admitted April 22, 1875. In attempting to jump on a train was struck by the axle and thrown against a post, causing a compound fracture of the leg, involving the fibula at its lower end. The tibia was also comminuted, and the soft parts badly lacerated. The posterior tibial artery was ruptured. The anterior tibial was uninjured. The patient was in good health, and, from the fact that the wheel did not pass over the limb, it was decided to make an attempt to save it. A protruding portion of the fibula was removed. The other fractures were adjusted, and the limb, which was warm, was placed in a fracture-box, with a bran dressing. Copious suppuration ensued. From time to time fragments of the bone were removed. Several large sloughs of deep cellular and muscular tissue were from time to time removed. Pain became in a few weeks a prominent symptom; separation of fragments and necrosis of others followed, and fistulous tracts occurred in all parts of the limb, greatly exhausting the patient. In September an exploratory operation revealed so much disease of bone that amputation was resorted to. Fifteen ligatures were used, and the stump dressed with carbolic charpie. Discharged, February 14, 1876, cured.

CASE 17. *Strumous synovitis; amputation of thigh; recovery.*—John H., aged 10, was admitted July 6, 1875, with strumous synovitis. He had always been a delicate child. One month ago a small, non-painful lump was noticed in the popliteal region. At first the tumor did not interfere with walking. The child was pale, anæmic, and emaciated. The right leg is the seat of diffuse swelling, extending from knee to hip, very tender to pressure. The patella is elevated by effusion. Fluctuation marked, not only over the knee, but well up on the thigh.

July 8.—A free opening was made in the inner side, giving vent to an immense amount of pus; then fistulous tracts appeared about the knee. The discharge gradually lessened, but intense pain continued with the least motion.

September 11.—An incision revealed serious necrosis of the femur; the tibia was also involved. Excision could not be entertained, and the thigh was amputated. Discharged, cured, October 25, 1875.

CASE 18. *Crushed leg; amputation; carbolized ligatures of catgut used; secondary hemorrhage; recovery.*—Archibald D., aged 56, was admitted September 15, 1875. A box weighing 800 pounds fell upon his leg, causing a compound comminuted fracture. The arteries were uninjured, but the tibia and fibula were so extensively comminuted that amputation was deemed advisable. Six carbolized catgut ligatures were used, which were cut off short; the stump was thoroughly syringed with carbolized water, and carbolized oil dressings, with bandage, were applied. For four days temperature was recorded as 104½°; it then fell to 102°, and soon after to 100°. The stump became much swollen the day after the operation, and there was slight sloughing of the flaps.

October 8.—Considerable hemorrhage; end of stump swollen and red. Pressure on the femoral, with elevation of the limb, and an application of Monsel's salt, controlled the flow of blood. Patient discharged, cured, January 19, 1876.

CASE 20. *Railroad crush of foot; sloughing of tissues; foot removed by incising simply the dead portion; discharged with ulcer of stump; readmitted three weeks later; amputation of leg; death from exhaustion.*—Thomas W. H., aged 40, was admitted September 29, 1875, with ulcer of stump. In January, 1875, he had been a patient in the hospital with a railroad crush of the left foot. Previous to the accident he had been drinking heavily, and was in such a wretched condition that immediate amputation could not be performed. Three days after admission mania à potu had developed; this was treated with chloral, morphia, stimulants, etc., and carbolic powder was applied to the foot, which sloughed and was removed by cutting through the tissues at the line of demarcation; being finally discharged by request with an unhealed stump. When readmitted with the ulcer there was great discharge, which was very offensive, while the skin about the ankle was red, swollen, and tender. The limb was elevated and poultices were applied. The patient had a murmur with the first heart sound; his knee-joints were swollen and generally rheumatic, and he had been drinking more or less for several years. Iron, quinine, and a good diet were ordered, and on October 2 amputation of the leg was performed by the Teale method. The arteries were found atheromatous.

October 11.—There was a slight erysipelatous flush.

October 17.—Patient somewhat weaker. There has been a gradual declination since the operation.

October 28.—Died of exhaustion. The temperature marked 103°, but at no time after the operation was it below 100°. No post-mortem was obtained.

CASE 21. *Necrosis of elbow-joint; amputation; recovery.*—Charles L. D., aged 30, was admitted November 8, 1875, with long standing necrosis; there were several sinuses, which discharged very freely; the necrosis was so extensive, involving the bones of the forearm, that excision was impossible. Teale amputation was made; the temperature reached 105° after the operation, then gradually fell. Discharged, cured, January 31, 1876.

CASE 22. *Railroad crush of arm; amputation of the shoulder-joint; recovery.*—Thomas H., aged 16, was brought in October 15, 1875. He had fallen asleep upon the track, and the wheels of a loaded car passed over the right arm, crushing it to the shoulder; the accident occurred at midnight, twelve miles from the city; he was admitted some eight hours afterwards. The upper part of the humerus was protruding: the axillary artery was torn across, but was blocked by clot. Amputation was made at the joint; six ligatures used. January 8, 1876, well.

CASE 25. *Elevator crush of foot; amputation: tetanus: death.*—Charles L., aged 12, a messenger-boy, was admitted November 13, 1875, with a most serious foot crush; the tarsus was entirely comminuted and the soft parts lacerated. Pirogoff amputation was made. On the ninth day there was some difficulty experienced in opening the mouth; four hours later trismus was marked. Fifteen grains of chloral were given every hour, with calomel and opium and a quarter of a grain of conia every two hours. 23d November, opisthotonos: immense relief followed injections of morphia, but at 2.30 P.M. died.

CASES 26, 27. *Exposure, followed by gangrene of both feet; natural amputation of both limbs; amputation of legs: recovery.*—Eba F. W., a seaman, aged 21, was admitted January 31, 1876. Two years ago, when on board ship in Alligator River, North Carolina, was cruelly treated and finally driven from the vessel: he took refuge in a swamp. The weather was then intensely cold, and the next day he experienced a numbness of both feet. He was in the swamp ten days without food, and was then found by some fishermen in an exhausted state. Some weeks afterwards both feet sloughed off. When admitted to the hospital the patient was very anæmic. The feet had separated at the ankles.

April 8.—The right leg was amputated at the middle third. Torsion was applied to the vessel, with the exception of the posterior tibial, where a pin was used.

May 14.—Amputation of left leg at a point to correspond with the other amputation and same form of operation.

June 18.—Well.

August 15.—Has remained in the hospital to date to have artificial legs applied to stumps.

CASES 29, 30. *Necrosis of tibia; amputation of leg; reappearance of necrosis: amputation of the thigh: recovery* (see Case 68, Amputation Hip-Joint).—Grace M., colored, aged 51, waitress, was admitted March 14, 1876. Has had disease of the leg for twenty-five years. Has had ulcers in various parts of the limb, which heal only to break out again. There is a depression four inches below the knee, at the bottom of which there is an ulcer three inches long. The depression, and nearly all the front of the leg, presents a cicatricial appearance. The tibia is fractured at this point by the progress of the disease, and the leg is held together by the fibula, which seems to be involved. The other leg is covered with scars, and there is some discharge of pus. There is no decided history of syphilis. Appetite good.

March 22.—Incisions were made, and the tibia was found to be entirely destroyed for several inches. Amputation two and a half inches below knee: rectangular flaps.

March 25.—Dressing changed; flaps uniting; no discharge.

March 30.—All but one ligature away: free discharge.

April 14.—Removed a quantity of slough; dressed with carbolized water twice daily.

April 24.—Still a great deal of discharge.

May 5.—Much improved; discharge slight.

June 6.—Improving in health, but the stump will not heal.

September 6.—An incision showed the knee-joint involved; in the cavity there was considerable pus; amputation; rectangular flaps.

September 8.—Considerable discharge.

October 12.—Applied a flaxseed-poultice.

October 19.—Stitches and ligatures removed.

December 10.—Very little discharge; doing well.

May 30, 1877.—Pain in groin, and tenderness on pressure over entire thigh.

June 16.—Under ether the patella, which was very much softened, was removed, also small sequestrum of shaft of femur.

June 20.—Discharging fetid pus.

July 13.—Up in a chair; improvement slow.

September 29.—Pain in stump.

October 5.—Discharged by request.

CASE 32. *Necrosis of the ankle; amputation of the leg; tetanus; death.*—John F., aged 22, was admitted May 3, 1876. Two months ago the patient was injured in right leg by a piece of lumber falling on him. A simple fracture of the leg resulted. Was treated three weeks in a fracture-box, after which a plaster dressing was applied for twelve days, and when it was removed there was a slough above the ankle; sinuses, then abscesses, have formed from time to time about the joint.

May 6.—An opening was made, and a large sequestrum of bone from the lower end of the tibia was removed. The tarsal bones were found in a necrotic condition, and amputation by the Teale method was performed at middle of the leg. The vessels were twisted, except the posterior tibial, which was secured by a pin. On the 12th twitching of the limb came on, with a sense of soreness about the jaws; one-fiftieth of a grain of atropia was injected every fourth hour. On the 14th the trismus was marked; bromide of ammonia, morphia, and whiskey were given freely. On the 15th the jaws were firmly locked, and there was some difficulty in respiration. In the afternoon the patient died suddenly.

CASE 33. *Scrofulous synovitis of four years' standing; amputation of thigh.*—Joseph W. H., aged 32, married, a laborer, of American parentage, was admitted to the hospital May 31, 1876. In 1872 the left knee began to swell. This difficulty increased to 1875, when abscesses opened, and sinuses were formed on both sides. On admission there was discharge of pus, with partial anchylosis of the joint. Free incisions were made by Dr. Hunt, and iron, quinine, etc., given. August 4, no improvement except in the general health of the patient having taken place, ether was administered, an Esmarch bandage applied, and the thigh was amputated by Dr. Morton. Teale flaps were made, which, after ligation of vessels, were loosely brought together with silver sutures; free drainage being allowed.

On August 6, primary union to a considerable extent.

An examination of the knee showed extensive ulceration of all the cartilages, and necrosis of the tibia as well as of the condyles of the femur, extending into the shaft of the latter bone.

CASES 36, 37. *Compound comminuted fracture of both legs; amputation of the right limb at the knee-joint, the left at the middle of the leg; secondary hemorrhage from the right stump; ligation of the femoral artery; recovery.*—John H. W., aged 27, was admitted August 10, 1876, at 2 A.M. The wheels of a loaded car had passed over both legs. The patient was in severe shock, with an irregular pulse. Coffee and morphia were given, and at 10 A.M. there was good reaction, when the right limb was amputated at the knee-joint; the left one about the middle. Oblique muscular cutaneous flaps were made. A long anterior and short posterior flap were used at the knee amputation. Tourniquet, silk ligatures, and dry dressings were used. Brandy, lime-water, and plain soda were ordered.

August 11.—Temperature 101°. Magendie's solution, eight minims were ordered this morning.

August 21.—There has been more or less diarrhœa for the last forty-eight hours, which was treated with chalk mixture. Lead, quinine, and opium were also given. Temperature to-day, 104°. Some sloughing of knee-flap.

August 29.—Hemorrhage from right stump, which was controlled by a tourniquet; some vomiting.

August 30.—Hemorrhage again occurred. The dressings were removed and the popliteal was religated higher up. Two hours later the bleeding again came on. The femoral was tied in the usual place, ether spray being used to anæsthetize the parts.

September 15.—Femoral ligature came away.

October 18.—There has been a portion of the tibia exposed in the left stump, from which some sloughing has occurred. Two pieces of bone were removed.

November 26.—Two days ago had a chill, after which high temperature. Has had malarial fever in times past. Sixteen grains of quinine given.

December 22.—Discharged, cured.

CASE 38. *Compound comminuted fracture of leg; gangrene; secondary amputation; recovery.*—Michael W., aged 15, was admitted September 1, 1876. Had his foot caught in an elevator and terribly crushed, the swelling being so great that no fracture could at the time be detected. Considerable suppuration and constitutional symptoms set in, requiring supporting treatment.

September 9.—Gangrene well developed.

September 18.—Foot nearly separated.

October 4.—Leg amputated; lateral flaps; lower third; silk ligatures.

December 8.—Discharged, cured.

CASE 39. *Amputation of the shoulder-joint for railroad crush; death.*—Henry D., aged 8, school-boy, was admitted September 28, 1876. Patient had been run over by a street car, producing a compound comminuted fracture of humerus in upper third, and extensive laceration of arm and elbow-joint. There was a good deal of hemorrhage, and the shock was marked. Shoulder-joint amputation made without much loss of blood, but patient became quite weak.

September 29.—Patient only partly reacted, and died suddenly.

CASE 40. *Bumper injury; immense swelling; detection of fracture after incisions made to relieve tension; secondary amputation of the leg; recovery.*—David C., aged 39, was admitted November 1, 1876. While coupling cars patient had his left leg caught between the bumpers. The heel of the boot somewhat saved the leg from the full force of the crush, but the limb up to the thigh was very tense, and no positive fracture could be detected. The skin was not broken. Later in the day the tension increased, and, as the limb was cold and with blebs, Dr. Morton divided the tissues from the anterior part of the foot to the knee on the outer side, and over the calf upon the inner side. The tissues were found infiltrated with blood. A fracture was found, which involved the tibia at its head; the fracture also entered the joint. The incisions gaped two and one-half inches. Morphia was given by skin freely. The limb placed in a fracture-box, and carbolized oil dressings applied. Quinine was given, and stimulants in moderate quantity were ordered.

December 26.—Abscesses in the ankle have opened, and the tarsal bones are found somewhat loose.

January 6, 1877.—The astragalus much necrosed and softened; the incision carried higher, showed necrosis of the tibia and fibula three inches above the ankle. Teale's amputation was made; Esmarch; ligatures.

January 14.—All ligatures are away save one.

January 20.—Last ligature is out.

March 9.—Cured; discharged.

CASE 44. *Compound comminuted fracture of the right ankle; amputation of the leg; also fracture of the left leg; recovery.*—William H. G., aged 49, was admitted November 30, 1876. He had fallen some thirty feet from the elevator of a hotel, thus causing a fracture of the tibia and fibula at the middle of the left leg, and a compound comminuted fracture of the right tarsus. The severity of the injury was seen in the right foot, and could only be estimated by a digital examination of the wound on the sole of the foot. The os calcis and astragalus were found comminuted to an enormous extent. There was marked shock; treated by Magendie's solution and hot coffee. Five hours afterwards reaction was established, and amputation of the leg at the middle third was then performed. Anterior and posterior flaps were made, and Esmarch and ligatures were used; no sutures; charpies and roller used. A silicate bandage was applied to the fractures of the left leg. Discharged, cured, March 10, 1877.

CASE 47. *Railroad crush; amputation at knee-joint; recovery.*—Michael M., aged 31, was admitted December 6, 1876, with his limb completely crushed at the middle third. Reaction was well established in three hours, and the limb was removed at the joint.

December 11.—Delirium tremens developed yesterday, but the stump looks well, although somewhat swollen. Bromide and morphia ordered, the latter by skin.

December 19.—Popliteal ligature removed.

February 6, 1877.—Discharged, cured.

CASE 49. *Compound comminuted fracture of the foot; Pirogoff amputation; delirium tremens; recovery.*—Gottlieb S., aged 51, a stone-cutter, was admitted February 1, 1877, with his left foot completely crushed by a heavy mill-stone falling upon it. Patient was a free drinker. Two hours after the injury a Pirogoff amputation was made.

February 6.—Active delirium since yesterday. Bromide of potassium was given every two hours, with whiskey, in small doses.

February 8.—Has required restraint, and is worse. Hydrate of chloral in twenty-grain doses given every three hours. Stump looks badly.

February 12.—Free from delirium for two days.

April 1.—Stump discharging freely. Bran dressing applied.

April 11.—Doing well.

June 18.—Discharged, cured.

CASE 50.—*Sacculated aneurism of the head of the tibia; forced flexion; compression; threatened gangrene; amputation of the thigh; recovery.*—Thomas H. F., aged 40, was admitted February 5, 1877, with a tumor the size of a small orange, situated posterior to the head of the tibia, which was the seat of expansive pulsation, a distinct thrill, and a loud systolic murmur. Forced flexion was used; then compression of the femoral. Gangrene threatening, the thigh was amputated. Discharged, cured, May 18, 1877.

(For full account see Case 4, in article on Aneurisms treated by Compression.)

CASE 51. *Railroad crush of leg; secondary amputation at knee-joint, the patella being retained; recovery.*—John D., aged 32, a miner, was admitted February 21, 1877. He was injured four days ago in the mines; suffered so much from shock that it was impossible to send him to the hospital earlier. On examination the limb was boggy and partly gangrenous; his general condition was very bad, with diarrhœa; tongue dry and fissured; pulse quick and feeble.

February 26.—Looseness of bowels continues, and delirium.

March 2.—Only two stools; weak; belly tympanitic.

March 7.—Retention of urine and weak. After this there was a general improvement.

June 2.—Amputation at knee-joint, the patella being retained.

June 13.—All ligatures away.

January 24, 1878.—Discharged, cured.

CASES 54, 55. *Exposure to intense cold; gangrene of both feet; amputation of both legs; recovery.*—George M., aged 68, was admitted March 21, 1877. Four days ago was exposed, and both feet were severely frost-bitten. Patient is feeble and pale.

April 4.—The sloughing has involved the entire sole of both feet, and in front extends to the instep, the tarsus being exposed and the ankle-joints involved.

April 14.—Right limb amputated by Teale method at lower third.

May 26.—Left limb removed opposite the right stump.

March 19, 1878.—Discharged, cured.

CASE 56. *Crush of leg; amputation at knee-joint; secondary hemorrhage; transfusion; death.*—Charles S., aged 15, was admitted March 31, 1877, with left leg crushed. Amputation at the knee-joint was made, the patella being retained; the condyles were removed; a pin controlled the popliteal artery.

April 4.—The pin was withdrawn, but reintroduced on account of bleeding; severe hemorrhages followed.

April 8.—Exhaustion was so great that transfusion was made, but a fatal result followed. (See Case 4, Transfusion.)

CASE 68. *Osteo-myelitis; amputation at the hip-joint; recovery.*—Grace M., aged 51, was a patient in the hospital in 1876 (see Case 29), and underwent an amputation of the leg; then of the thigh, just above the condyles.

October 15, 1877.—Readmitted for osteo-myelitis of the entire femur. The stump is swollen, open, and exceedingly painful, and so is the femur its entire extent on pressure.

October 20.—An exploratory incision was made, revealing necrosis as far as the great trochanter. Amputation at the hip-joint was then made by Dr. Morton, the abdominal tourniquet was used, and sixteen ligatures were applied: a grain of opium was given every six hours.

November 5.—Main ligature came away.

December 4.—Stump almost healed.

October 11, 1878.—Discharged, cured. This patient was allowed to remain longer in the house than the case actually required. Has since continued quite well.

CASE 74. Ellen C., aged 25, was first treated in the hospital in March, 1873, for intense neuralgia with tibial necrosis, which was so extensive as to demand amputation of the leg. A painful stump resulted, and two months later the limb was removed at the knee-joint. The wound united kindly, but the same neuralgic condition reappeared. Ellen remained in the house until August 6, when an artificial limb was applied, upon which she walked very well, and was nearly free from the pain. During the following winter she was admitted into hospital in West Philadelphia, and then she was suffering so much that an amputation higher up was made without any good result.

December 14, 1877.—Ellen was again admitted into the Pennsylvania Hospital suffering with the stump, in which large painful nodules could be felt.

December 29.—The stump was opened, and the sciatic reached and dissected high up and divided. Unusually large nodules were found on all the terminal nerve-trunks. At the same time the stump was amputated just below the great trochanter. Great relief followed this amputation. In April, 1878, discharged for misconduct.

CASE 75. *Crushed leg: amputation; gangrene: erysipelas; death.*—Peter M., aged 31, was admitted January 31, 1878, with a comminuted fracture of his right leg, a large sign having fallen on him. A flap amputation at the middle of the leg was made.

February 3.—Small bleb on the under flap observed.

February 6.—Stump sloughed: sutures all removed: three ligatures off.

February 7.—Chill; temperature 105°; quinine sulph., sixteen grains daily.

February 8.—Erysipelatous blush.

February 10.—Blush extending; pain in groin.

March 1.—At 5.30 A.M. became unconscious; pupils dilated; skin cold; pulse frequent and small; breathing stertorous; temperature 110°. Died at 7 A.M.

Autopsy.—Brain pale; slight serous effusion; no plugging of arteries; examination otherwise negative.

CASE 76. *Compound comminuted fracture of leg: resection of protruding bones: sloughing of tissues: secondary amputation; recovery.*—Jos. D., aged 31, admitted April 29, 1878, was knocked down and kicked or trodden upon by a companion; sustained a compound comminuted fracture of the tibia and fibula of right limb; some comminution of the tibia about the seat of fracture. The lower end of the upper fragment projects through the wound. It was found impossible, even under ether, to make a good approximation of the bones. Three large fragments were removed and the protruding bone was cut off. Slight hemorrhage only followed this operation. The patient is weak and sometimes delirious. Ordered quinine, ten grains daily; whiskey; bromide and chloral.

May 4.—Foot and leg much swollen, and a livid erythematous blush and crackling under the skin; over the outer ankle there is a bluish-black patch; evening temperature 105°.

May 5.—Temperature 103½°; leg about the same.

May 6.—Temperature 106°.

May 20.—Temperature nearly normal; tissues at ankle have sloughed.

June 1.—Inflammation returned a few days since; immense amount of discharge, which is producing great weakness.

June 21.—Amputation of leg below knee.

June 30.—Convalescence has been rapid.

August 27.—Discharged, well.

CASE 78. *Syphilitic disease of tibia; amputation at knee-joint; exhaustion; death.*—Mary S., aged 32, was admitted June 10, 1878. Originally a healthy woman; contracted syphilis in 1864. In 1871 had periostitis; then nodes appeared. These for a long time were kept in abeyance. In 1876 the nodes became very large and painful. In March, 1877, the skin ulcerated, and numerous fragments of dead bone came away. She formerly weighed 168 pounds; was fair in color. She is now emaciated, and very anaemic.

June 10.—Upon admission the tibia is exposed and dead for eleven inches; has great pain at night, and there is very offensive discharge.

June 14.—Amputation at the knee-joint; antero posterior flaps. The patella in this case was not removed.

June 17.—Severe chill, but stump looks very well.

June 18.—Stump painful; vomiting; quinine in large doses.

June 25.—Gradual exhaustion since last report, and died to-day.

Post-mortem.—Heart normal; but as all the organs were healthy, except slight fatty condition of the liver, the cause of death might be said to be "heart failure" from exhaustion, no lesions of any organs being detected.

CASE 79. *Necrosis following injury; amputation of the thigh; secondary hemorrhage controlled by pin; recovery.*—John R. J., aged 45, was admitted June 27, 1878. When fourteen years of age he fell and struck his right leg with some force. One year afterwards an abscess formed and opened, from which fragments of bone came away. Since then abscesses have formed from time to time. For four years he has had open sinuses, with very offensive discharges. On admission (which was for hemorrhage from one of the old sinuses) a sinus leads to the bone, in which there is a loose portion of bone two inches long, one inch broad.

July 3.—Number of fragments of necrosed bone removed.

October 9.—Since last report there has been no change, except an improvement in general health; and, as a large amount of the femur is involved and hopelessly so, amputation was performed this day by antero-posterior skin-flaps.

October 14.—Hemorrhage occurred. Ten P.M. stump opened, and a deep vessel secured. This did not control the hemorrhage, when a pin was used. A rapid recovery followed. Discharged, cured, November 22, 1878.

CASE 94. *Syphilitic disease of the bones of the foot; several operations; subsequently Syme amputation; recovery.*—James E., aged 46, was admitted October 5, 1878, with ulcerous disease of the toes, involving the joints. In 1860 he contracted syphilis, which was followed by secondary symptoms. Four years ago an abscess formed over the great toe, and was followed by necrosis. He had at various times, from February, 1876, to the time of admission, operations for the relief of this necrotic condition, but in each instance, with the removal of the part, a return of the disease manifested itself. The ulcer of the foot discharged a very offensive pus, and the parts were swollen and painful. In order that the best chance should be given, it was determined to amputate by a Syme method, thus avoiding the division of any bony structure, for, as the tarsus was involved, fistulous tracts extended in several instances into this part. The Syme was made November 2. After this there were a series of abscesses, although drainage-tubes were left in the wound, and every precaution taken, even to slitting the flap posteriorly, where a drain was inserted.

April 30, 1879.—Was discharged, well. Since then has walked five miles in a day, the weight resting directly upon the stump.

CASE 96. *Comminuted fracture of right femur about middle and fracture of tibia and fibula, with laceration of tissues.*—Frank M., aged 9, admitted November 13, 1878, about 5.30 P.M. His right leg was crushed between the bumpers of cars; two hours afterwards was brought to the hospital. Hemorrhage was slight. On examination there was a large lacerated wound of the leg superficial in character; the bones were severely fractured, but its character was not compound; below the knee the leg was quite cold, and no pulsation could be felt in the arteries, the femur was also fractured. As there was considerable shock and no attempt at reaction, an immediate operation was out of the question.

November 15.—Nausea; face reddish-orange color; temperature 100°, and symptoms of acute gangrene of leg. Quinine, three grains, and tinct. opii camph. to allay pain. Limb in position with sand-bags.

November 16.—Quinine increased to six grains; stimulants.

November 17.—Line of demarcation forming; temperature 103½°.

November 19.—Limb placed in fracture-box with bran and deod. pulvis.

November 24.—Increased stimulants one ounce a day; skin better color, but great irritability.

November 27.—Nature having divided the soft tissues to the bone, Dr. Morton divided the tibia about its middle: the fibula was found easily separated at its head. Oxide of zinc was applied and carbolized charpie. Temperature 103⅗°.

December 5.—Very much better.

January 18, 1879.—The surfaces of ulceration having been well covered with healthy granulations, Dr. Morton amputated the thigh, using all the skin which remained about the knee, and divided the femur five inches above the articular surface.

January 19.—Temperature 99⅘°; pulse 150, pain readily controlled by morphia.

January 21.—Stump dressed for first time; general union.

February 19.—Discharged, well. The comminuted fracture of the thigh united well, with some overlapping and deformity, for it was impossible to apply any dressing to retain the fragments in position.

CASE 97. *Cannon-ball crush of thigh; primary amputation on the field; osteomyelitis; hip-joint amputation; recovery.*—Boasso Dominico, aged 24, single, a musician, a native of Italy, was admitted November 20, 1878. During the Franco-Prussian war was struck by a cannon-ball; the thigh was then amputated at the junction of the middle and lower thirds. He was in the French hospital about one month, being discharged in July, 1877. During this period he had a stump abscess; following this he had no trouble until some months prior to his seeking admission to the Pennsylvania Hospital. Upon entering, necrosis was found, involving the remaining portion of the femur; fistulous tracts were on the inner and outer side; from these a large quantity of offensive pus was constantly being discharged, and the stump was exceedingly painful to touch. Was very pale and anæmic; appetite poor. Was ordered the compound phosphates, quinine, and a regulated diet. Hypodermics at night were required to induce sleep.

December 14.—For some days there has not seemed to be any decided improvement, so that it was decided to amputate at the hip-joint. Dr. Morton made a square anterior skin-flap, and a rather larger posterior one in the same manner. The femoral artery was then isolated and tied; the muscles anteriorly were divided with but little loss of blood; the muscles posteriorly were then cut close to the pelvis; some thirty vessels required ligation, and one pin was passed in deeply, to control hemorrhage from apparently rotten but vascular tissue on the inner face of the stump. The tissues about the ligation of the femoral were so soft that, for greater safety, a pin was passed under the femoral an inch and a half above, so that, in case of any hemorrhage, a ligature could be thrown around it and thus the artery could be compressed at once. The flaps were secured by silver sutures and strips of plaster. During the evening there was oozing, and the stump was opened and a number of small arteries were ligatured.

December 21.—Pins removed; no symptoms worthy of note occurred; the patient's convalescence was slow. The stump was dressed from the first with carbolized oil on charpie, and the dressings held in position by a towel.

March 1.—Stump firm; during the past month has had an intractable and curious skin affection, which resembled pemphigus, but due apparently to the exhaustion and drain. This eruption showed itself particularly on the arms, thigh, and face. Under a change in the hospital, to one of the upper wards, he at once improved. Previous to this, treatment seemed unavailing.

The specimens of bone showing osteo-myelitis after amputation at the hip-joint (from Grace M., Case 68, and from the above, Case 97, Boasso D.) present the following conditions, which have been kindly communicated by Dr. Longstreth:

"The compact layer of bone is very much thinned, and the cancellous portion is rarefied. The cartilage covering the head of the bone is smooth and shining, but has lost its transparency and become somewhat milky and opaque.

"The medullary tissue presents a yellowish-red color; is much less consistent than normal; in some places shows blood-red points, like hemorrhages; in others, yellow spots, like purulent matter.

"The microscopic examination of the cartilage shows that this tissue was involved in the inflammatory changes. The cartilage cells, in the layer next the bone, show granular and fatty changes, and in places are undergoing multiplication by division; the cells of the superficial layer are much less affected. The hyaline tissue around the cartilage spaces is also granular, and the granular change affects areas concentric with the cells of the tissue. Beyond these areas the cartilage tissue appears normal.

"Both specimens present about the same conditions, but are more advanced in the case of Grace Mitchell."

The following interesting example of immense tumor of the thigh was admitted into the hospital, and operated upon after the above series of amputations had been tabulated:

Osteo-enchondroma: amputation at the hip-joint.—Meyer R., aged 51, married, a native of Germany, was admitted into Dr. Morton's ward May 23, 1879, with a tumor of the left thigh, which extended from the knee to near the groin, which had been growing for ten years. The tumor presented hard, uneven, and in places softer nodules; the superficial veins were much enlarged; the skin greatly discolored. Pain at times had been a prominent symptom. The mass was rather larger at the upper part of the thigh, where it measured twenty-nine inches in circumference. The patient's general health had been fair, but on account of the great weight of the limb he had led a sedentary life, and the frequent accessions of pain had given him many sleepless nights, and he was considerably broken down. He was placed on the best diet, and at nights small doses of morphia were administered to produce sleep.

June 2.—After anæsthesia the abdominal tourniquet was adjusted, and an anterior oval skin-flap was made. The femoral vessels were then secured, and the muscles were divided down to the bone. The posterior skin-flap was then made, and the muscles divided close to the pelvis. The head of the femur was disarticulated, and the vessels secured so rapidly that little or no blood was lost. The stump was not closed until 5 P.M., four hours after the operation, during which time no bleeding had occurred. Carbolized oil upon charpie was applied, and a laudanum dressing to the abdomen. The patient did remarkably well until the 5th, when symptoms of peritonitis developed. On the afternoon of that day the temperature rose to 102½°, and in a few hours to 105½°, with marked tympanitis and pain. Gradually the respirations became labored, and on the morning of the 6th he died.

An examination of the tumor demonstrated that the growth was of the variety known as osteo-enchondroma. The specimen (see figure) shows a tumor growing mostly at the posterior part of the bone, and extending from just above its condyles, and involving its substance above its upper third. The tumor involves the anterior aspect of the thigh about five inches, rather less than one-third its length, and at this part rises above the line of the bone about one and one-half inches. On the posterior surface it is connected with the bone for about ten inches of its length. The breadth of the tumor is greatest at the lower third of the bone, and measures seven inches. The tumor projects six inches posteriorly to the line of the femur, and at this part projects upwards and backwards, and attains in length, measured on a cross-section of the bone, thirteen inches. The surface of the tumor is markedly lobulated but smooth, and covered by a thick fibrous membrane. Its consistence at its lower part is firm and elastic, whilst at its upper part, especially over the large, prominently projecting nodules, it is moderately soft and distinctly fluctuating. The condyloid extremity, and the cartilage covering it, as well as the upper half and head of the bone, present a normal appearance. A section made through the tumor, as well as the bone (antero-posteriorly), shows the interior

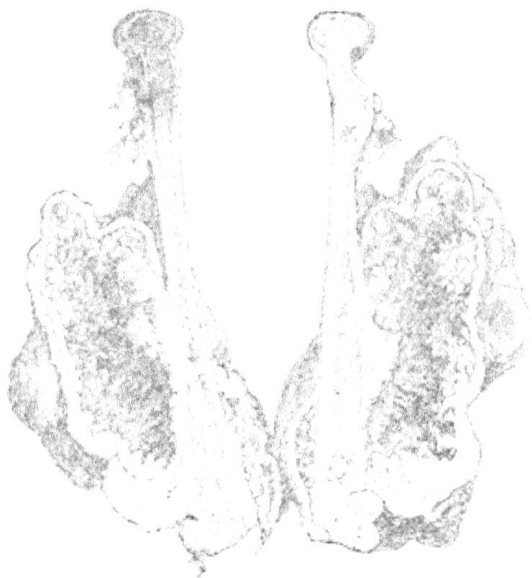

Osteo-enchondroma of the femur. (From a drawing made by Dr. H. M. Wetherill.)

of the tumor to be hollowed out into a large cyst, which is incompletely divided into several smaller cavities. In the fresh condition the fluid contained within the tumor was found to be thick, gluey, whitish, and opaque, but streaked with blood. The lining of the cavity is seen to be rough and shaggy, with numerous small rounded nodules, mostly of a whitish color and translucent, whilst the general surface is of a brownish-red color, evidently colored with blood. The general mass of the tumor, as well as the walls of the cyst, are composed of a clear hyaline, cartilage-looking material, which, at the lower (older) portions, is traversed by whitish-looking narrow bands, having in many places a calcareous consistence. The walls of the cyst (newer tissue) do not show this condition. The medullary cavity of the bone shows its structure quite altered, and in part occupied by small lumps of cartilage-like material. The compact substance of the bone, along its anterior aspect, is unchanged, while at the posterior aspect it has in places been absorbed, or very much rarefied, being replaced by the growth occupying its substance, or causing its atrophy by pressure.

Amputations from 1875 to (May inclusive) 1879.

During this period 108 amputations have been performed on 100 patients. Of this number 17 died. 6 of the patients were females. 59 of the amputations were primary. Of this number, 45 were cured, 13 died, 1 was removed by friends. 11 were secondary amputations. Of this number, 10 were cured, 1 died. 38 were amputations for chronic bone and other diseases. 34 were cured, 3 died, 1 was removed by friends. 34 of the amputations were of the upper extremity; 1 died. 74 of the lower extremities; 13 died.

	Cured.	Died.	Improved.
Of the 2 hip-joint amputations	2	...	...
" 16 thigh "	14	1	1
" 6 knee joint "	3	3	...
" 39 leg "	33	6	...
" 11 foot "	8	3	...
" 5 shoulder-joint amputations	3	2	...
" 13 arm "	10	2	1
" 8 forearm "	8	...	...
" 8 wrist and hand "	8	...	...
108	89	17	2

Hip-joint.—2 chronic, cured.

Thigh.—3 immediate, cured; 1 immediate, died; 1 secondary, cured; 10 chronic, cured (1 improved).

Knee.—2 immediate, cured; 2 immediate, died; 1 secondary, cured; 1 chronic, died.

Leg.—14 immediate, cured; 4 immediate, died; 5 secondary, cured; 14 chronic, cured; 2 chronic, died.

Foot.—5 immediate, cured; 2 immediate, died; 1 secondary, died; 3 chronic, cured.

Shoulder-joint.—3 immediate, cured; 2 immediate, died.

Arm.—6 immediate, cured (1 improved); 2 immediate, died; 1 secondary, cured; 3 chronic, cured.

Forearm.—7 immediate, cured; 1 chronic, cured.

Wrist and Hand.—5 immediate, cured; 2 secondary, cured; 1 chronic, cured.

Tables showing the Results of Amputations from 1830 to 1879.

Amputations.

Males	963
Females	48
	1011

	Cured.	Died.	Total.
Single amputations—males	682	225	907
" " females	30	16	46
Double amputations—males	50	6	56
" " females	2	0	2
	764	247	1011

NOTE.—In cases where double amputations have been performed and the patient has recovered, the result is mentioned as "*two cures*" in the general résumé, whereas when death occurs, it is put down as "*one death.*"

For the amputation tables by Drs. Norris and Morton, from 1830 to 1875, see *American Journal of the Medical Sciences*, volumes 22, 26, 28, 60, and 69; also, *Pennsylvania Hospital Reports*, 1868.

Results of Single Amputations.

	Cured.	Died.	Total.
Hand	58	...	58
Forearm	145	18	163
Arm	124	32	156
Shoulder-joint	21	9	30
Hip-joint	4	4	8
Thigh	89	46	135
Knee-joint	24	14	38
Leg	194	103	297
Foot	53	15	68
	712	241	953

Results of Double Amputations.

	Cured.	Died.
Forearm . . .	2	...
Arm	1	...
Thigh	2	...
Knee-joint . . .	2	2
Leg	14	3
Foot	2	1
	23	6

Results of Amputations, classified according to Ages, showing the Rate of Mortality.

Age.	Males.		Females.		Total.		Age.	Males.		Females.		Total.	
	Cured.	Died.	Cured.	Died.	Cured.	Died.		Cured.	Died.	Cured.	Died.	Cured.	Died.
1½	...	...	...	1	...	1	37	3	2	...	...	3	2
2	...	...	...	...	...	...	38	13†	8	...	...	13	8
3	1	...	1	1	2	1	39	14	1	1	...	15	1
4	3	...	1	...	4	...	40	21	5	1	1	22	6
5	7*	2	...	...	7	2	41	5	3	...	...	5	3
6	6	1	...	...	6	1	42	12	2	...	...	12	2
7	7	...	...	1	7	1	43	8	...	...	...	8	...
8	9	2	...	...	9	2	44	2	1	1	...	3	1
9	14*	1*	1	...	15	1	45	10	4	1	1	11	5
10	21*	2	...	...	20	2	46	8	3	...	...	8	3
11	12*	2	1	...	13	2	47	...	2	1	...	1	2
12	20	3	...	...	20	3	48	6	8	...	...	6	8
13	23	2	2	1	25	3	49	2	6*	...	...	2	6
14	19	5	2	...	21	5	50	7	5*	...	1	7	6
15	27	3	1	...	28	3	51	3	1	3*	...	6	1
16	27*	3	...	...	27	3	52	2	2	...	1	2	3
17	16	4	1	...	17	4	53	1	1	...	...	1	1
18	23	3	1	...	24	3	54	1	2	...	...	1	2
19	26†	10	1	1	27	11	55	2	1	...	...	2	1
20	16*	3	...	...	16	3	56	3	...	...	1	3	1
21	26*	7	2	...	28	7	57	...	2	...	1	...	3
22	31‡	13	...	...	31	13	58	5	...	...	1	5	1
23	24*	10	1	...	25	11	59	...	1	...	...	...	1
24	27*	7	1	...	28	7	60	3	...	1	...	4	...
25	17	8	1	...	18	8	61	1	...	...	...	1	...
26	22	5	...	...	22	5	62	2	1	...	...	2	1
27	16†	4	...	...	16	4	63	1	...	...	...	1	...
28	19	5	...	1	19	6	64	...	1	...	1	...	2
29	13	5	...	...	13	5	65	...	1	...	...	...	1
30	20	8*	...	1	20	9	66	1	...	...	...	1	...
31	11	7*	...	...	11	7	67	1	...	...	...	1	...
32	17*	6	2	1	19	8	68	3*	2	...	...	3	2
33	15	4	1	...	16	4	70	...	1	...	...	...	1
34	9†	9	1	...	10	9	75	1	...	...	...	1	...
35	14	13*	1	...	15	13	81	...	1	...	...	...	1
36	8	3	...	...	8	3							

* One double. † Two double. ‡ Three double.

Table showing the Mortality in Single Amputations according to Ages.

	Cured.	Died.	Total.	Mortality Per Cent.
From 1 to 10 years there were . .	70	11	81	13 +
" 10 " 20 " " " . .	218	40	258	15 +
" 20 " 30 " " " . .	220	75	295	25 +
" 30 " 40 " " " . .	131	62	193	32 +
" 40 " 50 " " " . .	63	36	99	36 +
" 50 " 60 " " " . .	24	14	38	36 +
Upwards of 60 " " " . .	10	8	18	44 +

Table showing Percentage of Mortality in Periods of Five Years, and also the General or Total Average for the Fifty Years.

	Number of Amputations.			Number of Deaths.			Rate of Mortality.
	Male.	Female.	Total.	Male.	Female.	Total.	Per Cent.
From 1830 to 1835........	34	...	34	12	...	12	35
" 1835 " 1840........	45	3	48	10	...	10	20 +
" 1840 " 1845........	49	...	49	13	...	13	26 +
" 1845 " 1850........	67	4	71	13	...	13	18 +
" 1850 " 1855........	112	1	113	26	1	27	23 +
" 1855 " 1860........	111	5	116	32	3	35	30 +
" 1860 " 1865........	155	8	163	33	2	35	21 +
" 1865 " 1870........	149	9	158	38	4	42	26 +
" 1870 " 1875........	146	10	156	36	4	40	25 +
" 1875 " 1879........	98	6	104	15	2	17	17 +

Average rate of mortality—males, 24+ per cent; females, 34+ per cent.

Of the 29 cases of double amputations performed.

In 6 cases, *both feet were amputated* for gangrene following frost-bite.

In 1 case, *amputation of the thigh* was performed for deformity and necrosis. The stump sloughed and the bone protruded after the operation. A reamputation was made.

In 1 case, *amputation of legs* for compound and comminuted fracture of feet and ankles, with fracture of femur.

In 1 case, *amputation of foot* for lacerated wound of foot. Ulcer of stump followed. Amputation of leg was performed.

In 1 case, *amputation of right foot and left leg* for compound fracture of both legs and feet.

In 2 cases, *amputation of foot and leg* for crushed feet.

In 1 case, *amputation of foot and forearm* for crushed foot and hand.

In 1 case, *amputation of knee-joint and leg* for crushed legs. (Compound and comminuted fractures.)

In 1 case, *amputation of foot and leg* for compound comminuted fractures of foot and leg.

In 1 case, *amputation of knee-joint and leg* for crushed leg and ulcer of stump.

In 1 case, *amputation of legs* for compound comminuted fracture of leg and ulcer of stump.

In 1 case, *amputation of leg* for railroad crush, irritable stump, and necrosis.

In 1 case, *amputation of knee-joint and leg* for compound comminuted fracture of leg.

In 1 case, *amputation of forearm* for compound fracture of forearm and necrosis of stump.

In 1 case, *amputation of leg and thigh* for compound crush of foot and necrosis.

In 1 case, *amputation of forearm and arm* for cog-wheel crush of hand, necrosis, and ulcer of stump.

In 1 case, *amputation of leg and thigh* for necrosis of tibia and necrosis of knee-joint.

In 1 case, *amputation of leg and knee-joint* for railroad crush of right and left legs.

In 1 case, *amputation of leg* for crushed foot by elevator and ulcer of stump.

In 1 case, *amputation of leg* for ulcers of stump.

In 3 cases, *amputation of leg* for chronic disease.

Table of Hip-joint Amputations performed at the Pennsylvania Hospital.

No.	Age.	Name.	Date of Operation.	Operator.	Side.	Disease or Injury.	Result.	Remarks, Cause of Death, etc.
1	32	Maurice Heston.	June 13, 1860.	Jos. Pancoast.	R.	Medullary sarcoma.	Cured.	Integumentary flaps, circular division of the muscles.
2	6	Thos. Saunders.	April 20, 1864.	A. Hewson.	R.	Railroad crush, thigh.	Died.	Flap operation. Continued shock.
3	28	Martha J. Reed.	July 26, 1865.	A. Hewson.	R.	Enchondroma.	Died.	Flap operation, shock. No reaction.
4	48	Jas. McGeehan.	April 24, 1866.	D. H. Agnew.	R.	Necrosis of the femur.	Died.	Integumentary flaps. Death from hemorrhage on the twelfth day.
5	21	Ed. D. Ulmer.	Feb. 17, 1868.	T. G. Morton.	L.	Osteo-myelitis; reamputation.	Cured.	Integumentary flaps.
6	19	Jno. Macomber.	May 3, 1872.	A. Hewson.	L.	Reamputation for malignant disease of thigh and stump.	Died.	Flap operation. Exhaustion fourth day.
7	51	Grace Mitchell, colored.	Oct. 20, 1877.	T. G. Morton.	L.	Osteo-myelitis; reamputation.	Cured.	Integumentary flaps.
8	24	B. Dominico.	Dec. 14, 1878.	T. G. Morton.	R.	Osteo-myelitis.	Cured.	Integumentary flaps.
9	58	M. Rosenbaum.	June 2, 1879.	T. G. Morton.	L.	Osteo-enchondroma.	Died.	Integumentary flaps. Symptoms of peritonitis and exhaustion.

THOMAS G. MORTON.

DR. MORTON'S SUCCESSFUL RE-AMPUTATIONS AT THE HIP-JOINT

BURNS AND SCALDS.

Of the 160 cases of burns and scalds reported, 43 were from petroleum, 13 from explosions of gunpowder,—4 of which were from blasting,—28 from grates, stoves, and furnaces, 47 were scalds proper,—*i.e.*, from hot water or coffee, steam, etc.,—26 were from miscellaneous causes, and in 3 cases the cause is not given.

Among the miscellaneous causes, the most peculiar are 1 from lightning-stroke, and 2 from striking parlor-matches whilst the victims were walking across the floor. In these latter cases both of the patients died.

Of the whole number reported 36 died, most of them in a short time after admission.

The old classification of burns is not particularly adhered to in the record, the terms superficial or deep, and a note as to extent, or if important organs are involved, being all-sufficient.

When bad cases of burns are brought into the house, after they are placed in bed, and covered with cotton or some protecting material, the most important thing to attend to is shock.

This is sometimes so profound as to blunt all external sensibility, the patients, whilst perfectly themselves mentally, complaining of very little or no suffering. The writer well remembers a little child who, doomed to death by fire, having its trunk and lower limbs burnt almost to a crisp, after being wrapped in cotton and put to bed, called for its toys, and laughed and played until a short time before its death, which occurred within a few hours after the injury was received.

The treatment of shock, as to internal remedies, does not differ from that in other cases. So far as external appliances and handling are concerned, we are often interfered with by the very nature of the injury.

After the sensibilities are restored through reaction the sufferings are very great, and no class of cases claim the sympathy and skill of the surgeon and attendants more than this.

It is fortunate, both for themselves and in a strictly hospital view, that most of the extremely severe cases die in a comparatively short time after admission. If it were not so, no institution could exist and escape contamination, for a bad burnt case presents nothing more nor less than a mass of dead and decomposing organic matter, which the enfeebled life-forces of the patient are trying to throw off. Separate and isolated wards would be desirable for these cases, but even then it would not do to convert such a ward into a veritable charnel-house by overcrowding it, for all would die. The justification for such wards is the comfort and well-being of other patients, for the writer has no doubt that the chances for recovery of the burnt one himself are better in a mixed general ward.

The remedies for burns, both popular and scientific, are truly legion. One primary object appears to have been aimed at, that of protecting the injured parts from the external air through some non-irritating or non-conducting and emollient substance. The far- and justly-famed carron oil (linseed oil and lime-water), huge poultices of flaxseed or slippery elm, dry cotton, flour, or starch dredged thickly over the parts, the ointment of the oxide of zinc or cosmoline (which last two, it will be observed, are great favorites), lime-water, cold or hot water, bicarbonate of soda, etc., have been and are used by all of us. At later stages anodyne and astringent washes contribute their share in the treatment, and various substances, such, for example, as atropia, are used to meet special indications.

As to the relative merits of the various remedies it is hard to give an opinion. The writer once took a case of extensive burn just within the range of a chance for life, and dressed about equal portions of the surface respectively with carron oil, raw cotton, dredged flour, cold water, and zinc ointment. The race to recovery was about even. The water was the cleanest, and the carron oil seemed to be the most soothing.

He also, in one case, as appears upon the notes, took the liquid from the recent blebs and tested it for acid, thinking that an explanation of the favorable action of alkaline solutions in treatment might be found. The liquid, however, was perfectly neutral, so that the idea of an irritating acid being present was not sustained.

It is after the sloughs have separated that the full extent of the injury is revealed, and if this has a wide surface range there is one very important point in treatment, that is, not to disturb the patient by too frequent dressings, for there is often an absolute repetition of shock at these times. We think, therefore, it is better to risk the odors and emanations, and to wait until the dressings have been softened by the discharges, so that they may be easily removed.

It is unjustifiable to inflict the tortures produced by the attempt to remove adherent dressings from extensive raw surfaces in burnt cases. In some extreme ones it would be justifiable to risk the depressing effects of frequent anæsthesia.

Much may be done, by attention to position and passive movements, to prevent contractions and deformities. These often, however, will take place where there is great loss of substance in spite of all efforts to prevent them. Where the skin alone is lost, and the consequent ulcerated surface has assumed a favorable character, we have an invaluable aid towards reconstruction by the process of skin-grafting. If the injuries are still deeper, and tendons, connective tissue, and muscles are involved, the patients may be subjected to subcutaneous or plastic operations, and to the use of various mechanical appliances. In the trial of these but partial success can be looked for in most cases, and many end in complete disappointment. In the higher orders, neither man nor nature appears to have the power to restore that which has once been completely destroyed.

Where there is sufficient tissue, however, to work upon, much better results than are looked for may reward the efforts of the surgeon. It is right to make the attempt to restore the functions of a joint, for example, even at considerable risk. Case 17 of the table is one strikingly illustrative of this point. Contractions had taken place at the knee to such an extent as to make the limb practically useless, and the patient was admitted for relief from this condition. The parts were so agglutinated that a rupture of the popliteal vessels was feared in any attempt to straighten the limb. The risk, however, was taken. Under ether the limb was forcibly straightened. The tissues ruptured with a tearing noise, and a great gap was left in the flexure of the joint. The vessels stood the strain, and the patient made a complete recovery. After leaving the hospital he persevered in the use of the limb and showed himself some time afterward. The powers of flexion and extension were almost perfect, and the position of the old cicatrices was occupied by soft and pliable tissue.

Burns and Scalds.

No.	Age.	Sex.	Nature.	Cause.	Treatment.	Result.	Remarks.
1	7	M.	Superficial, of hand and abdomen.	Explosion of powder, 4th July.	Ung. zinci oxid.	Cured.	In hospital 11 days.
2	14	M.	Superficial, of face and eyes.	Explosion of powder, 4th July.	Powder removed; sol. atropia to eyes, poultice to face.	Cured.	In hospital 9 days.
3	22	M.	Superficial, of face.	Exposed to very high heat in petroleum-refinery.	Ung. zinci oxid.	Cured.	In hospital 7 days.
4	30	M.	Superficial, of face and right eye; left eye perforated; prolapse of iris; compound fracture of left hand.	Premature explosion while blasting.	Rest; atropia to eyes.	Cured.	(See Fractures.)
5	38	F.	Superficial, of face and breast.	Poured coal oil on kindling and applied a lighted match.	Ung. zinci oxid.	Cured.	In hospital 6 days.
6	20	M.	Deep, of left foot; superficial, of right leg.	From molten iron dropping on him at a foundry.	Cosmoline.	Discharged by request.	In hospital 1 day.
7	18	M.	Superficial, but extensive,—from axilla to pelvis,—of right side.	Lighted candle fell on him while in bed and set fire to shirt.	Cosmoline and ung. zinci oxid.	To out department; relieved.	In hospital 10 days.
8	33	F.	Superficial, of face.	Hot coffee thrown on face.	Cosmoline.	Cured.	In hospital 18 days.
9	...	F.	Slight, of face and right arm.	Overturning of soup-pot.	Ung. zinci oxid.	Cured.	Did not remain in house.
10	55	M.	Superficial, of face and arm; deep, of right foot.	Water, causing an explosion among hot cinders in a puddling furnace.	Ung. zinci oxid., etc.	Greatly improved.	In hospital 27 days.
11	21	M.	Slight, of hands and wrist.	Pushed into an election bonfire.	Cosmoline.	Improved.	To out department.
12	14	M.	Superficial, of right cornea and brows.	From explosion of powder.	Wet dressings and atropia sulph. to eyes.	Cured.	In hospital 7 days.
13	28	F.	Superficial, of chest, back, right side, thigh, and arm.	From a kettle of hot water.	Cosmoline, ung. zinci oxid.	Cured.	In hospital 19 days.
14	69	F.	Fatal, from neck to knees.	Dress took fire from a stove.	Cosmoline.	Died.	Died same day as admitted.
15	31	M.	Deep, of small of back.	Sleeping on top of a brick-kiln.	Cosmoline, poultice, sloughs removed, skin-grafts, ung. zinci oxid., etc.	Cured.	In hospital 80 days.
16	25	F.	Very extensive, but superficial, involving the whole of both legs, the back, buttocks, and arms.	Dropped a coal-oil lamp on a stove and set fire to dress.	Cosmoline, etc.	Died.	In hospital 5 days.
17	35	M.	Partly deep and partly superficial, from ankle to knee of right leg.	Kicked over a coal-oil lamp.	Cosmoline, zinc ointment, poultices, operation, splints, skin-grafts.	Greatly improved when discharged.	In hospital 7 months, 1 week.
18	30	F.	Involves nearly all the body.	From clothes catching fire.	Cosmoline.	Died.	Died same day as admitted.
19	19	M.	Superficial, of arms and face.	Explosion of coal-oil lamp.	Cosmoline.	Improved; eloped.	In hospital 3 days.
20	18	F.	Superficial, of arms, knees, and parts of body.	Overturned coal-oil lamp.	Cosmoline.	Improved.	In hospital 9 days.
21	13	M.	Superficial, of hands and face.	Pouring coal oil into a lamp while burning.	Cosmoline.	Cured.	In hospital 19 days.

22	36	M.	Superficial, of foot.	Hot cinders.	Ung. zinci oxid.	Relieved.	In hospital 2 days.
23	40	M.	Scald, hands and thighs.	From kettle of hot water.	Poultice and ung. zinci oxid.	Relieved; to out department.	In hospital 15 days.
24	22	M.	Superficial, of hands, arms, and face.	From breaking of flask while distilling ether.	Cosmoline and ung. zinci oxid.	Cured.	In hospital 16 days.
25	55	F.	Deep, of three-fourths of surface.	Clothes caught from lighted paper.	Palliative, morphia, etc.	Died.	Died same day as admitted.
26	30	M.	Superficial, of hands and face.	Opening door of furnace.	Cosmoline and ung. zinci oxid.	Cured.	In hospital 20 days.
27	45	F.	Severe scald of lower extremities.	From kettle of boiling water.	Ung. zinci oxid.	Cured.	In hospital 55 days.
28	17	F.	Scald of shoulder and neck.	Spilling of hot tea.	Cosmoline and ung. zinci oxid.	Cured.	In hospital 12 days.
29	22	F.	Deep, burn of back and legs.	From clothes taking fire.	Cosmoline.	....	Was removed by friends, apparently in a dying condition.
30	25	M.	Number of burns, some deep, others superficial.	Buried in ruins at a fire.	Poultice to deep burns, ung. zinci oxid. to superficial.	Cured.	These injuries, combined with a comminuted fracture of right clavicle. In hospital 41 days.
31	17	M.	Deep, of outside of right leg.	Stocking caught fire in a foundry.	Poultice till slough separated, earth-dressing, skin-grafts.	Greatly improved.	Discharged to out department. In hospital 27 days.
32	16	F.	Deep, scalds of arms, neck, and face.	Overturned a pot of boiling water.	Ung. zinci oxid. and varied dressings, skin-grafts, nitrate silver, etc.	Improved.	Discharged for misconduct.
33	27	M.	Deep, burn of trunk and extremities.	Clothes caught from gas-burner.	Cosmoline, poultices, stimulants internally.	Died.	In hospital 8 days.
34	11	F.	Deep, of almost whole body.	Clothes took fire.	Palliative.	Died.	Lived 5 hours after admission.
35	38	F.	Scald of both legs.	From overturning a kettle of boiling water.	Ung. zinci oxid., lime-water, carbolic acid, etc.	Cured.	Also had delirium tremens. In hospital 41 days.
36	29	M.	Of face, with lacerated wound of hand.	From premature explosion of powder while blasting.	Water-dressing to hand, poultices, oxide zinc, nitrate silver, etc.	Cured.	In hospital 65 days.
37	10	F.	From head to foot.	Clothes took fire.	Palliative.	Died.	Lived a few hours.
38	40	F.	Scalds of back.	Overturned kettle of boiling water.	Poultice, ung. zinci oxid.	Cured.	In hospital 20 days.
39	...	M.	Scald of face.	From bottle of coffee being heated, with the cork tight; explosion.	Earth-dressing.	Cured.	In hospital 8 days.
40	30	F.	Deep, of buttocks and thighs.	Thrown against hot furnace by her husband; clothes took fire.	Poultice, ung. zinci oxid., blue wash, etc., skin-grafts.	Cured.	In hospital 5 months, 13 days.
41	24	F.	Deep, of most of body.	Overset a coal-oil lamp.	Palliative.	Died.	Lived 12 hours after admission.
42	...	M.	Of cornea, face, and hands.	Premature explosion of gunpowder while blasting.	Atropia to eyes, slippery-elm poultice, cold-water dressing, ung. zinci oxid.	Improved.	Was removed by request of friends. In hospital 11 days.
43	21	M.	Superficial, of great part of trunk and lower extremities.	From escaping steam in a sugar-refinery.	Earth and cotton.	Cured.	In hospital 21 days.
44	30	F.	Superficial, of left elbow.	Fell against stove.	Ung. zinci oxid.	Cured.	In hospital 16 days.
45	...	M.	Severe, of face, neck, and arms.	From holding a light in a vat of paper-pulp mixed with kerosene (?).	Clay-dressing, etc.	Burns greatly improved.	Patient died of tetanus. (See Tetanus, Case 5.)
46	29	F.	Superficial and deep, of thighs, buttocks, and right side.	Clothes caught fire from range.	Zinc ointment, poultices, etc.	Improved; removed by friends.	In hospital 89 days.

Burns and Scalds.—(*Continued.*)

No.	Age.	Sex.	Nature.	Cause.	Treatment.	Result.	Remarks.
47	36	M.	Deep and superficial, of buttocks and legs.	Clothes caught fire from barrel of oil exploding.	Palliatives, zinc ointment, etc.	Removed by friends.	In hospital 4 days; died at his home.
48	19	M.	Deep, of arm.	Caught between hot rollers.	Wet-earth dressing, carbolized oakum, etc.; amputation on fifth day.	Died.	Died of tetanus 11th day after operation. (See Tetanus.)
49	13	M.	Of face.	Explosion of powder, 4th July.	Wet-earth dressing.	Improved.	Removed by parents. In hospital 3 days.
50	13	M.	Of face and eyes.	Explosion of powder, 4th July.	Eyes syringed, grains of powder removed, wet clay to face, etc.	Improved.	Removed by friends. In hospital 7 days.
51	24	M.	Superficial and deep, of face and arms.	Explosion in oil-works.	Cotton, powdered clay, etc.	Improved.	Discharged by own request. In hospital 8 days.
52	40	M.	Superficial and deep, of face, arms, and shoulders.	Explosion of benzine.	Cosmoline, zinc ointment, etc.	Improved.	Discharged at own request; all healed except inside of arm. In hospital 17 days.
53	6	F.	Superficial, of right shoulder.	Hot water.	Ung. zinci oxid., etc.	Cured.	In hospital 9 days.
54	6	M.	Superficial, of foot.	Hot water.	Laudanum and water.	Cured.	In hospital 9 days.
55	23	M.	Superficial, of back, elbows, and knee.	Escape of steam.	Cosmoline and ung. zinci oxid.	Cured.	In hospital 9 days.
56	50	F.	Superficial and deep, of arms and back.	Hot water.	Simple cerate, tinct. ferri chlorid., gr. xxv. t. d.	Cured.	In hospital 19 days.
57	...	M.	Superficial, of chest and mouth.	From kettle of hot metal.	Ung. zinci oxid.	Cured.	In hospital 8 days.
58	9	M.	Superficial, of side, neck, and shoulders.	Escape of steam.	Cosmoline, morph. sulph., quinine, etc.	Cured.	In hospital 31 days.
59	10	F.	Superficial and deep, of legs and back.	Clothes caught at kitchen fire.	Carron oil, cosmoline, etc.	Improved.	Removed by friends. In hospital 7 days.
60	63	M.	Severe, of arm and hand.	Explosion of powder.	Cosmoline, stimulants, etc.	Cured.	In hospital 41 days.
61	30	M.	Deep and superficial, of face, with lacerated wound.	Burning lamp thrown in his face.	Cosmoline.	Cured.	In hospital 34 days.
62	34	M.	Deep, of both legs.	From kerosene lamp exploding.	Stimulants, poultices, anodynes, etc.	Died.	Died of delirium tremens. In hospital 14 days.
63	12	M.	Deep, of axilla, side, and arm.	Clothes caught from burning candle.	Poultice, carbolic oil, skin-grafts, tonics, etc.	Cured.	In hospital 11 months.
64	4	M.	Fatal, of more than half the body.	Clothes caught fire.	Wrapped in cotton.	Died.	
65	5	M.	Deep, of arm and side.	Clothes caught fire.	Poultice, cosmoline, etc., chloral.	Died.	Died of tetanus. In hospital 14 days. Case 12 of Tetanus.
66	30	M.	Scald of hands.	From escaping steam.	Cosmoline.	Cured.	In hospital 18 days.
67	30	F.	Fatal, of trunk and limbs.	From clothes catching fire.	Cosmoline, palliatives, etc.	Died.	Died same day.
68	40	F.	Scald of face.	From hot lard while cooking.	Cosmoline, etc.	Cured.	In hospital 10 days.
69	37	M.	Superficial, of face and hands.	From burning fat.	Chloral dressing, etc.	Cured.	In hospital 12 days.
70	25	M.	Superficial, of face and hands.	From boiling pitch.	Cosmoline, etc.	Cured.	In hospital 21 days.
71	25	M.	Superficial, of arm.	From burning lard.	Cosmoline.	Cured.	In hospital 9 days.
72	35	M.	Superficial, of face and hands.	From escaping steam.	Salicylic acid and moss-dressing.	Cured.	Patient had erysipelas. In hospital 23 days.
73	7	F.	Fatal, of whole body.	Lit a fire with coal oil; clothes caught.	Palliative.	Died.	Died same day.

74	11	F.	Severe, of abdomen and arms.	A sister of above patient, and burnt at same time.	Salicylic acid and moss dressing, raw cotton, etc.	Nearly well when discharged.	In hospital 4 months, 14 days.
75	26	M.	Fatal scald.	Fell into a vat of boiling water.	Palliative.	Died.	In hospital 1 day.
76	35	M.	Scald of legs.	Fell into vat while boiling soap.	Cotton and salicylic acid, etc., skin-grafts.	Improved.	Ulcers were nearly well when left hospital. In hospital 8 months.
77	16	F.	Severe burns of thighs.	From clothes catching fire.	Salicylic acid, cotton, etc., moss and chloral.	Cured.	In hospital 2 months.
78	24	M.	Superficial, of face, arms, and hands.	From coal oil catching while refining.	Cosmoline, etc.	Cured.	In hospital 14 days.
79	42	F.	Scald of side.	From overturning boiler of hot water.	Cosmoline, etc.	Cured.	In hospital 15 days.
80	45	M.	Superficial, of face.	From explosion of gunpowder while blasting.	Ung. zinci oxid., etc., atropia for eyes.	Cured of burns.	One eye was destroyed. In hospital 21 days.
81	25	M.	Superficial, of trunk and leg.	Struck by lightning.	Cosmoline, etc.	Cured.	Patient was at sea, and standing at foot of mast when this was struck and splintered. The burn extended from axilla to scrotum of one side, thence to opposite side and down to the foot. In hospital 11 days.
82	20	M.	Superficial, of arms and thighs.	Explosion of coal-oil lamp.	Oxide of zinc, etc.	Cured.	In hospital 17 days.
83	21	M.	Superficial, of face and arms.	From fire on canal-boat.	Cosmoline, etc.	Cured.	In hospital 4 days.
84	6	F.	Fatal, of face and trunk.	Clothes caught fire.	Palliative.	Died.	Died same day.
85	13	M.	Deep, of right arm and chest.	From coal-oil lamp.	Carron oil, poultice, opium, skin-grafting, etc.	Much improved.	In hospital 6 months.
86	26	M.	Superficial scald of foot and knee.	From boiling water.	Cosmoline.	Cured.	In hospital 18 days.
87	21	F.	Superficial, of face.	Fell against stove; an epileptic.	Ung. zinci oxid.	Cured.	In hospital 18 days.
88	43	M.	Superficial scald of right leg.	From boiling coffee.	Ung. zinci oxid. and chloral.	Cured.	In hospital 34 days.
89	8	F.	Superficial, of face and arms.	From explosion of coal-oil lamp.	Ung. zinci oxid., tonics, etc.	Cured.	In hospital 28 days.
90	40	M.	Deep, of hands, neck, and left ear.	From explosion of coal-oil lamp.	Ung. zinci oxid., etc.	Improved, discharged by request.	In hospital 6 days.
91	13	F.	Fatal, of face, trunk, and limbs.	From explosion of coal-oil lamp.	Stimulants and palliatives.	Died.	In hospital 4 days.
92	27	F.	Fatal, of trunk and limbs.	From pouring coal oil on fire.	Palliatives.	Died.	In hospital 1 day.
93	22	F.	Superficial, of face and arm.	From pouring coal oil into a lighted lamp.	Carbolated zinc ointment.	Cured.	In hospital 12 days.
94	2	M.	Fatal, of trunk and limbs.	Clothes caught fire.	Palliative.	Died.	In hospital 3 hours.
95	7	F.	Fatal scald of trunk and limbs.	From boiling coffee.	Carbolated zinc and palliatives.	Died.	In hospital 2 days.
96	25	M.	Scald of leg.	From boiling water.	Ung. zinci oxid.	Cured.	In hospital 42 days.
97	22	M.	Slight burn of hand.	....	Ung. zinci oxid.	Cured.	In hospital 3 days.
98	25	M.	Superficial and deep, of legs.	Lit a coal-oil lamp with candle; explosion.	Chloral, zinc ointment, starch bandages, etc.	Improved, to out-department.	In hospital 107 days.
99	22	F.	Superficial, of face and left hand.	Coal-oil lamp explosion.	Zinc ointment.	Improved, to out-department.	In hospital 9 days.
100	65	F.	Fatal, of trunk and limbs.	Clothes caught from stove.	Palliative.	Died.	In hospital 3 hours.
101	54	M.	Superficial and deep, of legs.	From falling of a lighted lamp.	Cosmoline, zinc ointment, blue wash, etc.	Cured.	In hospital 106 days.
102	32	F.	Superficial, of hand.	....	Zinc ointment.	Cured.	In hospital 10 days; was also treated for old ulcer of leg.

Burns and Scalds.—(Continued.)

No.	Age.	Sex.	Nature.	Cause.	Treatment.	Result.	Remarks.
103	9	F.	Superficial and deep, of trunk and limbs.	Clothes caught from stove.	Stimulants, zinc ointment, cosmoline.	Improved, to out department.	In hospital 29 days.
104	5	M.	Deep, of right arm.	Clothes caught from stove.	Stimulants, cosmoline, etc.	Improved, to out department.	In hospital 2 days.
105	14	F.	Fatal, of trunk and limbs.	Clothes caught fire.	Palliative.	Died.	Died same day.
106	10	F.	Fatal, of trunk and limbs.	Clothes caught from stove.	Stimulants, palliative.	Died.	In hospital 2 days.
107	25	M.	Scalds.	Hot liquid spilled over legs.	Salicylic acid, cotton, moss.	Cured.	In hospital 55 days.
108	18	F.	Superficial, of right knee.	Fell while carrying a lighted coal-oil lamp.	Zinc ointment.	Cured.	In hospital 11 days.
109	50	M.	Fatal, of trunk and limbs.	Explosion at gas-works.	Palliative.	Died.	Died same day.
110	23	M.	Scalds of trunk, limbs, and face.	From boiling liquid at chemical-works.	Palliative.	Died.	In hospital 1 day.
111	34	M.	Scald of face and left arm.	From hot water being thrown upon him.	Ung. zinci oxid.	Cured.	In hospital 21 days.
112	53	M.	Scald of eyes.	From hot water thrown on his face.	Atropia and cold water.	Cured.	In hospital 6 days.
113	13	F.	Superficial burns of hands and feet.	From explosion of gunpowder.	Opiates, ung. zinci oxid.	Cured.	In hospital 10 days.
114	34	M.	Deep, of right arm.	From coal-oil lamp exploding.	Quinia and morphia, ung. zinci oxid.	Improved.	In hospital 38 days. Admitted 10 days after the accident.
115	28	M.	Superficial, of buttocks.	Sat down on a barrel of some strong acid at the wharf. Did not sit long!	Chloral dressing.	Cured.	In hospital 10 days.
116	30	M.	Badly scalded in face.	From hot water being thrown on him.	Chloral dressing, morphia, stimulants, etc.	Cured.	In hospital 45 days. Patient became delirious, and during treatment jumped out of window and fractured his skull. He fell twenty feet.
117	45	M.	Superficial, of face, neck, and chest.	Kindling fire with coal oil.	Zinc ointment, etc.	Cured.	In hospital 2 weeks.
118	21	M.	Deep, of both arms and face.	Bursting of a steam paper-dryer.	Morphia, chloral, etc., chloral dressing.	Died.	In hospital 11 days. Died of tetanus.
119	32	M.	Superficial, of leg and foot.	By sulphuric acid.	Oxide of zinc, chloral, etc.	Cured.	In hospital 33 days.
120	1	F.	Superficial, of face and hand.	From explosion of coal-oil lamp.	Oxide of zinc, etc.	Cured.	In hospital 25 days.
121	23	F.	Superficial and deep, of arms and breast.	From explosion of coal-oil lamp. Mother of preceding case.	Zinc ointment, bromide potass., etc.	Cured.	In hospital 35 days.
122	3	M.	Scalds of back and thighs.	Fell into a tub of hot water.	Zinc ointment, opiates, etc.	Cured.	In hospital 42 days.
123	42	M.	Burns of hand and arms, deep.	Clothes took fire from coal oil.	Ointments, chloral, skin-grafts, etc.	Improved.	In hospital 63 days.
124	25	F.	Burns of face and breast, etc.	....	Zinc ointment.	Improved, to out-patient department.	In hospital 7 days.
125	20	F.	Almost whole body.	From clothes catching while asleep.	Palliatives.	Died.	Died in 2 hours from time of accident.
126	23	M.	Scalds of legs.	Fell into pot of hot water.	Zinc ointment, morphia, etc.	Cured.	In hospital 5 months.
127	2½	M.	Burns of body and limbs.	Clothes caught from stove.	Palliative.	Died.	In hospital 1 day.
128	16	F.	Superficial, of legs and right hand.	Clothes caught from upsetting of coal-oil lamp.	Zinc ointment.	Cured.	In hospital 18 days.
129	13	M.	Superficial and deep, of leg, side, and hand.	From discharge of loaded pistol, while in his pocket.	Zinc ointment, skin-grafts, etc.	Improved.	In hospital 73 days.
130*	29	F.	Superficial, of arms and chest.	From explosion of coal-oil lamp.	Zinc ointment, etc.	Cured.	In hospital 14 days.
131	25	F.	Deep, of dorsum of foot.	From hot soapsuds running into slipper, ten days before admission.	Zinc ointment.	Cured.	In hospital 13 days.

132	26	F.	Superficial, of arm.	From upsetting of a kettle of hot water.	Zinc ointment, lime-water, etc.	Cured.	In hospital 48 days.
133	33	M.	Extensive scalds.	From escaping steam in railroad accident.	Palliative.	Died.	In hospital 7 days.
134	3	M.	Scalds of back and buttock.	Fell into tub of hot water.	Zinc ointment, etc.	Cured.	In hospital 20 days.
135	30	M.	Burn of face.	From escaping steam from bursting of steam-chest.	Zinc ointment, atropia to eyes.	Discharged by request.	In hospital 1 day.
136	18	F.	Scalds of both feet.	Upset pot of hot coffee.	Zinc ointment.	Cured.	In hospital 28 days.
137	35	F.	Deep and superficial, of arm, chest, and abdomen.	Explosion of coal-oil lamp, setting fire to clothing.	Zinc ointment, morphia, quinine, etc., astringents to ulcers.	Cured.	In hospital 5 months, 20 days.
138	4	M.	Fatal, of neck, arms, and body.	Clothes caught fire.	Palliative.	Died.*	In hospital 1 day.
139	37	F.	Scalds of foot and leg.	From hot water; slipped while carrying it.	Bicarbonate of soda at first, without any special relief, afterwards zinc ointment.	Cured.	In hospital 14 days. The ointment gave more relief than the soda.
140	56	F.	Fatal, of body and limbs.	Clothes caught from burning wick of candle which she attempted to snuff.	Zinc ointment and palliatives.	Died.	In hospital 4 days.
141	21	F.	Fatal, of body and limbs.	Trod on parlor-match on the floor; clothing caught fire from it.	Restoratives and palliatives.	Died.	In hospital 1 day.
142	24	F.	Fatal, of body and limbs.	Clothes caught from those of 141, whom she was attempting to rescue.	Restoratives and palliatives.	Died.	In hospital 1 day.
143	21	F.	Deep, of hands and back.	Clothes caught from stove.	Zinc ointment, splints, etc.	Improved.	In hospital 41 days.
144	12	M.	Scalds of trunk and legs.	From bucket of hot water being spilled on him.	Cotton and salicylic acid.	Cured.	In hospital 19 days.
145	15	M.	Burns of hands and face.	From coal-oil lamp.	Zinc ointment, etc.	Cured.	In hospital 64 days.
146	5	F.	Burns of face and forearm.	Explosion of gunpowder.	Dry cotton, zinc ointment, etc.	Cured.	In hospital 3 months, 10 days.
147	16	M.	Scald of right thigh.	From hot coffee.	Zinc ointment.	Improved.	In hospital 6 days.
148	48	F.	Deep, of trunk and arms.	Clothes caught from stove.	Dry cotton, salicylic acid, etc., zinc ointment, chloral ointment.	Improved.	In hospital 76 days.
149	30	F.	Scalds of left arm and shoulder.	From boiling water.	Cotton and salicylic acid, skin-grafts to ulcers.	Improved.	In hospital 46 days.
150	29	M.	Superficial burns of forearms and hands.	Was washing paint from hands with benzine near fire; benzine caught.	Dry cotton, zinc ointment.	Improved.	In hospital 1 day.
151	13	F.	Fatal burns.	From upsetting of coal-oil lamp.	Palliatives.	Died.	In hospital 1 day.
152	21	M.	Burns of right arm and hand and left leg.	Explosion on board of vessel.	Zinc ointment.	Cured.	In hospital 20 days.
153	22	M.	Burn of conjunctiva.	From molten iron.	Atropia, mucilage, etc.	Improved, to out-patient department.	In hospital 10 days.
154	14	F.	Fatal burns.	From coal oil.	Palliative.	Died.	Died in 5 hours after admission.
155	31	M.	Scalds of arm, face, and side.	Fell into vat of hot liquorice.	Zinc ointment, morphia, etc.	Cured.	In hospital 37 days.
156	21	F.	Fatal burns.	From lighting fire with coal oil.	Palliative.	Died.	Died same day.
157	18	F.	Scalds.	From hot water.	Zinc ointment, etc.	Improved.	In hospital 18 days.
158	3	M.	Burns of leg, hand, and face.	From upsetting coal-oil lamp.	Zinc ointment.	Cured.	In hospital 37 days.
159	21	F.	Of left leg and arm.	Clothes caught.	Zinc ointment, etc.	Cured.	In hospital 27 days.
160	15 mos.	M.	Of both hands, left forearm, and thigh.	From coal oil.	Zinc ointment, etc.	Cured.	In hospital 27 days. The last three were of the same family, mother and two children.

* In this case, Dr. Hunt, thinking that some explanation might be found of the action of alkalies (such as soda, lime-water, etc.) in relieving pain, tested the serum of the blisters of the burn for acid, but found it absolutely neutral.

William Hunt.

ANEURISMS AND LIGATIONS OF ARTERIES.

ALTHOUGH the total number of the ligations performed in the hospital is comparatively small, yet included in the list will be found some of the most interesting cases on record. In the forty-three ligations, the subclavian artery was tied on two occasions, the axillary once, the brachial four times, the common carotid nine times, the common iliac once, the internal iliac once, the external iliac five times, and the femoral nineteen times. 31 of the ligations were for the relief of aneurismal disease; in 11 the operation was resorted to either for secondary or traumatic hemorrhage; it was performed once for elephantiasis of the lower extremity. Of the entire number of 43 ligations, 6 of the carotid, 1 subclavian, 2 of the external iliac, and 4 femoral operations terminated fatally; congestion of the brain, secondary hemorrhage, and serous effusion in the chest in the former, and peritonitis in the latter, being the immediate cause of death.

Four instances are recorded in which pulsation returned after the ligation:

In Case 17, a carotid ligation, pulsation returned on the twelfth and disappeared on the thirty-sixth day.

In Case 19, an orbital aneurism, the thrill and noise in the patient's head returned, with pulsation of the eyeball, ten hours after the operation. These symptoms continued for several days, and then finally subsided.

In Case 25, an external iliac ligation for aneurism of the profunda femoris, a slight pulsation returned soon after tying the ligature, but disappeared within twenty-four hours.

In Case 27, where the internal iliac was tied for a gluteal aneurism, pulsation returned on the twelfth day; it continued for six weeks, and then ceased to be perceptible.

Suppuration of the Sac occurred in three cases. In Cases 16 and 20—the former an inguinal, the latter a femoral, aneurism—this result followed, but no serious consequences resulted. In Case 21, a subclavian ligation for axillary aneurism, hemorrhage after hemorrhage occurred, with the discharge of the contents of the tumor; the patient finally recovered after ligation of the subscapular artery and the amputation of the arm, which became gangrenous.

Separation of the Ligature.—The earliest period of separation of the carotid ligatures was upon the twelfth, and the latest on the twentieth day. The subclavian ligature came away on the eighteenth day. The iliac ligature upon the common trunk separated on the thirty-fifth day; those on the external iliacs came away, the earliest on the fourteenth and the latest on the thirtieth day. The internal iliac ligature was removed on the twenty-second day. Of the femoral ligations, the ligature came away, the earliest on the eighth, and the latest on the twenty-sixth, day. It is probable that the ligature is often separated at a much earlier period than the surgeon is led to believe, from the disinclination on his part to meddle with it until it presents itself near the surface, being thrown off by the suppurative action.

The common iliac ligation, in 1842, Case 13 in this series of cases, represents the sixth case on record of successful ligation performed upon this great vessel. The external iliac ligation, in 1811 (Case 7), performed by Dr. Dorsey, deserves a special notice from the fact that it was the first time the operation had been attempted in this country, and it was followed by recovery of the patient. The first successful case in England had only occurred five years before, in the hands of Mr. Abernethy, after repeated failures. The ligation of the internal iliac, in 1867 (Case 27), represents the ninth case on record of the operation upon this important artery.

Considering the magnitude of many of the arteries ligated, as well as the very serious character of the cases requiring this operation, the results have been satisfactory. Of the carotid ligations, six of the nine cases terminated fatally; of the external iliacs, three in five died; of the eighteen femoral ligations, four died.

Peculiarities of Operation.—In nearly all the operations the ordinary silk ligature has been used. As early as 1820, Dr. Parrish used a catgut ligature upon the anterior tibial artery. In four of the later cases, however, viz., two of the femoral, one external iliac, one brachial ligation, carbolized catgut was used. Deep abscess and sloughing occurred in one of these femoral ligations, followed by phlebitis and a fatal termination; the external iliac ligation also died; the other two did well.

In two cases the aneurismal sac was punctured by a small trocar, and several feet of horse-hair were introduced; both patients died of sloughing of the sac and hemorrhage. In no case has the injection of coagulating fluids been resorted to, nor has electro-puncture been performed.

In several of the cases of popliteal aneurism, systematic flexion has been attempted, but only in one instance has this means proved effectual. In two instances digital compression was used; one (a popliteal aneurism) resulted in a cure after thirty-six hours; in the other the digital compression was attempted for the cure of an aneurism of the right external iliac. This not proving sufficient, mechanical compression was substituted and found to answer admirably, resulting in a complete cure, after anæsthesia and a compression of five hours and a half.

In 1848 a case of inguinal aneurism was admitted, which was judged to be a fair case for digital compression; for nearly four days continuous pressure was made upon the artery above the tumor by the aid of assistants, and afterwards by instrumental compression, without any benefit. The external iliac was subsequently successfully ligated.

Ligations of Arteries.

CASE 1. *Wound of the brachial artery: ligation: recovery.*—J. S., aged 24, was admitted August 3, 1802, on account of hemorrhage, after an ordinary bleeding operation, for which the basilic vein had been opened, and the artery at the same time was wounded. No union had taken place in the wound: the artery was ligatured with silk above and below the wound; ligatures separated on the fifteenth day.

October 20.—Discharged, cured.

CASE 2. *Traumatic aneurism: brachial ligation: recovery.*—Mary E., aged 25, was admitted August 29, 1802. Had been bled by a farrier in the basilic vein: hemorrhage was profuse: the wound readily healed. A week later an aneurism appeared. Dr. Physick tied the artery above and below the wound with "bobbin." Cured, September 19.

CASE 3. *Popliteal aneurism: left femoral ligation: recovery.*—H. B., aged 35, admitted February 4, 1803. Dr. Physick tied the femoral. The ligature separated on the twenty-eighth day. Cured, March 5, 1803.

CASE 4. *Popliteal aneurism: femoral ligation: recovery.*—H. B., aged 36 (see Case 3), readmitted June 7, 1804, with popliteal aneurism on right side. Dr. Physick tied the femoral. Discharged, cured, July 13.

CASES 5, 6. *Crushed ankle: amputation of the leg: hemorrhage: femoral ligated: recovery.*—J. Z., aged 31, admitted July 31, 1806, with crushed ankle, for which amputation was performed.

August 10.—There was a violent hemorrhage, and again on the 11th and 12th: then an acupressure-needle was used: bleeding occurred.

August 25.—The femoral was tied: hemorrhage recurring two days later, the femoral was ligatured higher up in two places and the artery divided. On October 1 the lower ligature was removed, and on October 4 the second.

May 2, 1807.—Cured.

CASE 7. *Inguinal aneurism: ligature of the external iliac artery: recovery.*—A. P., aged 30, admitted August 14, 1811. The aneurism was just below Poupart's ligament, and measured four by five inches; the skin was thin and discolored. The external iliac artery was tied high up, and on the fourteenth day the ligature separated. Cured, November 16, 1811.

CASE 8. *Traumatic aneurism at the ankle: ligation of the anterior tibial artery: recovery.*—J. T., admitted July 25, 1820. Dr. Parrish tied anterior tibial artery with catgut in two places, and then divided the vessel, August 5, 1820.

October 7.—Cured.

CASE 9. *Popliteal aneurism: ligation of the femoral after the sac was opened: death from hemorrhage.*—G. S., aged 30, admitted September 18, 1820. There was an obscure fluctuation in the tumor, but no pulsation.

September 23.—Dr. Hewson opened the tumor, which was found filled with clotted blood: arterial hemorrhage followed: the femoral was tied in two places: the vessel was divided. Arterial blood still continuing to flow from the sac, it was stuffed with lint. On the 27th the patient died from hemorrhage.

CASE 10. *Varicose aneurism: ligation of the primitive carotid: death.*—R. R., aged 59, admitted April 14, 1836, with a traumatic varicose aneurism.

May 26.—The common carotid ligated. Congestion of the brain proved fatal within twenty-four hours.

CASE 11. *Medullary osteo-sarcoma, involving the head of the tibia: tumor simulating aneurism: femoral ligation: supposed cure: return of the patient thirteen months after operation with a tumor of greatly increased size, occupying same situation: amputation of the thigh: recovery.*—J. R., aged 32, admitted August 10, 1838, for aneurism of the anterior tibial artery. The upper part of left leg was the seat of a pulsating tumor. The case was kept under observation for a month prior to the operation, during which period the tumor became larger and its pulsations stronger, so that on the 22d of September the femoral artery was ligated. The ligature came away on October 9, and on the 31st the note was recorded, "Tumor much less in bulk, and no movement felt."

On September 7, 1839, a year after his dismissal, R. returned with a pulsating tumor occupying the upper part of the leg.

October 16.—Amputation of the thigh was performed.

December 18.—Stump sound.

The tumor was thirteen and a half inches in circumference. The popliteal vessels and nerves lay over the back part of the tumor, but no branch communicated with the interior. There was no deviation from the ordinary size and structure of the popliteal vessels. The bony shell was from four to five lines thick where it joined the shaft of the tibia, and consisted of a globular expansion of the external table of the tibia, while the internal table, the intermediate spongy structure, and the medulla were continuous with the cavity of the tumor, and the matter contained in it, which was composed of a soft, solid, brainlike substance.

Case 12. *Inguinal aneurism; ligation of external iliac artery; recovery.*—J. E., a mariner, aged 28, was admitted July 17, 1841.

July 24.—Ligation of the external iliac artery was performed.

August 24.—Ligature removed.

November 24.—Discharged, cured.

Case 13. *External iliac aneurism; ligation of the common iliac; recovery.*—J. J., aged 39, United States, a laborer, was admitted August 22, 1842. The patient stated that originally he had felt something give way in the right groin whilst lifting a heavy stone. Soon after a hard tumor made its appearance, which became as large as a walnut in the course of a month, and continued to increase until the end of the fourth month. Pulsation began about the fourth week. About the beginning of the fourth month numbness and pain were experienced in the tumor, and extending along the anterior portion of the thigh.

August 29.—The common iliac was ligatured half an inch above the bifurcation. The ligature came away on the thirty-fifth day, and on October 8 patient was discharged, well.

The following year, on the 13th of November, this patient was readmitted for hemorrhage, consequent upon the rupture of an aneurism, which formed rapidly at a point lower down, and entirely beyond the seat of the original disease of the previous year. Hemorrhage terminated the patient's existence November 24, 1843.

Case 14. *Ligation of the common carotid for hemorrhage; death.*—C. La S., a clerk, aged 46, was admitted October 15, 1844, on account of a gunshot wound of and behind the right jaw. Eight days after the injury hemorrhage occurred, and the common carotid was ligated. The ligature separated on the twelfth day, but on the thirty-first day the patient died from recurring hemorrhage.

Case 15. *Popliteal aneurism; ligation of the femoral artery; recovery.*—J. S., a negro, aged 59, was admitted April 9, 1846, with aneurism of the popliteal.

April 22.—Femoral was tied; ligature separated on twenty-sixth day. Discharged, cured, June 6.

Case 16. *Inguinal aneurism; compression; failure; ligation of the external iliac artery; recovery.*—J. S., aged 24, Ireland, admitted August 29, 1848, with aneurism of the right groin. Digital compression was kept up continuously for nearly four days and nights; instrumental compression was then applied, and abandoned only after a faithful trial.

October 11.—The external iliac was tied; the sac suppurated; the ligature separated on the seventeenth day.

December 21.—Discharged, well.

Case 17. *Carotid aneurism; ligation common carotid; recovery.*—F. W., colored, aged 41, was admitted October 12, 1848.

October 21.—Carotid tied below the omo-hyoid.

November 1.—Pulsation observed in the tumor.

November 9.—Ligature separated on the twentieth day.

December 9.—Discharged, cured.

Case 18. *Popliteal aneurism; ligation of femoral artery; recovery.*—G. R., colored, aged 40, admitted December 21, 1852. Ligature separated on the twenty-first day. Discharged, cured, January 5, 1853.

(In August, 1867, this patient was readmitted with inguinal aneurism, occupying the profunda and femoral arteries; the external iliac was tied, but death followed from peritonitis. See Case 25.)

CASE 19. *Aneurism of the ophthalmic artery: ligation of the right common carotid: recovery.*—Mrs. P., aged 36, admitted December 3, 1864. Nine months before, when two months pregnant with her last child, now eight weeks old, she was awakened from sleep by what she described as the "report of a pistol" over her head, but was persuaded that it was only a dream. A slight purring noise in her head subsequently developed. During pregnancy all the symptoms of aneurism of orbit developed. Seven weeks after confinement the eye was very prominent, strabismus, pupil dilated, vision imperfect: the eyeball pulsated strongly: a distinct aneurismal thrill was perceptible all over the head: the bruit was loudest at the right temple: pressure upon the carotid controlled all sound and pulsation.

December 4.—Ligated the common carotid above the omo-hyoid.

December 21.—Ligature withdrawn, being the seventeenth day.

January 30, 1865.—Well.

1879.—Continues well.

CASE 20. *Aneurism: ligation of the femoral artery: recovery after suppuration of the sac.*—J. D., a mariner, aged 26, admitted March 9, 1866, with an aneurism six inches in length, involving the lowest portion of the femoral artery.

March 28.—Femoral artery ligated. The ligature separated on the sixteenth day.

August 18, 1866.—Discharged, well.

CASE 21. *Axillary aneurism: ligation of subclavian between the scaleni: suppuration of the tumor: hemorrhage on the forty-third day: gangrene of the limb: continued hemorrhage: ligation of the subscapular artery in the axilla: amputation of the arm at the upper third: hemorrhage on the sixty-seventh day from the circumflex humeri arteries: removal of humerus at the shoulder-joint: recovery.*—F. McK., aged 51, a stonemason, was admitted November 7, 1866. The tissues covering the tumor were so attenuated that bursting seemed imminent.

November 14.—Ligatured the left subclavian in its second part: separated on the eighteenth day. Suppuration of the tumor followed, with a violent hemorrhage on December 27,—the forty-third day. Between this date and the 8th of January, a period of thirteen days, ten hemorrhages occurred in succession. A ligature was then thrown around the subscapular artery. On the sixty-seventh day sloughing of the tissues about the head of the humerus opened the arteries, and almost fatal hemorrhage occurred. The limb became gangrenous. The desiccated member was subsequently removed by cutting through the dead tissues close to the living, and the bone a little lower down. Subsequently the remaining portion of the humerus, which was destroyed in its entire extent, was excised at the joint. The patient was discharged, cured, March 9, 1867, one hundred and fifteen days after the ligation.

CASE 22. *Femoral aneurism: ligation of the femoral artery: recovery.*—W. M., colored, aged 42, a barber, was admitted July 14, 1866. Femoral ligation was performed. Ligature separated on the twenty-first day. Discharged, well, September 27, 1866.

CASE 23. *Brachial aneurism: ligation of the brachial: recovery.*—1866, male. Aneurism following punctured wound of arm. Cured by ligation of brachial artery.

CASE 24. *Aneurism of innominate artery; ligation of the common carotid: death.*—J. G., a mariner, aged 51, was admitted March 3, 1867. Ligation of the common carotid was performed March 19. The patient died on the twelfth day, from serous effusion in the chest.

CASE 25. *Aneurism of the profunda and femoral arteries appearing fifteen years after a ligation of the femoral for popliteal aneurism: ligation of the external iliac artery: death.*—G. R., colored, aged 55, admitted August 12, 1867, with a strong pulsating tumor at and just below Poupart's ligament. Fifteen years previously he was a patient in the hospital with popliteal aneurism (Case 18), the femoral having then been successfully ligated. Was well until the spring of 1867, when, after a severe strain in lifting a piano, he perceived a tumor, which increased with rapidity. The tumor pulsated, and was about the size of a large orange. It filled up the groin and encroached upon the brim of the pelvis.

August 13.—Ligation of the external iliac was performed. Peritonitis developed the third day after the operation. The case terminated fatally.

Post-mortem.—General peritonitis.

Dissection of thigh.—Corresponding to point of ligation (operation of 1852) the femoral had been entirely obliterated for at least one and a half inches; that portion was converted into a thick fibrous cord; the femoral below this was dilated, thickened, and atheromatous. Above the obliterated portion, the artery for some distance was healthy, and laid upon and formed the back wall of the aneurismal tumor; but higher up, and opposite the profunda, the anterior portion of the femoral was much dilated, and formed a part of the front and inner wall of and communicated with the aneurismal tumor. The dilatation of the profunda extended down that vessel for an inch and a half, involving its entire calibre, and formed the outer and under walls of the sac. The aneurismal disease had been developed at the origin of the profunda, and implicated a part of the main vessel in its anterior wall, the posterior portion being undilated throughout its entire course. The profunda in its entire extent had become aneurismal, the sac being almost filled up with laminated clot. The ligature was applied two inches above the origin of the epigastric, which, as well as the circumflex ilii, was much enlarged. A small clot had formed just above the ligature, and none below it, although above and below a diffused clot, covering and closely adhering to the side of the vessel, was well marked. The inner coat of the artery at the point of ligation was puckered up, and, by projecting into the vessel above and below, produced a teat-like eminence, evidently due to the effect of the tightening of the ligature.

CASE 26. *Arterio-venous aneurism of the thigh following a penetrating wound; ligation of the femoral artery and vein; recovery.*—W. E., aged 21, admitted August 8, 1867. While sitting on his door-step, with the knees well drawn up, he had received a stab from the right side, the knife entering the postero-lateral aspect of the thigh, and taking a course upwards and inwards, emerged an inch below the apex of Scarpa's triangle. Syncope from profuse hemorrhage ensued; the bleeding was arrested by pressure. The anterior and posterior tibials pulsated rather feebly; the limb was warm. On the eighth day there was swelling of the thigh; marked pulsation, with the characteristic thrill of arterio-venous communication, which had developed within the preceding forty-eight hours. The knife had unquestionably passed into the artery and vein.

August 15.—The femoral artery was ligated first, and without any trouble, about half an inch above the original wound made by the knife. The incision was then prolonged until the wound of the artery was reached. This was signalized by a fierce gush of blood, which was arrested by pressure below. The femoral was again ligated; considerable hemorrhage continuing from the femoral vein, it was also tied. The femoral ligature separated on the seventh day; the others were all away by the thirteenth. The patient was discharged, cured, October 1, 1867, fifty-three days after the accident.

CASE 27. *Supposed aneurism of the gluteal region; ligation of the left internal iliac artery; discharged, improved, but died subsequently of malignant disease.*—J. M., aged 24, was admitted September 28, 1867, with a tumor involving the left buttock. The patient was spare in weight, about 135 pounds. His left hip had been painful from rheumatism, as he thought, since the preceding May, so that he had steadily-increasing pain, and was obliged to walk with a cane. In June, 1867, a swelling appeared on the dorsum of the ilium, and lameness and pain became prominent symptoms.

In July he fell and struck his hip. Following this a gradual increase in the tumor was observed. The disease had the appearance of a deep abscess, but distinct pulsation and bruit were detected; the former being quite expansive, the latter could be heard all over the tumor. Pressure diminished very much the bulk of the tumor, which rapidly filled up with the circulation unobstructed. The tumor measured five and a half inches transversely and six and a half in the vertical diameter. A careful examination per rectum gave no evidence of any intra-pelvic tumor, although the bruit was communicated through the abdominal parietes.

October 16.—Internal iliac ligated.

The patient was discharged December 14, fifty-nine days after the ligation. The tumor, which was much reduced in bulk, presented evidences of deep suppuration, and was entirely free from pulsation.

April 8, 1868.—Patient died.

Post-mortem.—A large tumor was found in the left iliac region pushing the bowels and bladder far to the right side. The substance of the growth was composed of callous tissue, from which a cherry-like substance exuded on section. The tumor was made up of cells filled with blood, in size ranging from that of a pea to some an inch in diameter. There was found no trace of any aneurismal tumor. The left ilium was almost wholly absorbed, the remaining portion being completely denuded of periosteum and full of perforations. The entire acetabulum was gone, but, with the exception of a little groove in the head of the femur, that bone was intact, being without periosteum, however, for one-third of its length.

NOTE.—The possibility of this tumor being cancerous was fully discussed prior to the operation. The principal fact in favor of malignancy was the negative one that aneurismal disease in this locality is extremely rare. The bruit and pulsation of the tumor were well marked; the expansion of the swelling with each impulse of the heart, the great diminution of the gluteal enlargement under pressure,—more than would be expected from an encephaloid growth; the normal condition of the abdominal cavity at that time opposed this view.

The cessation of pulsation, the subsidence of the tumor, the relief from pain, and the recurrence and the final disappearance of the pulsation also gave, seemingly, the unequivocal signs of an aneurismal disease.

That the case was not one of simple aneurism is fully conceded, yet it is still probable that a false aneurism existed in the tumor. The fall the patient received had probably ruptured some vessels feeding the growth or in the tumor itself, for the rapid increase at this time, the addition of excessive pain and pulsation, were evidences in favor of this view.

CASE 28. *Pistol-shot wound of the neck, involving the external carotid artery or one of its branches; ligation of the common carotid; recovery.*—A. S., aged 28, was brought in September 4, 1868, with a pistol wound. The ball entered the right side of the neck, posterior to and below the angle of the jaw, and between it and the mastoid process. Large quantities of blood were vomited, showing continued internal hemorrhage. As the patient was evidently sinking from loss of blood, the common carotid was tied in the apex of the superior carotid triangle. A speedy convalescence followed. On the twenty-eighth day the ligature separated. Discharged, well, October 12.

CASE 29. *Railroad crush of arm, with secondary hemorrhage; ligation of the subclavian artery; death.*—W. D., aged 37, was admitted September 21, 1868, with a complete crush of his right arm, involving all the tissues close to the shoulder-joint. The exposed axillary artery was tied. On the seventh day an attack of alarming hemorrhage, controlled only by pressure, reduced the patient almost to a dying condition. Another hemorrhage came on during the afternoon. The subclavian was tied, but reaction did not occur, and the patient died the same evening.

CASE 30. *Incised wound of the arm, dividing the brachial artery, basilic vein, and median nerve; ligation of axillary artery; recovery.*—C. B., aged 66, was admitted November 8, 1868, with a severe wound. The artery, vein, and median nerve were severed. Artery was tied. Discharged, cured.

CASE 31. *Popliteal aneurism; ligation of the femoral artery; recovery.*—P. C., a laborer, aged 25, was admitted on the 31st of May, 1869. The femoral artery was tied. Discharged, cured, July 28.

CASE 32. *Lacerated wound of the face and upper jaw; hemorrhagic diathesis; persistent bleeding; ligation of the common carotid artery; mediate transfusion; death from rupture of the internal jugular vein.*—W. P., aged 27, of a highly hemorrhagic diathesis, was admitted October 15, 1869, on account of a continued severe bleeding from a wound of the upper lip and jaw, caused by falling upon a piece of broken china. The carotid artery was tied October 20. He became comatose, without, however, having lost much blood since the operation. Mediate transfusion of eleven ounces of defibrinated blood was made, in two hours. The patient was semi-conscious, and on the next day was very much better. The patient improved until November 3, when suddenly the internal jugular vein gave way, and death ensued two days later.

CASE 33. *Traumatic aneurism of the femoral artery: ligation above and below the sac; death.*—W. McN., aged 23, was admitted January 8, 1869, having received a stab six weeks previously in femoral artery.

January 9.—The femoral was tied well up in Scarpa's triangle; the tumor was opened, and large masses of clot turned out. A ligature was placed on the vessel below the wound. On the 18th secondary hemorrhage occurred, which was controlled by pins. Another slight hemorrhage, and rigors and exhaustion and death, February 11, 1869.

CASE 34. *Severe lacerated wound of the hand; secondary hemorrhage; ligation of the brachial artery; recovery.*—I. H., aged 26, was admitted November 29, 1872, with deep incised wounds of the left hand, produced by the explosion of a mineral-water bottle, which he was filling. Serious hemorrhage occurred December 11; and, as it was impossible to tie in the wound, the brachial artery was tied in the upper portion of its course. Discharged, cured, December 31, 1872.

CASE 35. *Elephantiasis Arabum of the right lower extremity; ligation of the femoral artery; recovery.*—J. P. T., colored, aged 30, was admitted December 1, 1873. The patient, a farm laborer, had been a slave in Virginia, where he was born, and always resided until after the outbreak of the war, when he came to this city. His father was a cripple from a fall, but was otherwise in good health. His mother was also healthy. He and a younger brother were the only ones out of fifteen children who suffered from enlargement of the limbs. He had never been in Barbadoes. Fourteen years before coming into the hospital he noticed the first symptoms of increasing size of the right leg. At first he had some pain, especially at night, but of late years he had been unable to do active work from inconvenience from the weight of the limb, and from a serous oozing that issued from some abrasions, and kept his foot cold and wet. His general health was excellent, the disease being mainly confined to the right inferior extremity, which was markedly ichthyotic from the middle of the thigh down to the instep. The skin and hypodermic cellular tissue was very much hypertrophied, and hung in large folds over the ankle-joint. Under some portions of the thick scaly epidermis there were effusions of pus. The left leg above the ankle was slightly enlarged, but the surface seemed natural. The skin of the abdomen showed impairment of nutrition and alteration of structure, which he said was the result of a scald when he was small.

December 12.—The femoral artery was tied at the usual place. The temperature of the limb, taken on the eighth day after the operation, was 98° between the toes, and 101½° on the surface of the calf, the thermometer remaining on one hour; the axillary temperature was 99°. The limb was enveloped in a poultice of flaxseed in order to remove the old epidermis. The ligature came away on the twenty-first day, and the small wound remaining speedily closed. The limb was now enveloped in a roller bandage, firmly applied, which was exchanged, on the 7th of January, for one of india-rubber, which was, however, removed at the end of four hours on account of producing numbness. The compression was resumed as a regular portion of the treatment, and the limb gradually and steadily decreased in size. The patient was discharged as improved March 21. There was a marked improvement after the separation of the ligature, but as there was all along firm compression, with the limb in the horizontal position, it was difficult to know how much was due to the operation.

CASE 36. *Popliteal aneurism; ligation of the femoral artery; recovery.*—P. C., aged 30, admitted May 12, 1873. Six weeks before admission he slipped and sprained his left knee. Two weeks afterwards he noticed a small, painful pulsating tumor under the knee. The leg was painful and swollen, measuring fifteen and a half inches in circumference, the sound limb being only thirteen inches. Digital compression of the main artery was kept up for eight hours, and then had to be discontinued.

May 17.—The femoral artery was tied.

May 27.—Ligature separated. Was discharged, well, on the 23d of June.

On May 28, 1876, this patient was seized with dyspnœa of a most severe type. He had been for a long time suffering from what he thought was dyspepsia. He died, June 28, 1876, of hemorrhage. On examination a thoracic aneurism was found, which caused death by opening into the œsophagus.

CASE 37. *Traumatic aneurism following a gunshot wound of the thigh: ligation of the femoral artery: death.*—T. M., aged 40, was admitted May 28, 1874. In attempting to withdraw a loaded pistol from his pocket it exploded, and the ball entered the thigh about one inch to the inner side of the line of the femoral artery, near the middle of Scarpa's triangle. A pulsating tumor was found under the wound, in which a thrill was detected, and no pulsation could be felt in either the anterior or posterior tibial arteries.

June 5.—Oozing of sanious pus: pulsation still well marked. Three days later secondary hemorrhage occurred.

June 10.—The femoral artery was tied.

June 26.—A secondary hemorrhage occurred, which was controlled by an acupuncture-needle. Died June 28.

CASE 38. *Popliteal aneurism: ligation of the femoral artery: death.*—W. P. G., aged 48, a tailor. On admission, October 15, 1874, the aneurismal tumor filled the popliteal space; the leg was much swollen, and its surface œdematous.

On the 21st of October the femoral was tied with carbolized catgut ligature: the ends were cut off close to the vessel. Died November 8.

Post-mortem.—Fetid pus was found surrounding the occluded vessel. The ligature was in position, the knot adherent, but so slightly held that it readily came away. The artery was completely closed, but was impervious for three inches above and five inches below the ligation. The superior profunda artery was larger in size than the femoral. The main vein filled with broken-down lymph, and its walls were thick and roughened from inflammatory deposit.

CASE 39. *Supposed intracranial aneurism: ligation of the common carotid artery: death.*—M. D., female, single, aged 23, was admitted October 13, 1874. The right eye was normal, the left eye presented extreme exophthalmos, the globe was congested, the pupil dilated, the eye motionless and stone-blind. There was complete paralysis also of the third, fourth, and sixth nerves, with partial palsy of the ophthalmic branch of the fifth, and great œdema of the upper eyelid. The disk was regular, the retinal veins and arteries blocked. A loud aneurismal bruit was audible on the left side of the head, particularly in the temporal region and directly through the eyeball. Pressure upon the corresponding carotid entirely controlled the bruit, which was intermittent and synchronous with the cardiac systole.

October 20.—The left common carotid was tied. The bruit ceased at once, and the exophthalmos speedily subsided. In the afternoon of the following day, during an attack of sick stomach, the patient suddenly became very pale and expired.

Autopsy.—The anterior portion of left hemisphere was very much softened at its under surface, with marked evidences of recent inflammatory disease. All the nerves and blood-vessels which entered the sphenoidal fissure were firmly glued together by recent lymph; all the tissues were strongly contracted, especially at the entrance of the left orbital cavity; the nerve-trunks and blood-vessels were so inextricably involved and matted together in a mass that it was found impossible to trace out individual structures, *but there was no trace of any aneurismal disease.* The ophthalmic, cavernous, and circular sinuses were greatly distended with firmly-clotted blood, and all around on the left side masses of fresh plastic deposit were noticed. The left internal carotid was normal. On the right side the venous sinuses were free from clots and in a normal condition, but the internal carotid of this side showed, at the place where this vessel gives off the middle and anterior cerebral vessels, a slight enlargement, which was not of sufficient size to interfere with the blood-current in the circular venous sinus. When the artery was opened the internal coat showed no alteration from the normal artery. The brain was exceedingly pale, and its venous trunks generally engorged.

CASE 40. *Pulsating tumor of left orbit and temporal region; frontal and temporal arteries and the tumor acupressed; ligation of the common carotid artery; death.*—E. F. B., aged 43, a jeweller, of very intemperate habits, was admitted on the 16th of December, 1875, on account of hemorrhage from vessels leading to a pulsating tumor of the left orbit, whose origin dated back twenty-nine years, when he had been struck violently with a snowball. The inner portion of the left orbit was occupied by a pulsating tumor about the size of a pullet's egg. Immediately above this main portion a smaller growth existed, which also pulsated; it was from the latter part that the hemorrhage occurred. On the 19th he had another bleeding; this, however, was promptly treated, and ceased under pressure; but on the 28th he again had a hemorrhage. The patient, on the 5th of January, was much improved, although he continued to lose blood at intervals of a day or two, which was always promptly met with pressure.

January 8.—The patient was etherized, and acupressure-needles were introduced under the frontal and temporal arteries, which were then constricted by ligatures. As this was insufficient to control the pulsation, the common carotid artery was then exposed and tied. At 11.30 P.M. breathing became labored, his skin quite hot, and the pupils markedly contracted; the pulse was 120, respirations 12, and the temperature 103°,—soon after rising to 104½°,—and the breathing became still slower. Nineteen ounces of blood were now taken, with a most salutary effect; the patient revived sufficiently to converse, the pupil became more dilated, and his breathing gradually grew better and fuller; but at half-past two he rapidly became worse, and died of serous effusion the same morning.

Autopsy.—The dura mater was tightly adherent to the bone and markedly congested. The arachnoid was opaque in places, and generally clouded; beneath it was a large collection of clear serum. The sinuses were full of blood, but contained no clots. The vessels of the brain were normal; the enlargement and anastomotic condition of the vessels were principally at the front and inner part of the orbit.

CASE 41. *Femoral ligation for secondary hemorrhage after amputation; recovery.*—Joseph H. W. (Case 36, Amputations) aged 27, was admitted August 10, 1876, for a serious comminuted fracture of both limbs. Amputation of the right limb was performed by Dr. Andrews at the knee-joint, the left at the middle of the leg. On the twentieth day hemorrhage occurred from the knee-joint stump; the popliteal was exposed and religatured; two hours afterwards, bleeding recurring, the femoral was exposed and tied in Scarpa's triangle.

September 15.—The ligature came away. Discharged, cured, December 22.

CASE 42. *Ligation of femoral for supposed popliteal aneurism; recovery.*—J. L., aged 26, a negro, admitted January 22, 1877. Exhibited a large tumor in right ham, said to have gradually developed after a strain four years before, but within two months it had commenced to grow more rapidly, flexing the knee and compelling him to stop work. There was no history of syphilis. The girth of the knee was about nineteen inches; the swelling was seven inches long, and extended more towards the inner aspect of the limb. There was no thrill, pulsation, nor pain in the tumor, and no murmur could be detected. The tibial arteries could not be felt at the ankle. A mitral systolic murmur was heard in the heart. An aspirating needle allowed a stream of blood to flow, but not in jets. No pus. The following day a harsh systolic murmur was heard in the tumor. Without pronouncing positively regarding the character of the tumor (for pressure on the femoral did not control the circulation immediately, although it soon afterwards made the tumor smaller and softer), the femoral was ligated in Scarpa's triangle. Patient did well. A carbolized cat-gut ligature was used and the wound closed. No further trouble was experienced, and the patient was discharged March 25, with only slight swelling remaining and with full use of the limb.

CASE 43. *Ilio-femoral aneurism; introduction of horse-hair; gangrene of sac; ligation of external iliac artery; death.*—G. B., colored, aged 58, from Woodbury, New Jersey, was admitted March 11, 1877. No specific or inherited taint. He first noticed, about nine or ten months before, a swelling in the right groin, which pulsated, and had increased to the size of an orange. It extended from one inch above Poupart's ligament down to the apex of Scarpa's triangle, and from the pubic symphysis to the anterior superior spine of the ilium. Pulsation was detected over an area of seven and a half inches in diameter. The disease evidently implicated the external iliac and femoral arteries just at their junction. Having decided to attempt operative interference, it was decided to introduce a quantity of clean horse-hair into the sac to favor deposition of fibrin and the consolidation of the aneurism. This would avoid the risks involved in tying the external iliac.

March 19.—A delicate trocar was thrust into the aneurism by Dr. Levis, and four long pieces of horse-hair introduced through it, which were left in the sac. Two pieces, however, became entangled and were drawn out with the trocar, leaving in only two pieces measuring sixteen and a half inches each. Three days later some oozing was observed from the puncture, and on the next day there were signs of suppuration, and the patient had a chill.

March 25.—The pulsation was less distinct, but the overlying skin was inflamed and threatened to slough. The danger of rupture of the sac and fatal hemorrhage was so imminent that it was resolved that the external iliac artery should be ligated. Accordingly, on March 25, six days after the introduction of the hair, an incision was made from the crest of the ilium to within half an inch of the pubic symphysis and half an inch above the crural arch, and a carbolized catgut ligature applied to the vessel indicated. Two days later there was vomiting and delirium, and on April 3 there were symptoms of peritonitis, and the aneurismal sac was in a suppurating condition. No autopsy was permitted, but the sac was opened, and the hairs found imbedded in a soft black clot.

CASE 44. *Traumatic aneurism of the palmar arch; ligation and acupressure; recovery.*—E. O. K., aged 24, was admitted November 14, 1873. Three weeks before, while bottling mineral water, an explosion took place, and a fragment of glass entered his left palm. The wound bled freely, but was controlled by pressure. A few days before his admission the wound opened, and several hemorrhages have occurred. The wound pulsated strongly, and the parts were tumid and painful.

November 15.—The wound was well opened, and the bleeding vessels were secured by ligatures and pins. No further trouble occurred.

November 27.—Discharged.

CASE 45. *Incised wound of the forearm, with division of the ulnar and interosseous arteries; ligation; recovery.*—M. B. was admitted May 24, 1873, with a wound of the forearm involving the ulnar and interosseous; free bleeding resulted. An incision was made upon the bleeding vessels, and both were ligated above and below the wounds. Discharged, cured, June 14, 1873.

Aneurisms treated by Compression.—Quite a number of the cases of aneurism have been subjected to compression, but from excessive pain in some cases, and inability to control the circulation as desired in others, this method, after varying periods of trial, was desisted from and deligation of the vessels resorted to. In three instances, however, good results were obtained: in two cases by mechanical and digital compression, and in the other by forced flexion of the limb. The following are the clinical notes of these cases:

CASE 1. *Aneurism of the external iliac; failure of digital compression; mechanical compression successful.*—J. W., aged 45, admitted into Dr. Levis's ward August 28, 1871. Had been accustomed to laborious work. Three months prior to admission had walked fifteen miles. This unusual exertion was followed the next day by pain extending down the thigh, which had remained permanently. He walked with body bending forwards and supported by his hand on his knee. The right leg and foot were œdematous. A pulsating tumor was found in the right inguinal region, which was evidently a fusiform aneurism, involving the external iliac, and extending two or three inches along the course of the femoral. Digital compression of external iliac against the brim of the pelvis controlled the pulsation, but caused intolerable pain, so that it could not be continued. An apparatus was then devised to make mechanical compression of the external iliac near the origin, which, when applied, completely checked the pulsation.

The instrument, as shown in the cut, consisted of a hollow pad for counter-pressure on the buttock, a rigid steel band for partially surrounding the pelvis, with a small convex pad and a screw adjustable in a slot in front of the encircling band.

September 28.—Etherization was continued for five and a half hours with the apparatus in position. When it was removed at the end of this period some slight pulsation returned in the sac, but none was detected in the femoral or in the arteries of the leg. On the seventh day pulsation had ceased, and the tumor was much condensed and shrunken. Phlebitis, with œdema, appeared for a few days succeeding the operation, but subsided under ordinary applications. He was discharged, cured, November 4, 1871. One year later the lump had disappeared, and a hard, cord-like body indicated the position of the external iliac artery. He was then in good health, and employed as a laborer in a foundry.

CASE 2. *Popliteal aneurism cured by flexion.*—W. H., aged 26, admitted into Dr. Morton's ward October 23, 1873. Five weeks before, while working in the coal mines near Hazleton, Pennsylvania, he sprained his left knee by having it caught by a truck. Swelling and pain followed, and he was obliged to leave his work. He walked with difficulty, although using a cane. A dense, pulsating, and movable tumor was found in the popliteal region, about as large as a chicken's egg. Pressure on the tumor or on the femoral artery checked the impulse. No pulsation was found in the tibial arteries, and the temperature of the limb was 4° (F.) less than on the opposite side. The diagnosis of rupture of the popliteal artery, followed by a false aneurism, was made, a clot in the vessel having obstructed the flow of blood into the anterior and posterior tibial arteries. Since pressure controlled the pulsation, the treatment by forced flexion was determined upon. On October 25, a pad of linen was applied to the aneurism, and the leg was firmly flexed upon the thigh, and retained in this position by a bandage attached to a belt which encircled the patient's waist. Two days later the pulsation had greatly diminished; the pressure was lessened. November 18 the apparatus was removed; the tumor was without pulsation and much reduced in size. On the 20th of November the tumor was dense; there was no pulsation, but there was a difference of 4° between the surface of the leg on each side (93° and 97°). The anterior and posterior tibials pulsated faintly on the 28th, and on the 29th of November he was well enough to return home.

In March, 1874, he wrote that "the lump has not altogether left, but I walk very well; when the weather is cold my foot and toes feel very cold, but am well enough to work in the mines."

CASE 3. *Popliteal aneurism; digital compression; recovery.*—Samuel B., aged 35, was admitted May 26, 1879. Has been working in a spinning-mill for sixteen years: using his left knee to push the mule back some five or six times in the minute. His first experience of pain was some six weeks since, and this in his instep. Gradually the limb began to swell, and he only discovered the tumor within a few days before admission. On examination a well-marked popliteal aneurism was found. The left limb measured about one-half to three-quarters more in circumference than the right.

June 1.—Iodide of potassium with tincture of veratrum viride ordered every second hour to prepare him for digital pressure.

June 2.—Bowels were opened by enema, and one grain opium ordered; diet light. Digital pressure was commenced at 6 P.M.

June 3.—Aneurism seems firmer. P.M.—Hypodermics of morphia given several times to relieve pain.

June 4.—Pulsation stopped in the tumor this morning at 6.30 A.M., but pressure was kept up two hours afterwards,—in all thirty-six and a half hours. Limb placed on a splint; iodide of potassium and veratrum discontinued.

June 5.—Doing well; swelling almost left the foot.

June 11.—Splint removed; no pulsations in tibials; tumor firm.

June 14.—Tumor smaller; a slight pulsation in the anterior tibial; foot still remains a dusky color.

June 20.—Discharged. Pulsation in dorsalis pedis.

CASE 4. *Sacculated aneurism of the head of the tibia; forced flexion; compression; threatened gangrene; amputation of the thigh; recovery.*—T. H. F., aged 40 years, was admitted February 6, 1877. He was a sailmaker, and his occupation otherwise compelled him to climb awning-posts. Years ago he received a gunshot wound of the inner side of the right leg, near the middle, and the ball had never been extracted. He walked lame for a long time, though he felt no pain until a week before admission, and he only relinquished work three days ago, from pain in the calf of the limb. The patient being absolutely deaf, it was difficult to obtain an accurate history.

Examination showed a tumor, the size of a small orange, situated posterior to the head of the right tibia, which was the seat of an expansive pulsation, a distinct thrill, and a loud systolic murmur. These signs disappeared under femoral compression. The left leg at the same point was eleven and a half inches, while that of the right was thirteen and a quarter. When the circulation was cut off by pressure, the measurement of the right fell to twelve and a quarter inches. No cardiac murmur was detected, and there was no history of syphilis.

The radial artery did not appear atheromatous, and the dorsal artery of the foot, occasionally, it was thought, could be felt pulsating.

The case seemed to be aneurism of the sacculated variety, and connected with the lower portion of the popliteal or one of the tibial arteries. Treatment by compression appeared suitable, as forced flexion exerted no influence, because the aneurism was below the joint, and opposite the head of the tibia. Dr. Hunt placed a pad near the apex of Scarpa's triangle, which was retained in position for twelve hours. During the first five or six hours the current was almost, if not entirely, arrested; at times no pulsation whatever could be detected in the tumor, then occasionally it was feebly felt. During the last half of the time pulsation, however, was marked.

Partial flexion of the knee was also employed, and the patient kept well narcotized. Subsequently the pulsation seemed diminished, but owing to the inflammation of the skin caused by the instrument no further pressure was applied. Forced flexion was continued without any change taking place. Twenty days later the compressor was again applied to the femoral, and the man enjoined to keep it screwed up as much as he could bear, but to relax the pressure occasionally when the pain could not be endured. This was the treatment for two days, when it was found that the thrill had disappeared, and that the pulsation was so feeble that at times it could no longer be certainly felt. The bruit remained. On the fifth day after the compression, there was swelling of the leg and great pain in the ham. All compression was discontinued, and the limb enveloped in anodyne lotions.

About the same condition lasted for four weeks. The swelling decreased a little, and the pain gradually diminished, while the tumor was the seat of such feeble pulsation that it was impossible to state whether any actually existed.

There seemed every reason to suppose that suppuration was about to occur in the aneurism, which had become filled with clot, but fluctuation did not present itself. At the end of this period blebs appeared on the surface of the leg, and there were developed rather sudden symptoms of incipient gangrene, beginning in the neighborhood of the ankle. Amputation was performed six inches above the knee-joint.

The posterior tibial region was filled with over a quart of soft and laminated black clot, which had pushed aside and forced itself into and around the muscles.

No structure could be identified as a vessel or aneurismal sac. The popliteal artery, as far as the middle of the space, appeared normal, and then passed into the mass of clot.

The head of the tibia and bone just below were somewhat roughened, and some periosteum was removed.

No bullet was found, nor were there any evidences of suppuration. This was evidently an ordinary sacculated aneurism, which had ruptured and become diffuse, this latter taking place when the limb rather suddenly became swollen and painful, during the use by the patient himself of partial compression.

After the amputation the patient did well, and was subsequently discharged, cured, May 18, 1877.

Compression is used in the hospital in all favorable cases, but in several instances we have been compelled to abandon this method of treatment. The period occupied for effecting the cure in the cases of compression has varied from ten to thirty-six hours.

Aneurisms not Amenable to Operative Treatment.

CASE I. *Aneurism of the common carotid.*—Perry G., aged 44, colored, a farmer, was admitted May 27, 1875. Eighteen years ago he had a chancre; two months after had pains all over the body, and iritis; three years after this developed a chancroid. One year ago, after having had pretty good health, was suddenly attacked with severe epigastric pain, which compelled a cessation of work for two weeks. Towards the close of December last he first observed a swelling over the left clavicle, with occasional severe pain. The tumor has steadily increased, giving rise to local pain, as well as severe neuralgia of the head upon the same side as the tumor. Loss of power in left arm has been a prominent symptom for some months, and the radial pulse has been weaker on the affected side. On admission a well-defined tumor was found just above the left clavicle, and occupying the position of the left common carotid. There is strong pulsation and a slight thrill. Auscultation over the tumor reveals a low, soft murmur, with systole of heart. No pulsation or thrill in left axilla; left radial pulse can scarcely be felt; right one normal; left arm weaker than the other. Left forearm measures 9½ inches; left arm, 9½ inches. Right forearm measures 10 inches; right arm, 10 inches. The larynx is pushed half an inch beyond the median line. Breathing is almost entirely diaphragmatic. Impulse of the heart is at upper part of sixth rib, one and a half inches to the inner side of the nipple-line, and is quite extended. Body is shaken with each pulsation. Auscultation reveals a soft, low, blowing murmur, heard at apex, with systole, and converged round towards axilla, and of the same note as that heard over tumor. Pulse 96, rapidly rising under least excitement to 116.

Treatment.—Quinine, five grains; digitalis and iron, thirty drops. After a few weeks twenty grains of iod. potassium three times a day was ordered by Dr. Hunt.

June 11.—Ergotin every second day, hypodermically.

June 17.—Tumor smaller, but more thrill. The induration of the neck has disappeared. Voice more clear, and no pain in the head, as formerly.

June 24.—Since last report has had severe attack of pain in left arm and side of head, and radial pulsation is weaker.

June 26.—Tumor has increased in size, and pain is more constant, and patient is discharged at his request.

CASE 2. *Aneurism of innominate and subclavian; death from rupture of the sac.*—G. F., male, aged 48 years, colored, was admitted into Dr. Morton's ward June 11, 1877. Mother and one brother died of phthisis. He had a history of syphilis, contracted nine years before admission. In October, 1876, he strained his right shoulder while lifting logs, and from that time had suffered from pain in the shoulder and root of the neck. Ten days before applying at the hospital he first observed a swelling just below the right clavicle, and soon after experienced numbness, and impairment of sensation in the arm and forearm, which became lowered in temperature, and was swollen. The tumor presented the appearance of an aneurism; there was marked thrill and pulsation, and over its surface measured five and a half by eight inches. The radial pulse was quite weak on the right side. After admission the tumor progressively increased in size, and on August 11 the pulse at the wrist disappeared; and the patient had been suffering from dyspnœa. From this time, although the aneurism increased much more rapidly, he did not suffer much pain, because it spread above the clavicle, and the pressure was relieved upon the parts below. The tissues now became condensed and thinned over the tumor, and on the 28th of October blood began oozing through the integument; and two days later the skin at that place gave way, and a large stream of blood was projected to a distance of three feet above the patient on the wall. Death ensued almost instantly.

The treatment had been by rest and iodide of potassium, which had been gradually increased to seventy-five grains per diem. Morphia was given to relieve pain, and digitalis and aconite were used to control the action of the heart. By careful attention to diet, and occasional anodynes, the patient was kept comparatively comfortable. No opportunity offered itself for the performance of an operation with any hope of success, as the aneurism probably involved at least a portion of the innominate as well as the right subclavian and the common carotid.

At the autopsy, made by Dr. M. Longstreth, the examination of the heart and lungs and other viscera disclosed nothing abnormal. The ascending aorta, however, was thickened, rigid, and roughened, but there were no calcareous plates and no loss of substance. The innominate presented the same appearances, but at the distance of an inch from its origin the vessel entered a large aneurismal sac, which was composed of the remnant of the innominate and first portion of the subclavian. The tumor had dilated upwards and backwards, pressing the carotid artery against the spine, and entirely destroyed three-fourths of an inch of its calibre; it was imbedded in the aneurismal wall as a thickened ridge. The tumor was pear-shaped, tapering upwards. It was four inches from its beginning in the innominate, five inches wide, and extended three inches in its antero-posterior diameter. The inferior and posterior walls of the sac were thick, and had the same general appearance as the aorta and innominate. The anterior wall was much thinner, except at its point of contact with the sternal end of clavicle, where a mass of firm fibrous tissue was found. In fact, the anterior wall was composed mainly of laminated clot and the skin, although on section a very delicate sac wall could be detected, except at the point of rupture. Near the summit were two very thin spots, and a little below them a ragged opening, from which the hemorrhage took place, and which was filled with recent clot. On the lower outer surface of the aneurism the continuation of the innominate into the subclavian could be distinctly seen as a rounded trunk. The subclavian, after leaving the sac, was about the size of the innominate, and continued that size until its entrance to the smaller sac, which was composed of the lower wall of the subclavian, which had given way at the junction of the second and third portions of the artery for half an inch. This second aneurism was large enough to contain a hen's egg, and extended down behind the first rib to the second. Costal pleura was separated, and the right pleural sac had a low rounded prominence on its anterior surface. The aneurism extended upwards behind the clavicle, and was in contact with the wall of the larger sac over a space as large as a half-dollar. The terminal and mediate portions of the subclavian were normal. The nerve-trunks were found on the posterior surface of the dilatation. The cavity of the smaller aneurism was filled with layers of laminated fibrin and with a soft black clot. The internal mammary artery passed to the inner side of the smaller aneurism and appeared normal. No traces could be found of the other branches of the subclavian. The clavicle had been forced upwards and forwards, and its upper surface was eroded for one and a half inches. The first rib at upper and inner border was roughened, and on its under surface was a semicircular excavation about one-eighth inch in depth, and extending three-quarters of an inch across the bone.

Aneurism treated by introduction of Horse-hairs.

Aneurism of the subclavian artery: introduction of horse-hairs: sloughing of sac: death.—J. M., aged 41 years, was admitted September 23, 1873, with a large aneurism of the right subclavian artery, recognized only about three weeks previously to his applying at the hospital. It had evidently existed much longer; he had pain in the right chest and down the arm for nine months. A pulsating tumor, with distinct aneurismal thrill, above the clavicle, and below that bone another swelling, in which the thrill, however, is almost absent. It was difficult to determine whether this lower tumor, below the clavicle, was a portion of a large aneurismal sac bulging out between the first and second ribs, or a dilatation in the continuity of the vessel as it becomes axillary.

October 13.—To-day nine horse-hairs, having an average length of thirty-three inches, were introduced by Dr. Levis through a capillary-tube into the aneurism, below the clavicle. During, and immediately after, the operation, the pulse at the right wrist became weaker, and the throb and thrill in the tumor became markedly less. Pulse during the rest of the day varied.—73 to 84.

October 14.—Pulse, 80; respiration, 28; temperature, each axilla, 99½°.

October 16.—Tumor seems harder; not so much thrill and throb.

October 24.—Pulse at wrist very weak, but can be counted when in a recumbent position; when up, cannot be felt; tumor still hard, and pulsates slightly, and has slight thrill; pain considerable.

October 26.—Has been spitting blood and bloody mucus since 2 A.M., and has been suffering from very severe pain; no pulse at right wrist. There is a loud stridulous sound heard with expiration at upper part of chest; pressure upon trachea or bronchus. Quinine, digitalis, and opium given.

October 27.—Bleeding ceased.

October 29.—Arm more swollen; blood expectorated.

October 30.—Pulse 88; still expectorating blood, but less in quantity, and black.

October 31.—Arm less swollen, less pain; almost complete dulness of percussion over the upper lobe of right lung.

November 1.—Bloody sputa increased.

November 5.—Considerable amount of black blood thrown up; marked swelling of the axilla six inches below axillary fold.

November 8.—Bright-red blood expectorated; arm more swollen.

November 15.—Severe pain and arm more swollen.

November 16.—Died.

Autopsy.—There was some œdema of the axillary and pectoral regions, as well as of the right arm.

On the removal of the pectoral muscle, the whole region around into the axilla and under the scapula was noticed to be fluctuating.

On opening the thorax, the upper lobe of the right lung was found solid, of a dark-red color. The margins of this lobe, however, were still crepitant, and of their normal color.

The lower lobe was collapsed.

The aneurism commenced in the second portion of the subclavian artery, at an inch and three-quarters from the bifurcation of the innominate. The wall of the vessel had given way upwards and downwards.

The upper dilatation was about an inch and a half in diameter, and reached above the clavicle, partially covered by the fibres of the scalenus anticus muscle. It contained a dense white clot.

The opening on the lower wall of the vessel was oval-shaped, about an inch in its greatest diameter. The margins of this opening were smooth and quite regular.

On dividing the upper lobe of the right lung there was found a mass of black clotted blood behind it and extending beyond the line of the ribs, especially in the axillary region immediately below the clavicle, nearly two inches. In the upper portion of this soft clot a dense white clot was found attached around the lower aneurismal opening in the subclavian.

In the posterior part of the fibrinous clot were found the horse-hairs.

The lung appeared to have been pressed forwards and inwards by the increasing mass of blood behind it. Its vessels and the bronchial tubes had been occluded by its condensation. The lung-tissue in contact with the clotted blood resembled the condition of a lung in hydrothorax.

It was found that the first and second ribs had almost entirely disappeared; from the third to the fifth the ribs had been stripped of their periosteum; on the inner surface of the sixth rib the process of stripping was seen going on, the pleura and periosteum were slit for about an inch, and the blood had forced its way underneath them and almost surrounded the bone for this distance. The third, fourth, and fifth ribs maintained their normal position, but the blood extended through the costal interspaces, so that the clot in the thorax communicated with the fluctuating mass found beneath the pectoral muscles, and the fluctuating tumor in the axilla was covered with the dense fibrous tissues of the pleura, intercostal muscles, and costal periosteum, which had been pressed outwards by the blood accumulating in the thoracic cavity. The intercostal vessels were found lacerated.

An experiment was made to determine whether or not the blood-current in the subclavian communicated with the black clot found in the thoracic cavity, but it was found that water poured into the aneurismal dilatation of the vessel was perfectly retained, none escaping into the thoracic cavity. On very careful dissection, this fact was proved, for it was found that the mass of fibrinous clot in the dilatation varied from a half-inch to two inches in thickness on the circumference.

Diligent search was made for coats of the dilated vessels, but it was impossible to trace them beyond half an inch from the margin of the opening. The cavity in the fibrinous clot was about two inches and a half in diameter, extending downwards and outwards from the opening in the vessel, having the horse-hairs imbedded in its posterior walls. In the fresh condition this cavity was filled with a recent black clot, mingled with a little white fibrin connected with the fibrinous clot in the dilatation above the clavicle, but not attached to the fibrinous wall of the lower dilatation.

The clot in the dilatation above the clavicle extended across the calibre of the vessel, and prevented the flow of blood into the axillary vessel.

It should have been mentioned that the upper lobe of the lung was attached to the thoracic wall at the level of the seventh rib from the spine around to the sternum. In the thoracic aorta, atheromatous disease was very marked; the calibre of that vessel was irregularly dilated, and in parts the coats were quite thin. The other arteries, with the exception of the innominate and right subclavian, were normal. The cords of the brachial plexus of nerves were intimately connected with the upper portions of the tumor, and were pressed against the under-surface of the clavicle.

The subclavian vein was found stretched over the upper dilatation, and its calibre was so encroached upon that the return-blood was markedly obstructed.

The other organs were normal.

Traumatic Arterio-Venous Aneurism.

Traumatic arterio-venous aneurism of the thigh.—J. W. G. B., aged 58, was admitted January 22, 1879, with arterio-venous aneurism of the right thigh, with excessive varicosity of the limb, due to the obstructed circulation. Sixteen years before he had been wounded on the inner side of the right thigh by a bullet from a small pistol, the ball entering the thigh about five inches above the knee, and was subsequently removed from the popliteal region at the head of the fibula. In its course across the thigh it evidently passed through the artery and vein. He at once noticed a peculiar rasping sound, sensation, or thrill, and this has never been absent. Gradually the veins of the limb began to swell, and immense numbers of small superficial vessels appeared, all tortuous, and with distinct thrill and pulsation. The entire limb became greatly swollen, and the skin of the leg over the tibia gradually became ulcerous, and he repeatedly lost large amounts of blood, which was only controlled by pressure. The distention of his limb and the feeling of tension was so great that elevation at first only gave him relief; then pain became a most prominent symptom, and for the relief of this he got in the habit of using opium, until, at this time, he used from fifty to sixty grains a day. On examination of the limb in the upright position, the veins were immensely turgid and tortuous, especially so about the inner part of the thigh, some of these being an inch and a half in breadth. The deep veins were so much enlarged that, at the seat of original injury, the parts felt almost aneurismal. The same note was made of the deep inguinal and femoral veins at the groin. When lying down the circulation was more equalized, and the sense of a distinct tumor very much lessened; all the veins pulsated strongly. There was a very distinct arterio-venous thrill; in fact, the entire limb has a movement communicated to it by the impulse sent through the unnatural opening, while, also, the peculiar purring could easily be heard low down upon the foot as well as upon every part of the limb. The large doses of opium seem to give the patient a measure of relief, but from the continued pain and confirmed habit it seemed unlikely that he could do without his daily dose for any lengthened period.

It was suggested to him that the only operation which offered him any hope of a benefit would be either tying the femoral above and below the opening into the vein, or ligaturing the external iliac artery; that no disturbance of the vein be undertaken, so that, in the closure and subsequent shrinking of the portion of artery between the ligatures, the vein-opening would become necessarily closed. The case was a favorable one for the operation. The organs generally were normal, except that the heart had a slight aortic, systolic murmur. Rest, good diet, and emollients produced a change for the better, and the patient, after a short residency in the hospital, left, at his own request, to go home to recuperate, in order to return for an operation.

Aneurisms, Vascular Tumors, etc.

CASE 1. *Aneurism of the scalp.*—Judson V., aged 22 months, was admitted June 1, 1873, suffering from a congenital aneurism of the scalp, which is supplied by several large vessels, the temporal artery being greatly increased in size. The tumor has lately grown rapidly, and some hemorrhage has occurred. The thrill is very distinct.

June 7.—The tumor was encircled by pins passed deeply, and ligatures thrown around them.

June 11.—Doing well. Child removed from hospital.

June 16.—Returned to have pins removed, in doing which there was so much hemorrhage that an additional pin was inserted.

June 20.—Left temporal acupressed, on account of hemorrhage, at 3 A.M.

June 25.—Pins all removed; wounds healthy. Discharged, cured, June 28, 1873.

CASE 2. *Varix involving the buttock.*—Elizabeth N., aged 9, admitted January 21, 1875. Aneurismal condition on the buttock has existed since birth. During the past six months the tumor has grown rapidly; several hemorrhages have taken place. On examination the parts present a fungoid condylomatous appearance. The mass was excised by Dr. Morton; ligatures and suture-pins. On the second day the pins were removed.

February 6.—Discharged, cured.

CASE 3. *Excision of a large vascular tumor of the face, in a child.*—A. D., a male child, aged 2½ years, was admitted into Dr. Levis's ward January 21, 1875, with an immense tumor, involving the entire cheek and lower lip upon the right side, and projecting into the cavity of the mouth. At his birth it was the size of a hen's egg, but it had now attained a great bulk, forming a pendulous mass, extending backward from the middle of the lip to the ear, and bulging about four inches forwards. The inferior maxilla had become deformed and the teeth displaced by the pressure and dragging of the tumor, which projected so much into the cavity of the mouth that it was almost impossible for the patient to put the teeth of the two jaws in contact, and he had therefore been obliged to subsist almost exclusively on a fluid diet.

The tumor is painless, soft to the touch, its surface purplish in color, and the slightest scratch upon it is followed by copious hemorrhage, from the recurrence of which the child at several times almost perished. The boy's general condition being fair, an attempt to remove the growth was thought advisable. Accordingly, on the 23d of January it was extirpated, partly by the knife and partly by tearing tissues, after the more vascular portions had been secured by pins, passed deeply and surrounded by ligatures. The loss of blood was excessive, although all sources of hemorrhage were at once controlled, but the child rallied fairly. Two days after the operation there was considerable febrile reaction. On the fifth day there were severe muscular spasms, followed by great difficulty in deglutition, tetanic symptoms became marked, and finally stupor set in, and death ensued, on the eighth day.

Table of Ligations of Arteries performed at the Pennsylvania Hospital from 1802 to 1879.

No.	Operator.	Date.	Sex.	Age.	Side.	Artery ligated.	Disease or Injury.	Result.	Cause of Death.	Ligature separated.
1	Coxe.	1802	M.	24	Left.	Brachial.	Traumatic aneurism.	Cured.	....	15th day.
2	Physick.	1802	F.	25	Left.	Brachial.	Traumatic aneurism.	Cured.	....	12th day.
3	Physick.	1803	M.	35	Left.	Femoral.	Popliteal aneurism.	Cured.	....	28th day.
4	Physick.	1804	M.	[illegible]	Right.	Femoral.	Popliteal aneurism.	Cured.		
5	Wistar.	1806	M.	31	....	Femoral.	Hemorrhage after leg amputation.	....	....	18th day.
6	Wistar.	1806	M.	31	....	Femoral, 2d time.	Hemorrhage after femoral ligation.	Cured.	....	10th day.
7	Dorsey.	1811	M.	30	Right.	External iliac.	Inguinal aneurism.	Cured.	....	14th day.
8	Parrish.	1820	M.	...	....	Anterior tibial.	Aneurism.	Cured.		
9	Hewson.	1820	M.	30	....	Femoral.	Popliteal aneurism.	Died.		
10	Randolph.	1836	M.	59	Right.	Common carotid.	Varicose aneurism.	Died.	Congestion of brain day after operation.	
11	Norris.	1838	M.	32	Left.	Femoral.	Aneurismal tumor of leg.	Cured.	....	17th day.
12	Peace.	1841	M.	38	Right.	External iliac.	Inguinal aneurism.	Cured.	....	30th day.
13	Peace.	1842	M.	36	Right.	Common iliac.	Inguinal aneurism.	Cured.	....	35th day.
14	Peace.	1844	M.	40	Right.	Common carotid.	Hemorrhage.	Died.	On 31st day, from hemorrhage.	12th day.
15	Norris.	1845	M.	35	Left.	Femoral.	Popliteal aneurism.	Cured.	....	26th day.
16	Fox.	1848	M.	24	Right.	External iliac.	Inguinal aneurism.	Cured.	....	17th day.
17	Fox.	1848	M.	49	Right.	Common carotid.	Carotid aneurism.	Cured.	....	20th day.
18	Norris.	1852	M.	42	Right.	Femoral.	Popliteal aneurism.	Cured.	....	21st day.
19	Morton.	1864	F.	36	Right.	Common carotid.	Orbital aneurism.	Cured.	....	17th day.
20	Morton.	1865	M.	26	Right.	Femoral.	Femoral aneurism.	Cured.	....	16th day.
21	Morton.	1866	M.	51	Left.	Subclavian.	Axillary aneurism.	Cured.	....	18th day.
22	Hewson.	1866	M.	42	Left.	Femoral.	Femoral aneurism.	Cured.	....	21st day.
23	Morton.	1866	M.	...	....	Brachial.	Punctured wound.	Cured.		
24	Hewson.	1867	M.	51	Right.	Common carotid.	Innominate aneurism.	Died.	On 12th day, from serous effusion of the lungs.	
25	Morton.	1867	M.	57	Right.	External iliac.	Aneurism of profunda and femoral.	Died.	3d day, from peritonitis.	
26	Morton.	1867	M.	21	Right.	Femoral.	Arterio-venous aneurism.	Cured.	....	8th day.
27	Morton.	1867	M.	24	Left.	Internal iliac.	Pulsating tumor of buttock.	Cured.	Afterwards died of malignant disease.	22d day.
28	Hunt.	1868	F.	28	Right.	Common carotid.	Gunshot wound of mouth and palate.	Cured.	....	28th day.
29	Morton.	1868	M.	37	Right.	Subclavian.	Railroad crush.	Died.	Hemorrhage.	
30	Morton.	1868	F.	66	Left.	Axillary.	Wound of brachial.	Cured.	....	8th day.
31	Hewson.	1869	M.	25	....	Femoral.	Popliteal aneurism.	Cured.		
32	Morton.	1869	M.	27	Left.	Common carotid.	Hemorrhage.	Died.	Rupture of jugular vein.	
33	Hunt.	1869	M.	23	....	Femoral.	Traumatic aneurism.	Died.		
34	Levis.	1872	M.	26	Left.	Brachial.	Incised wound and secondary hemorrhage.	Cured.	Secondary hemorrhage.	
35	Morton.	1873	M.	30	Right.	Femoral.	Elephantiasis Arabum.	Cured.	....	21st day.
36	Morton.	1873	M.	30	Left.	Femoral.	Popliteal aneurism.	Cured.	....	10th day.
37	Hewson.	1874	M.	40	Right.	Femoral.	Gunshot wound of femoral artery.	Died.	Gangrene and exhaustion.	
38	Levis.	1874	M.	48	Left.	Femoral.	Popliteal aneurism.	Died.	Phlebitis.	
39	Morton.	1874	F.	23	Left.	Common carotid.	Supposed intracranial aneurism.	Died.	Cerebral anaemia.	
40	Morton.	1876	M.	43	Left.	Common carotid.	Pulsating orbital tumor.	Died.	Serous effusion.	
41	Andrews.	1876	M.	27	Right.	Femoral.	Hemorrhage after amputation.	Cured.	....	15th day.
42	Levis.	1877	M.	26	Left.	Femoral.	Supposed popliteal aneurism.	Cured.		
43	Levis.	1877	M.	58	Right.	External iliac.	Aneurism groin, sloughing after introduction of horse-hairs.	Died.		

THOMAS G. MORTON.

VARICOCELE.

In six cases, admitted for treatment, the disease was in each instance upon the left side. The youngest subject was 19 years of age. the others being respectively 24, 26, 29, and 31. The enlargement of the veins had existed for periods varying from four to seven years. All of the five cases operated upon were thoroughly cured. One was discharged in ten days. The duration of treatment in the others was respectively 18, 20, 28, and 38 days. In two cases silk ligatures were used; one ulcerated through on the twelfth day, the other having produced sufficient inflammation, was cut and removed on the sixth day after the operation. Silver wire was used in one case, and not having cut through, was removed on the eighteenth day. In one instance an acupressure-pin was introduced, and removed on the eighth day. A carbolized catgut ligature was used in one case; the result was very successful.

Operation.—A strong needle armed with silk or wire is carried under the veins well up on the scrotum. It is made to emerge, after engaging all the trunks, at or very near the place of entrance. The two ligature ends are carried through a perforated metal disk, or button, and then tightly compressed. It has not always been found necessary for the ligature to ulcerate its way through, but when sufficient inflammation has occurred the ligature is cut away. The wire ligature usually requires a longer period than the silk to produce the desired effect. This is partly due to the fact that the silk can be drawn more tightly. After the operation a full anodyne is given, preferably by rectum, and if the scrotum be very painful, laudanum and extract of lead is applied. On the second or third day a poultice is a very soothing application. Not unfrequently after this operation there is a considerable amount of pain in the course of the cord, with retention of urine; in such cases flaxseed poultices should be placed on the groin and abdomen, and the use of a catheter is often indispensable.

Varicocele.

Date.	Age.	Side.	Method of Treatment, Operation, etc.	Result.	Duration.
1874. Aug. 22.	24	L.	Silk ligatures carried around the veins, then through a button by a curved needle.	Cured.	18 days.
1876. June 6.	19	L.	Silver wire carried around the veins by a curved needle, then brought out of the same opening, and twisted on a cork over a metal button. Wire removed on eighteenth day.	Cured.	38 days.
1876. June 14.	31	L.	Rest; iodine locally and a supporter. Declined operation.	Improved.	10 days.
1876. Nov. 4.	24	L.	A carbolized catgut ligature was carried around the veins and tightened over a disk.	Cured.	20 days.
1877. May 24.	29	L.	Silk ligature was carried around by curved needle, then through a metal disk. Cut and removed on the sixth day.	Cured.	10 days.
1877. Aug. 20.	26	L.	An acupressure-pin was made to include the veins. A silver wire was then thrown around and tightened. On the eighth day the pin was removed.	Cured.	28 days.

THOMAS G. MORTON.

TRANSFUSION OF BLOOD.

THE operation of transfusion of blood has been performed in the hospital on nine occasions; six of these operations have been made during the past five years. Of the latter, four were for excessive prostration from loss of blood; of these, two were instances of secondary hemorrhage occurring at the end of the first week after amputations; in another, for shock and hemorrhage in a severe railroad crush of the leg and thigh; and once for impending collapse during the removal of a very vascular vaginal tumor, where much blood had been lost before the excision of the growth was attempted. In pure anæmia, transfusion was employed once; the operation was delayed too long, for the patient, at the time of the transfusion, was evidently sinking.

In a case of laudanum-poisoning, transfusion was followed by a marked improvement. It is greatly regretted that the operation was not resorted to much earlier, before the patient's forces were used up by the protracted efforts made to keep up respiration, etc. At 10 A.M. on the morning of November 15 the patient, a man of thirty years of age, was admitted. It was supposed that the laudanum had been taken some time during the preceding night, for he was found insensible at nine the same morning; the pulse was 140 and weak, pupils tightly contracted, body cold and blue, and total unconsciousness. The usual remedies were given, with vigorous external applications, until 6 P.M., when transfusion was performed, previously having drawn a pint of blood from the patient's arm; respiration improved, there were some movements of the arms and legs, the surface of the body assumed a better color; this continued for several hours, when respiration became more labored, and at ten o'clock, four hours after the injection, he died. In a similar case, after all reasonable efforts had been made without any decided improvement, I should advise a free bleeding and then a transfusion, with a repetition of the operation at short intervals if improvement in respiration occurred.

In a paper on transfusion of blood, which was published some time ago, I made the statement that[*] "an examination of the records of transfusion shows that the operation has been more frequently and successfully performed in cases where death has been imminent from simple loss of blood;" and after a careful examination of the subject since then, with an experience of fourteen

[*] *A Series of American Clinical Lectures.* New York: G. P. Putnam's Sons, 1877. Vol. iii, No. 1.

operations which I have performed. I am able to endorse fully these views, and also that "in hemorrhage associated with railroad injury we often have shock, not only as a result of loss of blood, but transmitted or direct shock, from injury of the nerves; and although we may be able partly to replace by transfusion the loss of blood, the injury to the nerve-centres may be irremediable; but as it is not always possible to form a correct estimate of the nerve-lesions where loss of blood has been excessive, transfusion may occasionally be resorted to, but, as might be expected, it usually proves unavailing."

CASE 1. *Compound comminuted fracture of the thigh; great loss of blood; transfusion; death.*—G. W., a policeman, aged 45, was admitted December 23, 1876. While engaged in trying to start a pair of balky horses, fell on the track of a street railway as a car approached, and the wheels passed over his left thigh, above the knee. He had bled profusely previous to admission, and was in profound shock. Pulse, which at first was imperceptible, soon could be counted, but was very thready. Examination of the limb showed a compound comminuted fracture of the lower third of the femur; the wound was four inches long, just above the knee, through which the fragments of the bone could be seen and felt; the thigh above, as far as the juncture of the upper and middle third, was "boggy"; the hemorrhage was from the femoral artery, which was ruptured at the upper part of the wound; there was no hemorrhage after being brought to the hospital. The femoral artery was at once tied in the wound; also hypodermics of whiskey and ammonia and ten minims of Magendie's solution of morphia were given at 12.15 A.M., which was fifteen hours after admission, and about sixteen hours after the injury. No reaction having occurred, transfusion was performed, his daughter, a healthy girl of 22, being the donor; the blood was defibrinated. An injection of four ounces was made into the right radial artery; two minutes and three-quarters were consumed in each injection of two ounces. The patient complained of some pain in his wrist and hand during the operation, but no other symptoms were observed during the injection. The entire operation, including the taking of the blood and its defibrination, etc., occupied about twenty minutes.

1 P.M.—There was no change in patient's condition, though he said, after the operation, that he could see clearer and felt better.

9.45 P.M.—Died.

CASE 2. *Vascular tumor of the vagina; great loss of blood; removal of the growth; profuse bleeding; exhaustion; transfusion; recovery.*—Eleanor P., aged 36, was admitted December 27, 1876. No history of any hereditary disease; naturally healthy; when eighteen years of age menstruated, and has so continued until five months ago; since then irregular; has had four children. Five months ago suffered from hemorrhages, shortly after which a vaginal discharge appeared, which has continued ever since. Four months ago pain came on in lower part of abdomen; has lost flesh, and health has declined. Six days ago was seized with violent hemorrhage from the vagina, losing, she thinks, a gallon of blood. On admission is pale, anæmic, but not emaciated; her appetite is good and bowels constipated. About two inches within the vagina there is a tumor the size of a hen's egg, irregular in shape and moderately hard; there is a fetid, scanty discharge of a bloody, purulent fluid. The mass was diagnosticated as cancerous. Patient was placed at rest, and ordered tonics and nutritious food; the urine showed no sugar nor albumen.

December 26.—After etherization the growth was found to be attached to the vagina, mainly upon its posterior aspect, and also slightly to the posterior lip of the uterus. An incision was carried around its base, and the greater portion of it was readily removed. The mass was soft and fungous. The base was then encircled with the écraseur and removed. Considerable blood was lost, and the patient suffered from extreme prostration. The vagina was packed with lint, dusted over with the dry Monsel's salt of iron. Brandy and nourishment were given at short intervals.

December 27.—Prostration is alarming, although there has been no bleeding since. Pulse at wrist almost imperceptible and irregular. Stomach very irritable, and rejects all food. Great pallor of face, lips, and gums.

5 P.M.—Having exhibited no tendency to react, indeed, having all the symptoms of impending collapse, three ounces of defibrinated blood were injected by Dr. Morton into a vein at the bend of the elbow. The blood was taken from a stout girl of eighteen. Ten minutes after the operation pulse fuller, 140.

5.30 P.M.—Seized with a violent series of rigors; vomited; gasped for breath. External warmth applied; mustard to abdomen, chest, and legs; enemata of brandy and beef-tea; small doses of carbonate of ammonia by mouth; one-quarter of a grain of morphia by the skin.

5.55 P.M.—Quiet.

10 P.M.—Much more comfortable. One drachm of brandy every hour; beef-tea every two hours.

December 28.—Much better. Some color in mucous membranes.

December 29.—Lint removed from vagina; parts syringed with permanganate of potassa. Morphia each day, one-quarter of a grain, by skin.

January 3, 1877.—Has been steadily improving; been taking quinia daily.

February 1.—Up and about the ward.

February 28.—Discharged.

CASE 3. *Profound anæmia; transfusion; death.*—James F., aged 26 years, was admitted December 14, 1876. During the summer he had been unwell, and was treated in the hospital for a general debilitated condition, with loss of flesh, and such excessive weakness as to confine him to bed, having a transient attack of jaundice, with a yellow skin, frontal headache, and attacks of faintness. His bowels were regular, and movements sometimes slate-colored, urine scanty and high-colored, and his feet showed a tendency to swell. In 1870 he had a chancre, but denied constitutional symptoms. Under treatment he improved; but the same general condition returning, he was again admitted, as above stated, when he was found to be rapidly losing flesh, with no appetite, severe headache, and a temperature of 103°. The urine was now amber-colored and acid, with a specific gravity of 1.012, and free from albumen, sugar, and bile. Splenic dulness was not increased, nor was there enlargement of glands; but he had tenderness on pressure in both loins, and the heart had a soft, systolic murmur. There was also a slight congestive dulness at the apex of the left lung.

On the 27th looseness of the bowels had developed, but the general condition was about the same.

On January 4, 1877, an examination of the blood showed extreme deficiency of red blood-corpuscles, but no marked increase of the white. The patient's general condition seemed to have improved. A few days later the breath became fetid, the face assumed a greenish tinge, and a few spots on the chest and face, showing slight bronzing, had developed. The ophthalmoscope simply showed pallor of the disk.

From this time the patient gradually became weaker, with all the signs of pure anæmia, and was transfused by Dr. Morton, at the request of Dr. Da Costa, in whose ward the patient was, on the 10th of January, 1877. Just previous to the operation the patient was delirious, and evidently sinking, with great restlessness, and extremely feeble pulse. The left radial artery was laid bare, and three and one-half ounces of defibrinated blood were injected with no marked effect. An hour after the patient became still more restless, and died within an hour and a half after the operation.

At the autopsy, five and one-half hours after death, the blood was found more fluid than usual, indeed it was actually watery; but neither gross nor microscopical examination of the bones, lungs, heart, liver, spleen, supra-renal capsules, nor kidneys showed anything to account for death, and it cannot be said that the cause of his disease was discovered. More than a week previous to the operation we urged an early transfusion; but, although the patient was carefully watched, there was apparently no sudden declination of the forces until the day the operation was undertaken, when a very rapid failure demonstrated that a fatal issue might occur within a very short time.

CASE 4. *Crush of leg; amputation at knee-joint; secondary hemorrhage; trans-*

fusion: death.—Charles S., aged 15, was admitted March 31, 1877, with a severe crush of the left leg. An hour after being brought to the hospital amputation at the knee-joint was performed, the patella being retained; the condyles of the femur were removed; oval flaps were made: a pin controlled the popliteal artery; the rest of the vessels were subjected to torsion.

April 2.—Some delirium.

April 4.—Stump first dressed. The pin was withdrawn, but, as the artery bled, it was re-introduced.

April 5.—Bleeding occurred. Silver wire was passed under the tissues to control the parts under the pin.

April 7.—Profuse hemorrhage from stump occurred, with great shock. The stump was opened, and the tissues were found to be sloughing. A tourniquet was applied.

April 8.—No reaction occurring, three ounces of defibrinated blood were injected into a vein at the left elbow. No improvement followed. Death took place three hours after operation.

CASE 5. *Compound crush of arm; amputation at shoulder-joint; secondary hemorrhage; transfusion; death.*—Robert McD., aged 35, a factory hand, was admitted May 17, 1878, with a bad crush of the right arm. Amputation at the shoulder-joint was performed. Patient continued to get along well until the morning of the 25th, when, at 3.30 A.M., there was a violent hemorrhage. Respirations were six per minute; no radial pulse; extremities cold. No reaction having taken place after all efforts by internal and external medication, transfusion to the extent of six or seven ounces of blood was made. Two ounces were injected into the median-cephalic. The blood not flowing freely, the canula was then removed, and some additional blood was injected into the left femoral vein. The operation itself occupied about two minutes, during which time artificial respiration was kept up. There was no reaction, and at 4.30 the patient died.

CASE 6. *Opium-poisoning; transfusion; death.*—Joseph G., aged 30, was admitted at 10.10 A.M., November 15, 1878, in an unconscious condition. Respiration stertorous, and three per minute, shortly afterwards six per minute; throat and nostrils filled with mucus; the whole surface of the body blue and cold; pulse 140 and weak; heart very feeble; pupils contracted to pin-points. Said to have taken laudanum some time during last night, and was not discovered until 9 this A.M. Upon admission into hospital an attempt was made to pass stomach-pump, without success. Then gave emetic of sulphate of zinc, twenty grains, which was repeated three times without any effect. About an hour afterwards passed the stomach-tube, and washed the stomach out thoroughly, removing some very black liquid, which smelled very strongly of laudanum, and afterwards injected hot coffee into stomach. He was given one-tenth of a grain of atropia hypodermically, and the pupils soon afterwards were largely dilated, without artificial respiration. By this time the respirations increased to eleven per minute. Was also given hypodermics of digitalis, whiskey, and ammonia, which for the time being increased the strength of the pulse, but without any effect on respiration. Was given an enema of turpentine and two drops of croton oil on tongue without effect. Faradaic electricity was used. Soon after this he had a well-marked opium convulsion, lasting from half to one minute, during which time respirations were suspended. By means of the continuous artificial respiration he was again restored, and continued to two o'clock, when they became very low. He was then bled from temporal artery and left arm, taking some sixteen ounces of blood. This, with cold water to chest, and aided respiration, increased his breathing to eleven again. He continued about the same for several hours, and at six o'clock pulse 130 and weak, heart very feeble, and respirations slow. Dr. Morton transfused five ounces of blood from a strong, healthy man, defibrinated, into the long saphenous vein. There was no marked change on the pulse, but the respirations gradually gained, and for some time he breathed without the aid of artificial breathing. The body and congestion of face have been somewhat relieved; before the transfusion this continued. He now had some slight movements of arms and legs, and the whole surface of body and face assumed a much more natural appearance. Two hours after transfusion he breathed some twelve times a minute, and without aid; 9.15 P.M., with aid, he breathed fifteen to eighteen, but very soon after he died.

Post-mortem by Dr. Chapman. No record obtained.

Operation.—The mediate operation has always been employed at the hospital. In hospital and private I have transfused now in fourteen cases, and have always used the defibrinated human blood, and have had no accident during the operation, and have never had any difficulty in obtaining blood from healthy young subjects, and for this reason have never had occasion to use milk. Indeed, it has always seemed a much easier task to get good healthy blood in an emergency than pure milk, and our experience has been entirely with defibrinated blood. It has been stated that blood subjected to the process of defibrination is unfit for use, but some years ago I submitted to Dr. J. G. Hunt some blood so treated, and requested his opinion on the subject. The doctor wrote me soon after as follows:

"I have examined the blood you sent me, and with some care. The fibrin among the broom-whisks really contains a slightly relatively larger amount of white blood-cells than in the blood which is defibrinated, and this we might expect from the character of the white cell, it being more adhesive than the red cell. Moreover, we know that blood during coagulation, probably by contracting, expels most of the white cells towards the outer parts of the clot, so that we might readily be deceived in an observation such as this. No change *in form* of either white or red corpuscle is apparent in either the defibrinated or in the fibrin blood. In defibrinating blood for transfusion there *is not* any appreciable loss of red corpuscles, neither is there any actual disturbance of the normal relative proportion of white and red, viz., about one white to three hundred and thirty red. In no way whatever is there any evidence that the blood after defibrination suffers any degradation in its nutritive or in its therapeutic properties."

I have consequently continued to use blood deprived of its fibrin, and can point to several cases of my own, now living, where by transfusion life has been saved. One great advantage in using blood so treated is that there need be not the slightest hurry or confusion during the operation; and, again, all the preliminaries are made away from the patient.

The subject of the injection of clots is a very interesting one, and, although I would not desire to transfuse clots, still I believe that the dangers are not so great as have long been supposed; and in sudden emergency I should have no hesitation in injecting pure blood, even if slightly clotted, for the lungs will arrest any such fragments of clot at once, and it is not likely that any harm will arise from their introduction. Some experiments, which I believe were quite original, upon dogs, made a few years ago, in regard to the injections of clot, in order to determine this point, are, I think, interesting enough to warrant the repetition of their details in this place.

I. In a medium-sized mongrel bitch that had been etherized, I opened the femoral vein in the left leg, and drew out about two ounces of blood into a warmed receiver; when this had been exposed about twelve minutes, and clot had formed, it was beaten so as to break up the clot, and one ounce was injected into the femoral vein of the other leg. Both veins were now tied and the skin drawn together by sutures. During the following night some blood was lost, bleeding being finally controlled by pressure, and the next day the animal seemed rather stupid, but took food, and in three or four days seemed as well as ever, save for lameness produced by the wounds. She continued in this good condition for a week, when she was killed by means of chloroform. Dr. Morris Longstreth, pathologist to the Pennsylvania Hospital, has kindly furnished the following report of the post-mortem examination: "Nothing abnormal was found anywhere in the body except in the lungs. The lungs, however, when first removed presented a uniformly pinkish-white color on the surface, and seemed crepitant throughout. There were no pleural adhesions. After the lungs had laid perhaps half an hour on the table, a thorough examination of them was made. The lateral and posterior portions of both lungs about equally presented whitish raised spots, varying in size from a large pea to a grain of wheat. Their color was not pure white, but merely whiter than surrounding lung; they were not projecting or pedicellated portions, but merely made prominent by the partial collapse of surrounding lung. Some few of these whitish places now showed very small red, almost hemorrhagic, centres on the pleural surface. There also now came prominently into sight much more numerous patches having a hemorrhagic appearance. Their size was in general about that of a grain of wheat. These hemorrhagic-looking points were not surrounded by any white area; they had not increased resistance, as determined by pressure of the fingers. On section of the whole lung the appearances just described were found most limited to the surface of the organ, although a very few whitish nodules were present below the surface, but yet lying more peripherally than centrally. The hemorrhagic points, if they existed, were concealed by the blood flowing over the surface of the section. The section of an individual white nodule showed its figure to be nearly spheroidal. In no instance did the figure appear wedge-shaped. The centre of the nodules presented no evidence of softening more marked than at their peripheries. They were by no means hard or fibrous in their consistence, although more resistant than the surrounding air-containing lung. The color was very light pinkish-white. They yielded very little, if any, serum, and nothing could be scraped from the surface of the section by the scalpel. Their immediate surrounding showed no hyperæmic (reactive inflammation) zone.

"The larger vessels of the lung contained no clot. The cavities of the heart contained no fibrous clots. The vein of hind leg was apparently obliterated at the seat of operation; above this point the vessel as well as the ascending vena cava were normal. . . . No microscopic examination was carried out."

It is plain that exactly what was expected had here occurred. The clots thrown in had gone from the heart to the lungs, and been there as it were filtered out, and the question now arose whether these small pieces would excite abscess or be themselves absorbed.

II. To determine this another dog that had been injected with one ounce of blood (which blood had been exposed for over fifteen minutes, and was so full of clots that the canula was once or twice choked by them during the operation) was killed at the end of two weeks. Dr. Longstreth says, "Nothing abnormal was found except in the lungs. On removal of lungs their surface presented whitish spots having irregular figures, and varying in size very greatly, none larger than a two-cent piece. These spots never became prominent above the surface, as in the former case. There were seen no distinctly red areas. On section of the lung, which was crepitant throughout, it was found that the whitish areas were limited to the surface of the lung, extending below the pleural surface very little more than a line or two in depth. As the knife cut through them it seemed to meet with slightly greater resistance, indicating increased firmness at these whitish spots. The immediately adjacent lung showed no hyperæmic zone, and the pleural surface was normal. The base of the right lower lobe of lung was distinctly more congested than that of the left side, and this appearance became more marked after perhaps half an hour than at time of removal; otherwise the lung at this part was normal. The whitish spots were not more numerous, and on section the cut surface was not granular, nor was the consistence of the tissue lessened in this lung. No clots were found in the heart cavities." It is evident that the lungs here were regaining their normal condition, and in this case at least no bad results would have followed the injection of clot. To make the matter more certain, however, I made one more experiment.

III. I injected a small and active mongrel dog with one ounce of clotted human blood, which had been exposed for about two and one-half hours. There were some symptoms of constitutional disturbance, but when, at the end of three weeks, the animal was killed he seemed perfectly well, and Dr. Longstreth reported that nothing abnormal was discovered in the lungs or in any other part of the body.

These experiments would seem to show that the injection of clot is not nearly so dangerous as it has been supposed, and more especially when it is remembered that under no circumstances would a patient receive as much clot as was injected into the veins of the subjects of these experiments.

The injection of air has also been looked upon as a most serious accident, but I am quite convinced that there is but little danger from this source provided only a small quantity be introduced. To test this I have repeatedly injected air, also, in a number of dogs, without any bad or unfavorable results.

Instrument for the Operation of Transfusion.—The instrument which we employ in the hospital, and which still appears well suited to the purpose when defibrinated blood is to be used, consists of a blood-receiver, syringe,

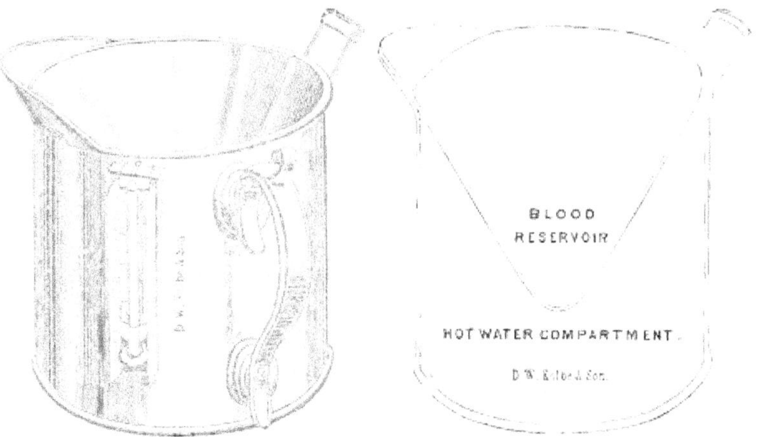

and canula. The receiver, a cylindrical vessel, made of German silver, is six inches in diameter and six inches in height, having at its upper edge a short metal tube closed by a screw top, and one-half inch in diameter, through which warm water is introduced into the lower part of the vessel. On the outside is a thermometer, the curved bulb of which lies in the chamber. The cover or top of this receiver is an inverted cone, gilded on the exposed surface, and dipping down into it so as to be surrounded by the hot water. The syringe is of glass, five inches long by one in width, metal bound, with hard rubber at each end, and holding two ounces; a tapering rubber nozzle, one and one-

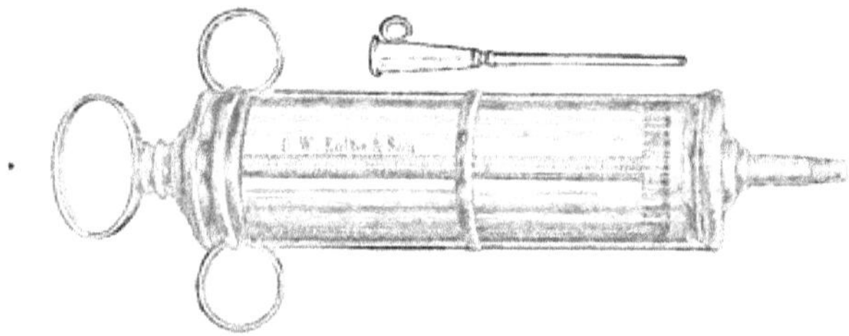

half inches long, projecting from one end, while at the other is a ring at the end of the piston-rod. Besides these there is a steel canula, two inches long and one-sixteenth of an inch in diameter, to which is attached a conical metal neck, into which the end of the syringe fits. The canula is fitted with a trocar. There should also be at hand, a tumbler, or bowl, surrounded by hot water, a fine piece of linen for straining the blood, and a half-dozen little bundles of broom-straw, which are easily made from a new clothes-whisk or broom. Lancets, forceps, and whatever is necessary for the opening of a vein will also be required.

The method of operating is as follows: The giver of the blood (a vigorous man should be chosen if possible) is bled in the ordinary way; that is, a bandage is bound firmly above the elbow, the hand is made to grasp some hard substance, and a convenient vein is opened near the bend of the elbow. The blood is drawn into the receiver, previously filled in its lower receptacle with warm water (temp. 110°), and lightly whipped with the broom-whisks, fresh ones being substituted when the first becomes coated with clot. This operation will take two or three minutes. The blood is then carefully strained through the linen into the tumbler surrounded by warm water, and the receiver being washed the defibrinized blood is replaced. While all this has been the work of an assistant the surgeon has exposed a vein in one arm of the patient, tied a ligature firmly below, and placed a knot loosely above the point where the opening is to be made. The vein is now lifted and nicked, and the canula inserted, and the tube allowed to fill with blood.

The syringe, into which blood has been drawn from the "receiver," is now fitted into the end of the canula and held with its end rather downward, so that any air may rise to the handle end. It should not be completely filled, so that the piston when all is ready, being drawn back a little, may extract any remnant of air from the tube. The injection should be made very slowly, and repeated as often as necessary.

If the pure blood is used, Aveling's apparatus is the best. * "This instru-

* *Morton on Transfusion of Blood; A Series of American Clinical Lectures.* New York: G. P. Putnam's Sons, 1877.

ment consists of an india-rubber bulb, without valves, holding about two drachms, to each end of which an india-rubber pipe of six or seven inches long is fastened; at the outer end of each pipe is a metallic stop-cock. Besides these there are two small silver tubes, one round and the other bevel-pointed, to enter the veins, and a scalpel and pair of fine forceps will also be necessary. The operation is performed as follows: The bulb, with its attached tubes, is placed in a basin of warm water, the cocks opened, and the tubes and bulb entirely filled with water, which can be easily done by alternately pressing and relaxing the grasp until it is clear that all air has been driven out. The two silver tubes are also put into the basin and filled with water. The patient

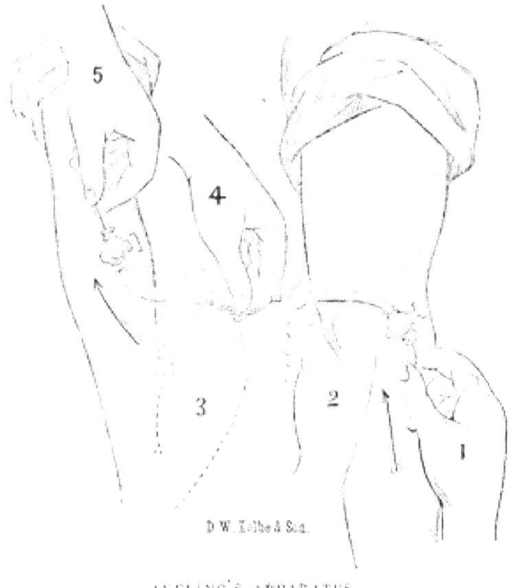

AVELING'S APPARATUS.

1. Assistant's hand holding tube in donor's arm. 2. Surgeon's thumb and forefinger acting as valve to prevent return of blood. 3. Same, while blood is being drawn into bulb. 4. Surgeon's hand alternately compressing and relaxing bulb. 5. Assistant's hand holding tube in receiver's arm.

is now brought to the edge of the bed, and a small fold of skin over one of the veins at the bend of the elbow is raised, transfixed, and divided. This brings the vein into view. Meanwhile, an assistant has lifted the bevel-pointed silver tube from the basin, carefully holding a finger over each end so as to retain the water. The surgeon now seizes the vein with the forceps and nicks it; the assistant, still holding his finger over the larger end, quickly introduces the tube into the upper portion of the vein, and, retaining his finger, closes the point of entrance into the vein with the fingers of his other hand, so as to prevent all escape of blood. If the donor's arm has not already been prepared, which is better, by another assistant, it now goes through the same operation; but here not quite so much care need be used, as a slight loss of blood can be of small consequence. The tubes having been adjusted, the ends of the rubber pipes are now fitted into the ends of their respective tubes, and the stop-cocks, which have been turned so as to prevent the escape of water, are opened. A connection is thus established between the giver and the receiver. The surgeon now presses the pipe on the side towards the giver of the blood. The bulb is then slowly squeezed, and two drachms of water thrown into the vein of the patient. While the bulb is firmly squeezed the surgeon's fingers leave the pipe on the side of the giver, and the pipe on the side of the receiver is compressed, by which a return of the water is prevented, and, the bulb being allowed to expand, two drachms of blood are drawn into it, when the previous operation is repeated, an injection of two drachms of blood being made, and this process is repeated as often as may be deemed necessary."

The amount of blood used must depend upon circumstances. The successful injections have ranged from over a pint to one or two drachms. A very small quantity of blood has often a wonderfully tonic and invigorating effect. Where the body is wasted by disease, and the blood-supply has long been bad, great care and a small amount must be used. Convulsions and other complications which have occurred during the operation are due to the overpowering effect of a quantity of blood suddenly driven in on a weakened heart, and depressed nervous centres. Should there be any subsequent tendency to a nervous chill, a hypodermic injection of morphia or Hoffmann's anodyne may be at once given, and warm bottles placed to the feet.

Transfusion.

No.	Date of Admission.	Sex.	Age.							Result.
1	Dec. 23, 1876.	M.	45	Hemorrhage after railroad crush.	Mediate.	Defibrinated.	4 oz.	R. radial artery.	Reaction.	Died.
2	Dec. 27, 1876.	F.	36	Secondary hemorrhage from vaginal tumor.	Mediate.	Defibrinated.	3 oz.	M. cephalic.	Reaction.	Recovered.
3	Jan. 10, 1877.	M.	26	Anæmia.	Mediate.	Defibrinated.	3¼ oz.	R. radial artery.	No reaction.	Died.
4	April 7, 1877.	M.	15	Secondary hemorrhage on seventh day after amputation for railroad crush.	Mediate.	Defibrinated.	3 oz.	M. cephalic.	No reaction.	Died.
5	May 17, 1878.	M.	35	Secondary hemorrhage on eighth day after shoulder-joint amputation.	Mediate.	Defibrinated.	2 oz.	Femoral vein.	No reaction.	Died.
6	Nov. 15, 1878.	M.	30	Opium-poisoning.	Mediate.	Defibrinated.	8 oz.	Saphena vein.	Reaction.	Died.

THOMAS G. MORTON.

HYDROCELE.

Within the last five years, 33 cases of hydrocele have been under treatment. Of this number, 9 were simply tapped, and 24 were subjected to the radical treatment.

In seven cases the injection consisted of carbolic acid diluted with glycerine, in proportion of one part of the acid to two or three parts of glycerine.

Tincture of iodine was used in seven cases. In two of these the inflammation excited was not sufficient to entirely close the sac, and subsequently a silk ligature was successfully used.

In ten cases the seton was applied. All made excellent recoveries. In four of these abscesses more or less marked followed, without, however, any serious consequences.

Goulard's solution of lead was applied over the scrotum, with advantage in one case, where a slight re-accumulation of fluid had occurred after the use of the seton.

The instruments used in many of these tappings are shown in the figures.

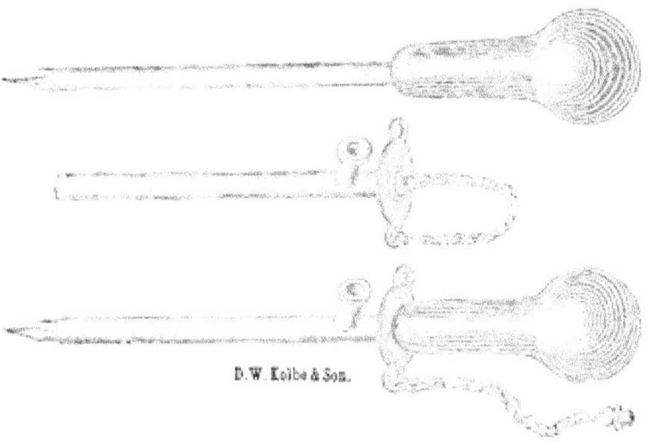

The silver canula has on its under surface an outlet near the hilt, to which a rubber tube is attached, by which, after the trocar has been withdrawn, and the plug, which hangs by a chain, is introduced, the fluid is conducted into a vessel. When an injection is used, a small glass syringe, with a gutta-percha tube, connects with the canula. In using Lugol's solution of iodine it is sometimes diluted with one part of water. The length of time necessary to produce sufficient inflammation with the seton varies according to the condition of the patient. In most cases, twenty-four hours is sufficient, but if at the end of that time the scrotum be not painful nor swollen, a longer time may be required. Should abscess develop, an incision may be made, and the parts dressed with carbolized oil; a tent should also be inserted.

Table showing the Results of the Cases operated on by the Various Methods, with the Average Number of Days under Treatment.

	Cases.	Days.
Injections of carbolic acid	7	20
Injections of iodine	5	23
Silk seton (two after iodine)	10	40
Silver seton	1	12
Excision of a portion of the tunica vaginalis testis, after seton	1	28
Simple tapping	9	5

In all the cases of hydrocele, with one exception, the fluid removed was like that usually found. The one case referred to, was a man, aged 74, who had been sent to the hospital supposed to be suffering with irreducible hernia. Four years previously he had injured his scrotum, by falling astride of a rail. Six ounces of a whitish, milky fluid were drawn off, which, on microscopic examination, showed spermatozoa in profusion.

It is probable that there had been an encysted hydrocele, or cyst of the testicle, which had ruptured into the tunica vaginalis testis, this allowing the spermatic fluid to be emptied into the unnatural situation. At first, a cyst probably formed between the tunica albuginea and the tunica vaginalis, and, on account of a communication with some of the seminiferous tubules, it must have contained spermatozoa. Finally this cyst ruptured, and poured its contents into the vaginal tunic, which became distended, and perhaps, indeed, secreted serous fluid in addition to the other, until the tumor, reaching the inguinal region, presented the appearance of a hernia.

Hydrocele.

No.	Age.	Injection used.	Tapping.	Remarks.	Result.	Days in Hospital.
1	44	Carbolic acid.	....	The injection consisted of carbolic acid ʒi to ʒiii of glycerine.	Cured.	11
2	36	Iodine.	....	....	Cured.	18
3	43	Iodine.	....	Right (double).	Cured.	37
4	..	Iodine.	....	Left.		
5	28	Carbolic acid.	....	Carbolic acid ʒi to ʒiii of glycerine.	Cured.	6
6	29	Seton (silver).	....	....	Cured.	12
7	26	Seton (silk).	....	Abscess followed.	Cured.	70
8	58	Carbolic acid.	....	Abscess followed.	Cured.	30
9	60	Seton (silk).	....	Abscess followed.	Cured.	38
10	31	Carbolic acid.	....	Carbolic acid ʒi to ʒiv of glycerine.	Cured.	41
11	30	Seton (silk).	....	Removed in 24 hours.	Cured.	60
12	37	....	Tapped.	An enormous tumor. Declined radical treatment.	....	4
13	60	Iodine first, re-accumulation, then seton.	....	....	Cured.	45
14	19	Seton (silk).	....	....	Cured.	68
15	35	Seton, re-accumulation, then excision of a part of the tunica vaginalis testis	....	No complications.	Cured.	28
16	74	....	Tapped.	Contained spermatozoa, and probably some cholesterin.	Improved.	3
17	34	....	Tapped.	No re-accumulation.	Improved.	9
18	35	....	Tapped.	....	Improved.	1
19	34	....	Tapped.	....	Improved.	1
20	33	Seton (silk).	....	....	Cured.	41
21	31	Seton (silk).	....	Removed in 72 hours.	Cured.	24
22	56	Seton (silk).	....	Slight abscess.	Cured.	33
23	70	....	Tapped.	....	Improved.	4
24	50	Carbolic acid.	....	....	Cured.	18
25	41	Tinct. iodine failed, re-accumulation, then seton.	....	....	Cured.	38
26	58	Carbolic acid.	....	....	Cured.	8
27	28	....	Tapped.	....	Cured.	10
28	21	Carbolic acid.	....	....	Cured.	27
29	36	Tinct. iodine.	....	....	Cured.	32
30	29	....	Tapped.	Strapped, no re-accumulation.	Cured.	15
31	16	Silk seton.	....	....	Cured.	28
32	50	Tinct. iodine.	....	Iodine repeated; the first injection failed.	Cured.	30
33	38	....	Tapped.	Declined radical operation.	Improved.	2

THOMAS G. MORTON.

DISEASES AND INJURIES OF THE EYE.

During the past five years, 79 cases of diseases and injuries of the eyes have been admitted into the wards for treatment. A very large number of minor cases treated in our reception ward, and then transferred to the care of the out department, have not been included in this report.

Of the 79 cases, there have been 25 cases of cataract, 6 of ectropium, 8 of iritis (syphilitic and rheumatic), 7 of corneitis and of conjunctivitis, 1 of gonorrhœal ophthalmia, 7 of granular lids and tarsal ophthalmia, 1 of scald of both eyes, 2 of corneal ulcer, 1 of sympathetic ophthalmia, 4 of symblepharon, staphyloma, obstructed duct, and pterygium, 2 of opacities of cornea, 9 of wounds of the eye and hernia of the iris, 6 of panophthalmitis.

Of the 25 cases of cataract, 22 were subjected to extraction. In 6 of these both eyes were operated upon at the same time. All of these double operations were successful except one; in this case one eye was lost, the other recovered with excellent vision.

Of the entire number, 17 extractions were made by the Graefe method, 5 by the Liebreich method. 2 needle operations (anterior) did well, but left the hospital to await absorption. 1 case declined treatment. 3 of the 17 Graefe operations failed, 14 were successful. All of the Liebreich did well.

The after-treatment in these cases is very simple. The eye is thoroughly closed by a piece of black adhesive plaster of sufficient width to go from just under the eyebrow to the edge of the upper eyelid. The patient is placed in a moderately darkened room, kept at rest, and given the best diet. Exceptionally, the eyes are covered with a soft white flannel bandage.

In syphilitic iritis, we use atropine solutions, from half a grain to eight grains to the ounce of distilled water. Internally, small doses of the biniodide of mercury, with large doses of iodide of potassium; occasionally, mercurial inunctions. In rheumatic iritis, instillations of atropia, the artificial leech, and oil of turpentine in emulsion or in capsules. Not unfrequently tonics are required early, and often stimulants. Quinine, in decided doses, often proves very beneficial. Morphia, when required to relieve pain, often answers best when given by the skin.

In granular lids we use atropia in moderately weak solutions, half a grain or a grain to the ounce of water, with repeated *but very superficial* scarifications.

All the cases of ectropium required two or more operations, and in several a series of operations were necessary to restore the lids. When the deformity, as is often the case, is the result of burns, as a rule only a bare improvement can be expected.

We have nothing new to offer in regard to corneitis, conjunctivitis, and wound of the eye; we rely principally upon anodynes, emollients, atropia, the artificial leech, very weak solutions of astringents (nitrate of silver or zinc), closure of the lids, moderate pressure in some cases, rest, good diet, tonics, and occasionally alteratives with cod-liver oil. In obstinate cases of strumous complications we have found that a linear application of a saturated solution of chromic acid just above the eyebrow often acts admirably. Now and then a thorough old-fashioned leeching of the temple will bring about a change and lead to speedy convalescence. We discard entirely all strong caustic applications.

In two cases, both males, who were laboring under religious delusion, each, successfully, accomplished the enucleation of an eye. In both cases the patients used the nail of the forefinger as a knife. In the last case, when the extracted globe was examined it was found that it had been removed quite as well as if by the usual method in the hands of a skilled surgeon. All the muscles were cleanly cut from the ball, and the nerve was divided as neatly as if it had been severed by the scissors, and the ball was in a perfect condition.

Ophthalmic Cases.

Sex.	Age.	Disease or Injury.	Operation or Treatment.	Result.	Days in Hospital.
F.	18	Iritis.	Atropia and tonics.	Cured.	[illegible]
M.	26	Corneitis.	Atropia and tonics.	Improved.	[illegible]
M.	22	Ectropium, traumatic.	Plastic operations.	Cured.	49
F.	60	Cataract.	Extraction (G.).	Cured.	49
F.	45	Cataract.	Extraction (G.).	Cured.	24
M.	8	Wound of cornea.	Atropia, cold, and pressure.	Cured.	[illegible]
F.	56	Cataract.	Extraction (L.).	Cured.	6
M.	69	Cataract.	Extraction (L.).	Cured.	[illegible]
M.	22	Wound of eye, lacerated (premature explosion).	....	Incurable.	69
M.	70	Cataract.	Extraction (G.).	Cured.	25
F.	24	Gonorrhoeal ophthalmia, double.	Atropia, scarification; 1 lost; 1 saved. Weak nitrate of silver applications.	Cured.	[illegible]
M.	53	Scald of eyes.	Wet dressings and atropia.	Cured.	6
M.	34	Gunshot wound of eye.	Enucleation.	Cured.	20
M.	20	Granular ophthalmia.	Scarification, atropia, tonics.	Improved.	11[illegible]
M.	70	Cataract, double.	Extraction; R. lost; L. cured.	....	36
M.	44	Ophthalmia, traumatic.	Enucleation.	Cured.	1[illegible]
F.	1	Ectropium from burns.	Plastic operations.	Improved.	20
M.	29	Ectropium, traumatic (explosion in tunnel).	Plastic operations, V incisions.	Improved.	16
M.	25	Ectropium from burns, double.	Plastic operations.	Improved.	105
M.	77	Cataract, double.	Extraction; R., Liebreich; L., Graefe.	Cured.	21
M.	17	Chronic granular lids.	Peritomy, copper wash, atropia, chloride zinc lotions, tonics internally.	Improved.	1[illegible]7
M.	40	Granular lids.	Atropia, scarifications, nitrate of silver, weak solutions.	Improved.	25
M.	63	Ulcer of cornea, traumatic.	Atropia and mild astringents.	Cured.	67
M.	27	Iritis, syphilitic.	Atropia, artificial leech, iodide of potassium.	Cured.	[illegible]
M.	40	Iritis, syphilitic.	Hyd. bichlor. and iod. potas.	Relieved.	1[illegible]
F.	20	Tarsal ophthalmia.	Nitrate of silver solution to edges of lids.	Relieved.	2

Ophthalmic Cases.—(Continued.)

Sex.	Age.	Disease or Injury.	Operation or Treatment.	Result.	Days in Hospital.
M.	34	Panophthalmitis.	Anterior section of eye.	Cured.	17
M.	23	Gonorrhœal ophthalmia.	Weak sol. nit. silver, atropia.	Cured.	34
M.	16	Corneal ulcer.	Atropia, closure of lids.	Cured.	21
M.	58	Iritis, syphilitic.	Ungt. hyd. inunctions, atropia and iodide of potassium.	Cured.	45
M.	24	Rupture of eye (mine injury).	Emollients and closure of lids.	Cured.	100
F.	65	Cataract.	Extraction (L.); sloughed.	Incurable.	15
M.	22	Gun-cap in the eye.	Removed; atropia.	Cured.	48
M.	26	Ectropium from burns.	Plastic operations.	Relieved.	80
F.	60	Sympathetic ophthalmia.	Enucleation.	Cured.	25
F.	57	Cataract.	Extraction (G.).	Cured.	18
F.	50	Panophthalmitis.	Enucleation.	Cured.	45
F.	49	Panophthalmitis.	Abscission.	Cured.	6
M.	48	Cataract.	Extraction (G.).	Cured.	20
F.	60	Cataract.	Extraction (G.).	Cured.	17
M.	27	Cataract, traumatic.	Extraction (G.).	Improved.	28
M.	28	Iritis, specific.	Atropia, iod. potassium, and inunctions.	Improved.	62
M.	32	Opacity of cornea.	Atropia, tonics.	Improved.	9
M.	48	Cataract.	Extraction (L.).	Cured.	15
F.	17	Iritis and sclerotitis.	Atropia, oil of turpentine, artificial leech.	Improved.	73
F.	18	Corneitis, strumous.	Atropia, tonics.	Cured.	83
F.	40	Corneitis, ulcerative.	Atropia, tonics.	Unimproved.	89
F.	20	Wound (laceration) of eyeball.	Atropia, tonics.	Cured.	12
M.	31	Occluded pupil.	Artificial pupil.	Improved.	6
M.	57	Cataract.	Extraction (G.).	Cured.	24
M.	47	Ectropium.	Plastic operations.	Improved.	47
M.	17	Lachrymal obstruction.	Canaliculus split, dilatation.	Improved.	33
M.	70	Cataract.	Declined treatment.	....	4
F.	13	Keratitis, interstitial.	Tonics, iodine, atropia.	Improved.	130
M.	20	Cataract, traumatic.	Extraction.	Cured.	28
M.	8	Cataract, traumatic.	Atropia, anodynes (to await absorption).	Improved.	18
F.	58	Cataract.	Double extraction.	Both cured.	23
M.	19	Cataract.	Double extraction.	Cured.	31
M.	65	Pterygium.	Excised.	Cured.	4
M.	61	Cataract.	Extraction (G.).	Cured.	55
M.	27	Symblepharon.	Plastic operations.	Cured.	19
F.	2	Cataract, double.	Anterior needle (to await absorption).	Improved.	37
M.	20	Conjunctivitis.	Atropia, 1/4 gr. zinc wash, scarification.	Improved.	55
F.	20	Conjunctivitis.	Atropia, 1/4 gr. zinc wash, scarification.	Improved.	25
M.	65	Cataract.	Extraction (G.). Destructive inflammation.	Incurable.	42
M.	15	Staphyloma of the cornea.	Abscission.	Cured.	28
M.	17	Cataract, traumatic (powder).	Needle, atropia 1 month, then extracted.	Cured.	101
M.	40	Opacity of cornea (explosion).	Iridectomy.	Cured.	20
M.	64	Cataract, double.	Extraction (G.).	Cured.	28
M.	23	Conjunctivitis, traumatic.	Piece of iron removed.	Cured.	10
F.	60	Cataract.	Extraction (G.).	Cured.	15
M.	58	Iritis, syphilitic.	Atropia, iod. potassium, mercurials.	Cured.	13
M.	21	Iritis, rheumatic.	Atropia, turpentine, iod. potassium.	Cured.	45
M.	19	Panophthalmitis (blow with black-jack).	Atropia, sealing of lids.	Cured.	19
F.	7	Hernia of iris, traumatic.	Atropia, sealing of lids, and pressure.	Cured.	48
M.	43	Panophthalmitis.	Atropia, sealing of lids, counter-irritation.	Cured.	50
M.	40	Suppuration of eyeball (mining accident).	Enucleation.	Cured.	165
F.	50	Granular lids.	Scarification, weak copper wash, atropia, canthi slit.	Improved.	20
F.	22	Granular lids.	Scarification, weak copper wash, chromic acid to brows.	Cured.	49

THOMAS G. MORTON.

EXCISION OF A PORTION OF THE RECTUM FOR MALIGNANT DISEASE.

THE subject of cancerous affections of the lower portion of the bowel is ably treated in another part of this volume, under the headings of cancer, stricture of the rectum, etc.; but as two cases were subjected to an operative procedure that had previously been done but five times in America, a more detailed description of these cases is here given.

As is well known, Faget, in 1739, excised an inch and a half of the rectum, and in 1826 Lisfranc revived the operation of extirpation of the rectum for carcinoma; but, though Dieffenbach and, at a later time, Billroth practised this method of treatment, it seemed to gain no followers in America. Up to the time of the first case operated upon in the hospital there had been in America, as far as could be ascertained by a search through a great deal of surgical literature, only five instances of extirpation of the rectum. One of these operations was performed by Bushe, two by Mott, one by March, and one, two months prior to the first of Levis's cases, by Briddon, of New York.

Since the publication of the cases operated upon by Dr. Levis many others have been recorded as occurring in the practice of surgeons all over the country, and it is perhaps not too much to say that some of the interest taken in this revived operation was due to the successful work done in the wards of this hospital.

The clinical record of the two cases is given, that the condition of the patients may be fully understood.

Excision of three inches of the rectum for the removal of a cancerous growth; recovery.—The patient, aged 60, stated that he had first noticed the existence of some rectal trouble about a year previous to his admission, which occurred December 29, 1876. The first symptoms were pain and the occasional passage of pus and blood, accompanied by constipation. On digital examination, Dr. Levis found a nodulated mass, about two and one-quarter inches in width, occupying the anterior rectal wall, and extending to a limited extent laterally, rather more, it would seem, to the patient's left than to the right. It extended about two and one-half inches up the gut, but did not involve the anus, which was free from disease, except that there were a few hemorrhoidal tumors, some of which, according to the patient's account, had been strangulated by a ligature a few weeks previously. It was easy to hook the point of the index finger over the top of the cancerous mass; the posterior wall was free from involvement, and there was no stricture, though, of course, the calibre was slightly lessened by the nodulated thickening of the anterior surface of the cavity of the rectum. The man's lungs, heart, and urine appeared normal, and there was no stricture of the urethra.

On January 6 excision of the rectum was done, and the whole cancerous mass removed. After a large metallic bougie had been introduced into the bladder, to serve as a guide to the position of and to steady the urethra, an incision was made from the base of the scrotum to the coccyx, encircling both sides of the anal aperture. The hand of the operator was then introduced behind the bowel, into the hollow of the sacrum, in order to tear the rectum loose from its posterior attachments. By means of the finger and a pair of serrated scissors Dr. Levis broke up the adhesions all around the rectum to the front, where it was more firmly attached, on account of the disease, to the prostate gland and neck of the bladder. The cancerous gut was next carefully dissected from these parts, exposing to view the prostate and the lower part of the bladder. While this was being done the vessels were carefully ligated as soon as divided, and double sutures passed through the skin into the rectum, above the proposed line of excision. These were not fastened, but left in position, to give perfect control of the parts. When the rectum, including the cancerous portion, had been thus carefully and thoroughly enucleated, the gut was drawn forcibly down by seizing the tumor, and the scissors employed to cut through the walls of the bowel. A section of the rectum, three inches in length, was thus excised, leaving behind a perfectly soft and smooth mucous membrane. The sutures were then shotted, and some extra ones applied to keep the gut in position, which was by this means securely stitched to the surrounding integument. The whole operation was completed with the loss of about one fluidounce of blood, because the ligatures, some half a dozen in number, were applied as each vessel was cut, and the operation suspended until the hemorrhage was thus controlled. The wound was then dressed with carbolized oil. The growth was subsequently examined microscopically by Dr. Longstreth, pathologist of the hospital, and found to be an epithelioma.

The patient reacted perfectly after the operation, and was treated with small doses of stimulants and anodynes, and large doses of quinine, until twenty-four hours had elapsed, when he returned to the tonic doses of iron and quinine that he had taken before the operation. His urine had to be removed by catheterization for ten or eleven days, and for a number of days was chocolate-colored, from the admixture of blood. His temperature on the evenings of the second and the fourth days after the operation reached 102° and 101¾°, but afterwards steadily declined, reaching 98½° on the morning of the tenth day. After this time it remained below 100°, with the exception of once, when it attained that height. The wound suppurated pretty freely, without any burrowing of pus, and there was slight tympanites for a few days, but the patient had not sufficient pain to require more than an occasional opiate at night. On the seventh day his bowels were freely opened for the first time by castor oil, and by the tenth day all the sutures were removed. These, by the way, in many instances had pulled loose long before. Fifteen days after the operation the patient was allowed to sit up, and his convalescence from the grave surgical procedure of excision of the rectum was secure.

The condition of the patient on March 1, 1877, was as follows: He had habitual constipation, and was obliged to have continual resort to laxatives to keep the fæces from being retained too long, for then the hardened masses gave pain when expelled. To accomplish this he used compound rhubarb pills, according to indications, and occasionally employed enemata of soap and water; by this means he had a passage every few days. If the contents of the bowels were very loose after an active purgative, he was apt to soil his clothes, but otherwise he had perfect control of defecation, and even seemed to exercise slight control over the escape of flatus. By care he had avoided an involuntary evacuation of fæces for weeks, but the call to stool had to be rigidly obeyed. He could not stand upon the order of his going, but had to go at once.

An examination of the parts showed some contraction at the anus, as would be expected from the cicatricial nature of that orifice, and from the anus to the lower end of the gut the cavity was lined with what had the appearance of mucous membrane. When the finger was introduced, it at times passed into a sort of cul-de-sac alongside of the inferior end of the rectum, but there was no difficulty in passing directly into the bowel.

The patient was discharged very happy, and went to his home, many miles distant. Some months afterwards it was reported that the disease had returned, but no subsequent history of the case was obtained.

About two months and a half later Dr. Levis operated upon a second case in a similar manner. Of this case the notes are not so full, because I had not personal charge of his after-treatment.

Enucleation and removal of the lower part of the rectum, followed by death on the fourth day.—The patient, aged 52, had a carcinomatous tumor, the size of a small hen's egg, which had existed about three months, situated at the right side of the bowel. An incision was first made along the right side of the anus, and the finger introduced to tear up the attachments all around the lower end of the rectum. The incision was then extended around the anus in such a manner as to encircle it, and the operation completed in very much the same manner as in the first case. The section of the tube removed was about one and one-half inches. The patient became jaundiced, and died on the fourth day. The autopsy, made by Dr. Longstreth, showed a slight pneumonic patch in the right lung, considerable lymph and pus in the pelvic cavity, and general peritonitis. The jaundice appeared to be the result of pressure from a few enlarged glands, probably not cancerous, near the common duct. There was no wound found in the peritoneum, the lowest point of which was three-quarters or one inch above the end of the excised bowel.

Such is the history of the two cases of extirpation of the rectum performed in the hospital. The first case rapidly recovered from the operation, in fact, had not a bad symptom, and obtained a new lease of life without even being annoyed with incontinence of fæces. The second died a few days after the operation, though the portion of rectum removed was less than in the former instance.

The ease with which the operation was performed in these cases, and the rapid convalescence of one patient, present the operation in a more favorable light than would at first be expected.

What, then, are proper cases to be subjected to excision of the lower end of the rectum? Lisfranc considered it improper to undertake excision if the index finger could not reach the upper limit of the disease. This is certainly true, and I should be inclined to say that the operation is contra-indicated if the surgeon cannot easily hook his finger above the upper edge of the mass and feel smooth, healthy mucous membrane beyond. If the surrounding tissues are involved in the carcinomatous disease, so much that there is little probability of the operator being able to pull down the intestine after the lower end has been enucleated and excised, the operation should not be attempted. Again, secondary glandular involvement and extreme exhaustion are strong contra-indications.

The suffering attendant upon carcinomatous disease of the anus and rectum, from the chronic constipation, the painful defecation, the continual tenesmus, and the exhausting discharges of pus and blood, render the patient a pitiable object, and almost any risk which promises alleviation is justifiable. There is no disease, unless it be cancerous stricture of the œsophagus, that is at all comparable in mental and physical distress to cancer of the rectum.

When the suffering is intense in cases where the adjacent viscera are implicated, Esmarch recommends even partial excision, and considers applicable the method of Volkmann and Simon, who scoop out with sharp spoons as much of the heterologous growth as possible. By this method a great portion can be extirpated without the occurrence of hemorrhage, and if cauterization be employed in addition, alleviation can be obtained for a long period, even as in cases of uterine cancer. At any rate, this, as a palliative measure, is as beneficial as colotomy, and withal is less repulsive to the feelings.

In women the position of the vagina in front of the rectum renders the operation not only less complicated, but more favorable as to prognosis, for the urethra and bladder are not concerned in the extirpation, and the surgeon is better able to determine the extent of the disease, and has also more room to work. The female rectum is, in fact, a much more superficial organ than the male. It is necessary in women, however, to save as much as possible of the vaginal wall, or, if it be removed, to form an artificial septum by proper suturing between the two cavities.

There have been proposed a number of methods of attacking the malignant growth, in order to have as little hemorrhage as is consistent with thorough eradication. Most operators prefer placing the patient in the lithotomy position, though the knee-elbow posture has been adopted at times. Lisfranc operated in the following way: Having encircled the anus by two crescentic incisions, he dissected the bowel loose from surrounding tissues, then split the rectum longitudinally, to expose the parts fully, and excised as much of the cylinder as was necessary. The splitting of the tube was done at the posterior part, in order to avoid the peritoneum and the larger vessels. In females the vagina affords opportunity for introducing the finger in front of the growth, and in males it is well to have a large bougie in the bladder. If the anus is not involved, the external sphincter may be preserved by making a single straight incision from the central tendon of the perineum to the coccyx, dissecting up the skin and the split sphincter on each side, and then extirpating the lower part of the rectum in the ordinary manner. Another method is to form a perineal flap, convex towards the scrotum, and to dissect this and the sphincter backwards over the coccyx, which procedure exposes the rectum in such a manner that the diseased portion can be removed.

During the operation the hemorrhage may be profuse from the hemorrhoidal, the transverse perineal, and the superficial branches of the internal pudic arteries. The cut vessels may be carefully tied as soon as divided, or the galvano-caustic knife or the écraseur may be employed in the various stages of the extirpation. The first method was adopted by Dr. Levis in the present case, and was eminently satisfactory, for scarcely one fluidounce of blood was lost. This is certainly at variance with the usual descriptions of this formidable and so-called bloody operation.

It is probable that the danger from hemorrhage is over-estimated, for when the bowel is enucleated by tearing it loose from surrounding structures with the finger, so that the hand is gradually insinuated into the hollow of the sacrum behind the rectum, there seems to be little arterial bleeding.

The last step in the operation consists in drawing down the amputated gut, which is done by loosening the cellular tissue, and by the effacement of the normal curves in the viscus, and attaching it to the integument. The sutures will probably tear out, but it perhaps gives a chance for portions to become united, and may thus hasten the cure.

The sequelæ of excision of the rectum most to be dreaded are pelvic suppuration, phlebitis, and peritonitis. Two of Lisfranc's early cases succumbed to the first of these causes, and cases have been recorded of a fatal issue attending the occurrence of phlebitis. Billroth considers the use of many drainage-tubes a very important item in the operation, to prevent burrowing of pus. Owing to the proximity of the peritoneum in all cases, and the great danger of wounding it in those instances where the disease is situated high up in the rectum, peritonitis is to be anxiously looked for in every patient, and its advent gives a very foreboding outlook. That it may occur, even when the tissue removed does not extend far up from the anal aperture, is evident from the history of the last patient operated upon. There was only an inch and a half excised, and no injury done to the peritoneum, yet peritonitis supervened and caused death.

The peritoneum passes from the bladder or uterus to the anterior surface of the rectal tube, but the exact distance from the anus at which this takes place must be an indeterminate quantity. Lisfranc, so says Vidal, gave the distance as six inches in woman and four inches in man, while Malgaigne states that two inches for the female and two or three for the male is the proper estimate. Blandin, according to the same authority as above, gives three inches in man and one and a half in woman. Vidal himself measured it in several subjects, and found that the mean was less than two inches, the women being below the men in every instance. Notwithstanding the discrepancy in these measurements, and the low mean at which the distance from the anus to the peritoneal investment is put, it seems to be a fact that at least three, if, indeed, not four, inches of the tube can be removed with comparative impunity. The importance, however, of respecting the immediate vicinity of this readily-inflamed membrane is seen when it is recalled that Vidal reports a fatal case, in which the autopsy showed a hole in the peritoneum large enough to admit three fingers. On the other hand, Maisonneuve had a case of recovery where the peritoneum was extensively wounded.

The great difference in these measurements depends, I think, on the manner in which they were made. The rectum is a tube that is subject to great distention, and hence does not extend from the anus upwards as a perfectly straight smooth cylinder. Consequently, it is easily appreciated that to determine the number of inches that may be excised without wounding the peritoneum, it is necessary to have the tube detached from its surroundings, and to have the reduplications and curves effaced. This is the condition in which the surgeon places the gut by dissection and traction before he cuts it off from its connection with the remainder of the alimentary canal.

To determine this point I made a number of measurements in the dead-house of the hospital, in the following manner: First, having placed the left hand in the peritoneal cavity, I carried my finger-nail to the point where the serous membrane crossed from the bladder or uterus to the anterior wall of the rectum. Then I introduced a graduated stick into the anus, and thrust it upwards until the end touched the finger-nail. This was done as carefully as possible, to avoid undue pressure. By this means the distance from the bottom of the peritoneal cavity to the verge of the anus was given; but this was not the length of rectum that extended from the lowest point of attachment of the peritoneum to the anus, for that was longer, on account of the folds in the tube. To obtain the latter measurement, I dissected out the rectum and a piece of the bladder or uterus, with the peritoneum still attached. This was laid upon a table, and a graduated rule pushed into the anus as before. The rectum was then smoothed out upon this, but not rendered tense, and the number of inches noted. The difference was so marked that this must be the cause of the great discrepancy between authors as to the number of inches of the rectum that are uncovered by peritoneum in front. Posteriorly it makes no difference, because the peritoneum does not come anywhere near the seat of operation.

A table of the results shows that in the eight cases measured the average was 3⅘ inches.

1.	Male,	21 years.	In situ.	1½ in.	Removed	4 in.
2.	"	28 "	"	2 in.	"	4 in.
3.	"	27 "	"	1¼ in.	"	3¾ in.
4.	"	23 "	"	1¼ in.	"	3½ in.
5.	"	26 "	"	1½ in.	"	3⅝ in.
6.	"	40 "	"	2 in.	"	4¼ in.
7.	Female,	45 "	"	1 in.	"	3½ in.
8.	"	68 "	"	1¾ in.	"	3¾ in.

The measurements *in situ* are, of course, of less importance, because, when the operation of excision is performed, the rectum is drawn down, and has its folds obliterated. They are given in the table because I believe that the conflicting opinions mentioned above are the results of inaccurate statements as to which method of measuring was adopted.

It will, perhaps, be interesting to discuss the results of the operation of extirpation of the rectum, to consider the subsequent condition of the patient thus deprived of his sphincter muscle, and to obtain some idea of the prognosis in regard to a return of the malignant disease.

In all cases of carcinomatous disease a return sooner or later is to be expected; hence it is impossible to state that a patient has been perfectly cured, though he may be greatly relieved of suffering, and live many months before the malignant growth recurs. If no operation is done the patient becomes exhausted and dies; but numerous instances are recorded of persons living several years after the removal of the disease by excision of the rectum. Owing to the difficulty of obtaining the subsequent history of cases, and to the fact that the term "cured" in most instances refers only to the immediate recovery after operation, the statistics of extirpation of the rectum are very unreliable. The many cases of successful operation recorded during the last two or three years show that the removal of the lower end of the rectum is not as formidable a procedure as would be expected.

It is impossible to arrive at exact data concerning the average period between excision and the return of the disease, or in what number of cases the malignant affection is truly cured. Instances are mentioned, however, where several years elapsed without any recurrence; and as epithelioma is the usual type assumed in this locality of the body, a better prognosis obtains than if scirrhus were the usual form of malignant growth found in the walls of the rectum.

The experience in the vast majority of cases, where several inches of the bowel have been removed, is that incontinence of fæces does not follow unless the contents be very fluid. Whether it be because the superior circular fibres of the muscular coat of the rectum act as a subsidiary sphincter, or because in the normal condition the rectum is empty and becomes full only as the desire to go to stool occurs, it matters not. Case after case have caused operators almost universally to know that absence of the sphincter ani muscle, and even of three or four inches of the rectal tube, does not necessitate a condition of involuntary evacuation of consistent stools; and in certain instances liquid fæces and even flatus have been under control.

A secondary result which sometimes occurs is cicatricial stricture; but this is a condition quite readily treated by dilatation, and is not by any means such a severe complication as the cancerous constriction likely to be developed if no operative interference be undertaken.

The operation of extirpation of the rectum is well illustrated by the two patients whose histories have been given in detail. It remains to be said that since the time when these patients were treated in the hospital great interest has been manifested in the subject by surgeons throughout the country, and many similar operations recorded. The success attending the operation has not been invariable, but enough so, I think, to show that in selected cases the operation can be done with a reasonable hope of alleviating the distress of the patient and making life endurable. It should be done in the earliest stages of the disease.

JOHN B. ROBERTS.

SUFFOCATIVE GOITRE.

GOITRE is generally regarded as an affection to be treated in the medical wards, but the following cases are deemed of surgical interest. Two of them were actually treated in the surgical wards, and the other was subjected to a surgical consultation to decide upon the advisability of performing tracheotomy. A short account of these cases may be interesting. The first patient was under the care of Dr. Morton, and his history is as follows:

Cystic bronchocele of many months' standing suddenly taking on rapid growth and causing death.—Joseph S., aged 40, a native of England, married, and by trade a cotton-carder, had had a tumor in the anterior part of his neck for eighteen months, which, however, had grown with great rapidity during the last three months, and had caused for four weeks previous to his admission dyspnœa of a severe character. On November 18, 1874, when the man was admitted, the neck from the centre of one sterno-mastoid muscle to that of the other measured six and one-half inches; the vertical measurement of the tumor was four inches. The tumor was freely movable and showed the presence of a cyst by fluctuation. Under treatment there was improvement, and on December 7, 1874, the tumor measured six inches transversely, and three and three-quarters inches vertically; and there was no longer any interference with respiration. The size of the tumor, however, was not constant, but, without any relation to the treatment, increased and decreased irregularly.

On February 18, 1875, the tumor had increased very much, and greatly interfered with respiration and deglutition; so much so that he could not lie down, and was scarcely able to take nourishment. He died in the afternoon of this day exhausted. The post-mortem examination of the tumor showed on the left side an enlargement of the thyroid gland, extending from the œsophagus to the middle line of the trachea, and from the level of the middle of the thyroid cartilage to within half an inch of the origin of the great vessels from the aorta. The upper portion of this tumor was a cyst the transverse circumference of which measured seven and one-half inches; the internal surface of this cyst was lined with hard nodules. Below the cyst, and lying in the space between the left carotid and the innominate and right carotid, and to the left of the trachea, which was pushed to the right by the tumor, was a fibrous mass, one and one-half inches in diameter. The thyroid isthmus was not much hypertrophied: the right lobe was one and one-quarter inches in transverse diameter. The trachea was pushed to the right, and about one inch below the larynx its calibre was reduced about one-half by the lateral pressure.

Here was an instance of an old tumor, of nearly two years' duration, suddenly increasing with great rapidity, and producing symptoms of dyspnœa and difficult deglutition.

The second case, which was treated in the medical ward, was not a cystic tumor, but was remarkable on account of the rapidity of its growth.

Acute bronchocele, producing death by asphyxia in three weeks.—W. C., aged 18, a painter, unmarried, born in Ireland, was admitted on July 5, 1875. He stated that he had observed, about the middle of June, that he buttoned his collar with difficulty. The following day his throat had increased in size so much that he could not button his shirt (?). The swelling rapidly increased until he had difficulty in breathing, when he sought treatment in the hospital. At the time of admission the tumor, which was evidently thyroidal, was the size of a cocoa nut, and his respiration was a good deal embarrassed, though he could eat his dinner, and afterwards walked in the yard. There was no cardiac murmur detected. In the afternoon a violent paroxysm of dyspnœa occurred, which soon abated; but he grew worse, and his labored respiration could be heard for a long distance through the house. The attacks of dyspnœa became more and more frequent, and tracheotomy was suggested, but deemed useless. The patient died cyanosed the morning after admission. The specimen shows enlargement of both lobes and of the isthmus, completely surrounding the trachea. The tumor is very soft and lobulated; on the sides it extends upwards to the base of the arytenoid cartilages, while the isthmus reaches to within a quarter of an inch of the notch in the thyroid cartilage. Between the isthmus and the highest point at each side there is a notch in the tumor which leaves exposed the surface of the thyroid cartilage. Both lobes are about equally enlarged. The circumference from the œsophagus on one side around to the other, including the trachea in the mass, is now six and three-quarters inches. It is said that, when removed, the tumor weighed four pounds. The specimen has been kept in chloral. The trachea was not laid open.

The first of these cases is an example of the rapid development which may occur in old bronchoceles, while the latter shows with what extraordinary rapidity the thyroid gland may increase and cause death by asphyxia in a few weeks. The second patient noticed that his collar was too tight—a point, by the way, likely to be noticed at the very incipiency of the disease—in the middle of June, and died on the 6th of July.

The next case also is interesting, because it is one of acute bronchocele, causing fatal dyspnœa; but the association of pregnancy with the suffocative goitre adds very much to the rarity of the condition.

Acute bronchocele, with cardiac hypertrophy, occurring during pregnancy.—The woman, who was a patient in Dr. Levis's ward, was aged 27, Irish, in about the sixth month of utero-gestation, and was admitted on December 30, 1875, on account of violent and threatening paroxysms of dyspnœa. Four months previously, when she was about two months pregnant, there had been noticed a slight prominence of her eyeballs, which was followed by a small swelling of the anterior part of her neck, to the right of the median line. About the same time her attention was attracted to palpitation of the heart when taking active exercise. As she progressed in her pregnancy the cervical tumor continued to increase in bulk, and about two weeks before admission, at the time of quickening, she was attacked with frequent paroxysms of dyspnœa that almost reached the point of asphyxia.

On admission there was slight prominence of the eyes and an enlargement of the thyroid gland, which was especially marked on the right side, where the tumor was about the size of a large egg. The first sound of the heart seemed rather heavy, but there was no murmur. The patient's respiration was labored, there was harshness of voice, and three or four times daily the dyspnœa became so aggravated that the woman was almost asphyxiated. During these paroxysms, which lasted three or four hours, she presented a picture of the utmost distress; she screamed outright, threw her arms about, and struggled for breath, while the turgid veins of the neck, the streams of perspiration running down her cheeks, and the lividity of her countenance added to the horror of the spectacle. The paroxysms not unfrequently terminated by vomiting, and during them she sometimes became delirious.

Dr. Cohen, by the laryngoscope, found tumefaction of the arytenoid cartilages, and saw that the glottis was not œdematous, but that, although vibrating spasmodically, it did not close as much as normal. There was, moreover, particularly on the *left* side, bulging inward of the tracheal wall.

It was decided to undertake some operation, and accordingly laryngotomy was performed, as tracheotomy was not possible on account of the enlarged thyroid extending upwards over the trachea. Although this was done, and although the operator could force his finger into the windpipe, thus dispelling the idea of stenosis of the tracheal tube, yet the patient's respiration was still much embarrassed. She did not have a recurrence of the violent paroxysms, but continued to breathe with difficulty, and finally died, January 1, 1876, twenty-nine hours after the operation, or about fifty-five hours after admission into the hospital.

The autopsy was made about sixteen hours after death. The lungs were not adherent, and presented evidences of slight congestion. The pericardium contained a large amount of effusion; the heart was greatly hypertrophied, weighing, without the contained clots, twenty and one-half ounces, but there was no valvular insufficiency. The liver was normal. In the situation of the spleen there was found a small, hard, granular body, rather triangular in shape, and with what appeared to be a hilum. Its weight was fifty-four and one-half grains. The supra-renal capsules were normal in size, but were somewhat hardened, and had indurated masses on the surface. The kidneys were of the natural size, and the capsule peeled off normally, but they were lobulated on the surface, and were somewhat congested.

The tumor of the neck was found to be the thyroid gland, which was hypertrophied and surrounded the trachea from its junction with the œsophagus on one side to a similar point on the other. The tumor at the sides reached as high as the top of the larynx, and with the enclosed trachea measured eight and one-half inches in circumference; the enlargement on the left side was about the same as on the right, though before death the tumor was more prominent on the right side. The isthmus was hypertrophied and bulged forwards between the ribbon muscles of the neck. The tracheal wall was slightly pushed inward by the tumor on the left side. The incision made in the operation was about one inch long, and was found to extend obliquely through the left side of the thyroid down to the cricoid cartilage. Its upper part divided the two left vocal chords, and it extended beneath the enlarged thyroid isthmus. The tumor was a simple hypertrophy of the thyroid gland, and contained no cysts. The recurrent laryngeal nerves ran beneath the gland along the sides of the trachea, and the tumor had to be dissected up to expose them. The left nerve was smaller than the right. The bronchial glands also were enlarged. The fundus of the gravid uterus was about two and one-half inches above the umbilicus. The fœtus was not removed.

These three cases of suffocative goitre differ from each other in several respects. The first two patients were males, while the last was a female, in whom the goitre was developed as a concomitant of pregnancy.

The connection that exists between the reproductive organs of woman and the thyroid gland has long been recognized. It is a common occurrence for goitrous women to find the tumor increased in size during menstruation and pregnancy. Guillot has reported[3] two cases of fatal asphyxia from bronchocele originally developed during pregnancy. Tait has recorded in the *Edinburgh Medical Journal* for May, 1875, twelve instances of enlargement of the thyroid body in pregnancy.

It may be profitable to discuss the cause of the dyspnœa in these cases, and

[3] *Archives Gén. de Méd.*, t. xvi. p. 513.

then to consider the line of treatment to be pursued in similar instances. In all of them death occurred, though the management of the condition varied. In the first, medical means alone were adopted; in the second patient's case, tracheotomy was suggested, but deemed useless; while the larynx was opened for the relief of the third sufferer without preventing a rapidly fatal issue.

What was the cause of the dyspnœa in these cases? Was it due to the growth pushing inward the tracheal wall, which in two of the patients was shown by the autopsy to be the fact? or did it occur because of pressure on the recurrent laryngeal nerves? In the woman it might be suggested that the difficult respiration was due to cardiac phenomena occurring as a concomitant of Graves's disease; but this view is hardly tenable, since the distinctive signs of cardiac dyspnœa were not present. That the case is to be classed under the head of Graves's disease I think there is little doubt, but the fatal dyspnœa is to be attributed to a thyroid rather than to a cardiac origin.

The three cases, then, may be looked upon as instances of suffocative goitre, in which the dyspnœa is due to some derangement of the respiratory act, resulting from the presence of the tumor itself.

The narrowing of the trachea was not sufficient to be a cause of great dyspnœa in the two cases where a post-mortem examination showed the extent of stenosis; hence it is very improbable that the diminution in lumen was of itself a factor in the production of asphyxia. Many cases of bronchocele die with suffocative symptoms where no great constriction of the tracheal calibre can be found. Another explanation, therefore, must be sought in attempting to arrive at a correct appreciation of the etiology of these cases.

Either pressure on the recurrent laryngeal nerves, or what Browne has designated as interruption in rhythm of the involuntary muscles of the trachea, must be considered as the real cause of death in these cases.

The hoarseness of voice mentioned in the history of the last case would point perhaps to pressure on the recurrent nerves, but it was hardly determined by the autopsy that such direct pressure actually existed. Tracheal interruption is said[a] to give rise to violent attacks of dyspnœa, with lividity of the countenance, terminating at times by vomiting. Laryngotomy or tracheotomy would be expected to give relief in cases of recurrent laryngeal pressure, though not in tracheal constriction causing muscular interruption.

Whichever of these explanations is the true one, it is of more immediate importance to us to consider the treatment calculated to give relief in suffocative goitre.

Tracheal pressure is most apt to be produced when the thyroid tumor is situated behind the sternum, and cannot bulge forwards. The same effect is produced in a less degree by the tense cervical fascia, for if the tumor develops with great rapidity the cervical fascia does not distend before it, and as a consequence pressure must be exerted on the parts which are in the vicinity, and

[a] Amer. Jour. of Med. Sciences, April, 1877, p. 370.

either tracheo-stenosis or nerve-pressure is induced. Hence in certain cases the dyspnœa is relieved by making a free incision through the cervical fascia, and permitting the tumor to protrude; and if the urgent symptoms are caused by sudden congestive increase of the glandular tumor, the patient may thus be rescued from impending death.

If the dyspnœa be dependent upon partial occlusion of the trachea, tracheotomy may be of service, especially if a long tube or a flexible catheter be introduced in order to pass below the point of constriction. The hypertrophied and vascular thyroid gland generally requires the incision to be made high up in the trachea or in the larynx, and therefore a long tube is required. If tracheal interruption be the cause of the symptoms, tracheotomy or laryngotomy can be of no service, as was exemplified in Dr. Levis's case, where there was, to be sure, a slight amelioration of the symptoms, but no marked relief.

Ligation of the thyroid arteries has been done in suffocative bronchocele, but its value varies with the portion of the gland involved, and it is only available in certain cases. It would not be available in sudden emergencies. The insertion of a seton has been recommended as giving relief to the respiratory distress, by producing suppuration and dispersion of the tumor.

In cystic goitre evacuation of the cyst and injection of an irritant is sometimes available, but in Case No. 1 this remedy was not deemed advisable.

The most radical procedure is certainly the complete removal of the enlarged gland, but this is attended with great danger on account of the important vascular and nervous structures of the neck, which are in proximity to the thyroid gland. Nevertheless there have been a number of successful cases reported, especially since Greene's brilliant results.

If the dyspnœa in the cases before us was caused by pressure on the recurrent laryngeal nerves, none of these operations would have been of much avail, for the nerve was probably compressed by the posterior portion of the lateral lobes of the gland. It is possible, perhaps, that removing the anterior portion of the tumor might have relieved the general tension sufficiently to have allowed respiration, and thus have preserved the patient's life.

Any strictly surgical treatment then would have been, as was tracheotomy in the last case, of little avail, unless it had been instituted at the very beginning of the disease, or unless the whole tumor had been dissected out regardless of hemorrhage and the dangers of wounding important nerves. At the stage in which two of the patients were admitted it is evident that no remedial surgical procedure would have been feasible.

Would any form of medical treatment have been of use? It was ordered that the third patient should be bled very freely if she became so cyanosed as to point to immediate asphyxia; and, in order to make a very rapid flow of blood, the temporal arteries were to be cut rather than the veins of the arm. This was to be done in order to lessen the internal congestion, relieve the engorged lungs, and so secondarily to diminish the dyspnœa. Arteriotomy was not done, however, and when the consultation was called it was decided to open the trachea. Phlebotomy, in a good many cases of bronchocelic asthma, has been followed by relief of the symptoms; and a case of Graves's disease is related by Trousseau, where a perfect cure followed venesection. In addition, however, ice was applied to the neck, and digitalis administered internally. Before this triple treatment was begun tracheotomy was indicated by the extreme oppression and the imminent asphyxia, and it had been decided in consultation to open the trachea if a rapid subsidence of symptoms did not occur. In three days from the time the treatment was instituted the boy could go up and down stairs without oppression, and finally the cardiac palpitation entirely, and the exophthalmos and the goitre almost entirely, disappeared. It would seem in Trousseau's case that the venesection relieved the distended thyroid gland, and that the subsequent continuous employment of tincture of digitalis quieted the heart's action and prevented a second congestive enlargement of the gland.

In our first case medical treatment of various kinds was employed for a long period without leading to any permanent benefit, while in the second and third instances no time seemed to be given for anything short of surgical procedures. In fact, in the second and third cases death occurred in a few hours.

The fatal results of these cases, in which different lines of treatment were adopted, points to the necessity of early treatment and constant supervision of all goitrous enlargements exhibiting a tendency to increase. If these cases had early been treated by surgical methods it is possible that death from asphyxia might have been averted. Especially does this seem probable in the last case, where in the early days abortion, combined with medical treatment, would possibly have prevented further growth of the tumor.

John B. Roberts.

ERYSIPELAS.

This paper, we think, is of great interest in its bearing upon the questions of contagion and upon the germ theory of disease, which are now so much discussed. The record utterly fails to confirm the theory of any high degree of contagion in the disease itself, and if the contagious elements are in the shape of specific germ spores, upon the diffusion of which the spread of the disease is dependent, then these spores certainly neglected their opportunity. Note the perfect freedom with which applicants for admission in the receiving-room have been taken into the house. Of the sixty-five cases, *sixty-two were admitted for the disease*, and only three cases in the list took erysipelas in the hospital. The forty-five cases marked M. W. were treated in the medical wards. The object of reporting them here is to show that they produced no noticeable effect on the general atmosphere of the building. Twenty cases were treated in the surgical wards, and both the medical and surgical patients were placed indiscriminately as to position in the beds assigned them. There is no known instance in this record of any communication of the disease from one patient to another, or to any of the attendants.

There were some surgical patients, no doubt, besides those in the reported list who had attacks of erysipelas, but, if so, they were very few in number.

These remarks are made as statements of absolute facts, and are in no way to be used in support of any carelessness or want of precaution as to the patients with erysipelas. On the contrary, the record might be quoted as showing the good effects of care and cleanliness in the handling and treatment of these cases. Any attendant who would neglect to wash his hands after dressing one of these patients, before going to another, whether having the same disease or not, would be guilty of a great wrong. We are almost as much in the dark as ever as to the true nature of contagion, and until we are more enlightened we must adopt empirical measures to prevent it. Erysipelas, we all know, sometimes fairly sweeps through a surgical hospital, and then suddenly disappears, leaving for a while no cases at all, and then a few gradually appear. It may be that, for the last six or eight years, we have been passing through a period of comparative exemption. However this may be, it is certain that there were enough cases in the house to propagate myriads of germs, which, if developed, were either still-born or quiescent.

This disease, so formidable in appearance, appears to be well under the control of treatment. This is sufficiently indicated in the table. Only five cases were fatal, and three of these died in the medical wards.

Erysipelas.

No.	Age.	Sex	Nature	Cause.	Treatment.	Result.	Remarks
1	18	F.	Of leg.	Followed sprain of ankle; 15 days before admission.	Laudanum and lead-water.	Cured.	In hospital 5 days. Admitted for the disease to the surgical wards.
2	30	M.	Of face.	Followed blow with piece of coal; 3 weeks before admission.	Emollients and tinct. ferri chlorid.	Cured.	In hospital 17 days. Admitted for the disease to the surgical wards.
3		M.	Of foot, with subsequent abscess.	Followed a cut of toe, while putting his coats at sea, some days before admission.	Elevation of foot, emollients, poultice; opened abscess.	Cured.	In hospital 19 days. Admitted for the disease to the surgical wards.
4	45	M.	Of scalp.	Followed removal of two sebaceous tumors; 1 week previous to admission.	Emollients, soda sulph.; tinct. ferri chlorid.	Improved.	In hospital 3 days. Returned, and remained 11 days, and was cured. Admitted for the disease to the surgical wards.
5	45	M.	Of face, larynx, pharynx, etc	From a beating while drunk.	Emollients, scarification of pharynx, tinct. ferri chlorid., etc.	Died.	In hospital 7 days. Admitted for the disease to the surgical wards.
6	49	F.	Of face.	Followed a blow on forehead; 6 days before admission.	Tinct. ferri chlorid., ung. zinci oxid.	Cured.	In hospital 4 days. Admitted for the disease to the surgical wards.
7	42	F.	Of leg.	Followed bruise; abscess occurred.	Slippery elm and tinct. ferri chlorid.; abscess opened.	Cured.	In hospital 11 days. Admitted for the disease to the surgical wards.
8	26	M.	Of foot, with small abscess.	From chafing of toe.	Laudanum and lead-water, ung. zinci oxid.	Cured.	In hospital 23 days. Admitted for the disease to the surgical wards.
9	24	M.	Of face. M. W.*	No account of injury.	Fever mixtures, iron, etc.	Cured.	Had albumen in urine; throat was involved. A severe case. Admitted for the disease. In hospital 4 months.
10	22	F.	Of scalp.	From scalp wound; 6 days before admission.	Tinct. ferri chlorid., slippery elm, etc.	Cured.	In hospital 11 days. Admitted for the disease to the surgical wards.
11	32	M.	Of whole of right arm. M. W.	From fall on elbow; 6 days before admission.	Emollients, tinct. ferri chlorid., Rochelle salts, etc.	Cured.	In hospital 18 days. Admitted for the disease.
12	24	M.	Of right hand and forearm.	Followed compound fracture of forefinger, for which he was being treated in the out department.	Emollients, ice, morphia, etc.	Cured.	Admitted to the surgical wards for the erysipelas. Returned to the out department. In hospital 18 days.
13	22	M.	Of face. M. W.	No history of injury.	Tinct. ferri chlorid., lead-water and laudanum, morphia, etc.	Cured.	Had albuminuria, delirium tremens, etc. Admitted for the disease.
14	28	M.	Facial. M. W.	....	Tinct. ferri chlorid., mucilage, etc.	Cured.	Had four relapses. In hospital nearly 3 months. Urine albuminous. Admitted for the disease.
15		M.	Facial. M. W.	Followed an attack of tonsilitis.	Tinct. ferri chlorid., slippery elm, etc.	Cured.	In hospital 10 days. Admitted for the disease.
16	...	F.	Facial. M. W.	....	Tinct. ferri chlorid., mucilage, etc.	Cured.	In hospital [illegible] days. Admitted for the disease.
17		F.	Facial. M. W.	....	Tinct. ferri chlorid., etc.	Died.	Albumen in urine. In hospital 7 days. Admitted for the disease.

* Medical ward.

Erysipelas.—(Continued.)

No.	Age.	Sex.	Nature.	Cause.	Treatment.	Result.	Remarks.
18	...	F.	Facial. M. W.	....	Tinct. ferri chlorid., etc.	Cured.	In hospital 10 days. Admitted for the disease.
19	62	M.	Facial. M. W.	....	Tinct. ferri chlorid., mucilage, etc.	Cured.	In hospital 12 days. Admitted for the disease.
20	20	F.	Facial. M. W.	....	Tinct. ferri chlorid., mucilage, etc.	Cured.	In hospital 17 days. Admitted for the disease.
21	22	F.	Facial. M. W.	....	Tinct. ferri chlorid., etc.	Cured.	In hospital 25 days. Admitted for the disease.
22	36	M.	Facial. M. W.	Followed an operation on face outside; 2 weeks before admission.	Tinct. ferri chlorid., etc.	Cured.	In hospital 18 days. Admitted for the disease.
23	36	M.	Facial. M. W.	....	Tinct. ferri chlorid., etc.	Cured.	In hospital 18 days. Admitted for the disease.
24	40	M.	Of foot. M. W.	....	Tinct. ferri chlorid., etc.	Cured.	In hospital 5 days. Admitted for the disease.
25	22	M.	Of leg.	Followed cut below patella; a few days before admission.	Cold to limb, tinct. ferri chlorid., etc.	Cured.	In hospital 4 days. Admitted for the disease to the surgical wards.
26	23	M.	Facial. M. W.	....	Liq. ammoniæ acetatis, tinct. ferri chlorid., etc.	Cured.	In hospital 24 days. Admitted for the disease.
27	57	M.	Of legs. M. W.	....	Tonics, stimulants, etc.	Cured of the erysipelas.	Patient also had albuminuria. In hospital 60 days. Admitted for the disease.
28	24	M.	Facial. M. W.	....	Tinct. ferri chlorid., sodæ hyposulph., etc.	Cured.	In hospital 16 days. Admitted for the disease.
29	23	F.	Of leg. M. W.	Thought to follow tight lacing of shoe.	Tinct. ferri chlorid. and quinine, lead-water and laudanum.	Cured.	Transferred on account of abscess to the surgical wards. In hospital 57 days. Admitted for the disease.
30	32	F.	Of left leg. M. W.	....	Tonics, poultice, carbolic oil dressing, etc.	Cured.	In hospital 17 days. Admitted for the disease.
31	21	F.	Of scalp.	Followed wound and fractured skull.	Tinct. ferri chlorid., cold, etc.	Cured.	This case had been in the hospital for the original injury, was discharged, and *took the erysipelas outside*, and was readmitted for the disease to the surgical wards. In hospital 21 days.
32	37	M.	Of leg.	Followed incised wound of knee, a few days before admission.	Tinct. ferri chlorid., etc.; incisions, etc.	Improved. Discharged for misconduct. Nearly well.	In hospital 6 months. There were sinuses, ulcers from sloughs, etc. Admitted for the erysipelas to the surgical wards.
33	23	M.	Of scalp.	Following lacerated wound.	Quinine and iron, etc., carbolic oil dressing.	Cured.	Was attacked in the hospital. In hospital 24 days.
34	35	M.	Facial. M. W.	....	Tonics, etc.	Cured of the erysipelas.	Erysipelas occurred in house. Patient was admitted for phthisis; had fistula, etc.
35	16	M.	Facial. M. W.	....	Tinct. ferri chlorid., etc.	Cured.	In hospital 5 days. Admitted for the disease.
36	27	F.	Facial. M. W.	....	Tinct. ferri chlorid., quinine, etc.	Cured.	In hospital 14 days. Admitted for the disease.

37	35	M.	Facial. M. W.	....	Tinct. ferri chlorid., quinine, mucilage, etc.	Cured.	In hospital 16 days. Admitted for the disease.
38	32	M.	Facial. M. W.	....	Tinct. ferri chlorid., etc.	Cured.	In hospital 51 days. Admitted for the disease; also had rheumatism.
39	25	M.	Facial. M. W.	....	Tinct. ferri chlorid., etc.	Cured.	In hospital 7 days. Admitted for the disease.
40	48	F.	Facial. M. W.	....	Tinct. ferri chlorid., etc.	Cured.	In hospital 14 days. Admitted for the disease.
41	64	M.	Of leg.	Admitted for an old ulcer. Erysipelas began in hospital.	Tinct. ferri chlorid., stimulants, etc.	Died.	Died of exhaustion from excessive suppuration and sloughing. In hospital 67 days.
42	48	M.	Of face. M. W.	....	Tinct. ferri chlorid., etc.	Cured.	In hospital 15 days. Admitted for the disease.
43	32	M.	Of face. M. W.	....	Tinct. ferri chlorid., chloral hydrat. to face, etc.	Cured.	In hospital 9 days. Admitted for the disease.
44	28	M.	Of right arm.	....	Abscesses opened, tonics, etc.	Improved.	Admitted for the sequelæ of erysipelas to the surgical wards. In hospital 77 days.
45	42	M.	Facial. M. W.	....	Tinct. ferri chlorid., etc.	Improved. Eloped.	In hospital 13 days. Admitted for the disease.
46	34	F.	Of foot and leg. M. W.	....	Tinct. ferri chlorid., etc.	Cured.	In hospital 14 days. Admitted for the disease.
47	21	M.	Facial. M. W.	....	Tinct. ferri chlorid., etc.	Cured.	In hospital 15 days. Admitted for the disease.
48	40	F.	Of scalp and face.	From a wound of forehead.	Tinct. ferri chlorid., etc.	Cured.	In hospital 12 days. Admitted for the disease.
49	35	M.	Facial. M. W.	....	Tinct. ferri chlorid., etc.	Cured.	In hospital 5 days. Admitted for the disease.
50	20	F.	Facial. M. W.	....	Tinct. ferri chlorid., etc.	Cured.	In hospital 16 days. Admitted for the disease.
51	51	M.	Facial. M. W.	....	Tinct. ferri chlorid., etc.	Cured.	In hospital 5 days. Admitted for the disease.
52	39	M.	Of left leg.	From irritation of cicatrix of old burn.	Poultice, etc.; rest.	Cured.	In hospital 15 days. Admitted for the disease to the surgical wards.
53	45	M.	Of face. M. W.	....	Tinct. ferri chlorid., etc.	Cured.	In hospital 3 months. Admitted for the disease.
54	28	M.	Of face. M. W.	....	Quinine and tinct. ferri chlorid., stimulants, etc.	Died.	In hospital 3 days. Temperature rose to 104°.9 and 106°. Urine albuminous. Admitted for the disease.
55	28	M.	Of face. M. W.	....	Tinct. ferri chlorid., etc.	Cured.	Admitted for the disease.
56	34	M.	Of face. M. W.	....	Tinct. ferri chlorid., quinia, etc.	Cured.	In hospital 15 days. Albuminous urine. Admitted for the disease.
57	26	F.	Of face. M. W.	....	Tinct. ferri chlorid., etc.	Cured.	In hospital 11 days. Admitted for the disease.
58	25	F.	Of right forearm.	....	Tinct. ferri chlorid. and quinine.	Cured.	In hospital 12 days. Admitted for the disease to the surgical wards.
59	45	M.	Of face. M. W.	....	Tinct. ferri chlorid.	Cured.	In hospital 7 days. Admitted for the disease.
60	28	M.	Of face. M. W.	....	Tinct. ferri chlorid., slippery elm.	Cured.	In hospital 7 days. Admitted for the disease.
61	26	M.	Of face. M. W.	....	Tinct. ferri chlorid., quinia, etc.	Cured.	In hospital 26 days. Admitted for the disease; also had catarrhal pneumonia.
62	36	M.	Of face. M. W.	Was struck in face 3 weeks since.	Tinct. ferri chlorid., etc.	Cured.	In hospital 1 month. Admitted for the disease.
63	40	M.	Of foot.	From old ulcer being irritated.	Tinct. ferri chlorid., mucilage to foot.	Cured.	In hospital 2 days. Admitted to the surgical wards for the disease.
64	30	F.	Facial. M. W.	....	Tinct. ferri chlorid., quinine, etc.	Cured.	In hospital 17 days. Admitted for the disease.
65	40	F.	Facial. M. W.	....	Tinct. ferri chlorid., etc.	Died.	In hospital 37 days. Had meningitis. Admitted for the disease.

CONCUSSION OF BRAIN AND SPINAL CORD.

Of the 47 cases recorded in the table, 37 involved the brain only, 5 the cord, and in 5 both portions of the cerebro-spinal axis were the seats of injury. One brain case was complicated with a linear fracture without pressure. The physiological and pathological manifestations of course more or less involved the whole of the cerebro-spinal tract, but the distinctive diagnosis is founded upon the immediate symptoms, both rational and physical, presented, when the patient is brought into the house. In nearly all of the cases the original diagnosis is verified by their progress to recovery, or by post-mortem examinations. Severe complications often coexist, which may or may not have anything to do with the brain or cord.

After being put to bed the patient is carefully examined. He is apt to be pale, has a cool and clammy skin, is disposed to vomit, and his friends are mostly able to tell that his stomach was emptied immediately after the injury. Evidences also of this fact will often be about his clothing and beard. Sometimes, especially in cases that have been picked up in the street, with no history of accident, the suspicion arises that profound drunkenness may be the trouble, and it is by no means easy always to positively distinguish this condition from either compression or concussion. Where there is any doubt, we direct that the patient be kept in the house long enough for symptoms of a debauch to pass over; for, as this class of cases are likely to be the subjects of judicial inquiry, it is a duty even to err on the safe side. Unpleasant instances of too hasty judgments in these cases are doubtless known in every large general hospital.

The condition of the eyes, the tongue, the mouth, the breathing, the heart's action, both by pulse and auscultation, the varying degrees of paralysis and the parts involved, if any, and the state of the bladder, as to fulness or emptiness, are rapidly but carefully noticed. There is no confirmation of the view often repeated in surgical writings, to the great confusion of medical students, that the pupil is dilated in concussion and contracted in compression, or *vice versa*. The pupil is simply sluggish in action, or altogether irresponsive to the stimulus of light, and may be dilated or contracted whether the case be one of compression or concussion. In fact, in the eyes of the same patient exactly opposite conditions as to the pupils may exist.

It is the belief of many that fatal cases of concussion will present some positive rough lesion on post-mortem examination. However this may be as to most of them, there is no doubt that some die from the jar or shake alone to which they have been subjected. This is proved not only by autopsies, but by the fact of the complete recovery sometimes of those who at first were apparently as seriously hurt as those who died. This certainly would not be if there had been any serious lesion to contend with.

In the treatment of the shock the use of opium should generally be avoided. Hypodermic injections of brandy or ether may be given. Sinapisms should be applied to the spine and feet and over the stomach. If the temperature is much reduced the patient may be placed in a hot bath for a few minutes, after which he should be carefully dried and put to bed.

The first urgent symptoms having been dealt with, the after-treatment is sufficiently indicated in the table. The reaction is sometimes followed by inflammation and effusion, and has to be dealt with accordingly.

Concussion of Brain and Spinal Cord.

No.	Age.	Sex.	Nature.	Cause.	Treatment.	Result.	Remarks.
1	24	M.	Slight concussion.	Struck on head by heavy beam. No external wound.	Rest.	Discharged by request.	In hospital 2 days.
2	26	M.	Profound shock.	Fell backwards from height of 5 or 6 feet.	Rest, purging, and bromide of potash.	Discharged, cured.	In hospital 12 days. There was some bleeding from the left ear, which proved to be of no serious importance.
3	36	M.	Medium shock. Was insensible on admission.	Fell from a wagon, striking back of head.	Rest.	Cured.	In hospital 2 days.
4	37	M.	Paraplegic, in arms and legs. Loss of control over bladder and rectum. No paralysis of sensation. (Brain and cord.)	Fell from scaffold, striking head and shoulders on pile of sand.	Rest.	Improved.	Remained only a day or two in hospital.
5	22	M.	Slight concussion, with partial facial paralysis.	Fell from top of a four-story building.	Rest, cathartics, bromide potass., quinia, morphia sulph., tinct. ferri chlorid., p. r. n.	Cured.	Injury was complicated by fracture of ribs of left side. (See Fractures.)
6	46	M.	Profound shock. (Brain and cord.)	Said to have fallen from great height and struck on feet.	Was moribund on admission.	Died.	Died 3 hours after admission. The only external injury was a small scalp wound.
7	30	M.	Profound shock. (Brain and cord.)	Fell from scaffold; struck a sign before reaching the ground.	Was moribund on admission.	Died.	Died half an hour after admission.
8	...	M.	Medium shock.	Fell over a banister on to marble floor.	Rest.	Relieved.	In hospital 2 days. Removed by request. There was a lacerated wound of face also.
9	26	M.	Medium shock.	Squeezed head between bumpers of two cars.	Cold and rest.	Relieved.	In hospital 9 days. Removed by request.
10	14	M.	Deep shock.	Injured on railroad.	Rest, etc.	Cured.	In hospital 22 days.
11	...	M.	Concussion, spinal.	Fell down shaft of mine.	Rest and palliatives.	Stationary.	A hopeless case. In hospital 55 days.
12	12	M.	Slight concussion of brain.	Fell from load of hay.	Rest, enemata, cold to head.	Cured.	In hospital 5 days.
13	62	M.	Profound shock.	Fell down stairs.	Palliative, cold to head, etc.	Died.	In hospital 1 day.
14	55	M.	Profound shock.	Fell down hold of canal-boat.	Palliative.	Died.	In hospital 1 day.
15	40	F.	Slight concussion of spine.	Fell down stairs.	Rest, etc.	Cured.	Had also lacerated wound of scalp; afterwards erysipelas. In hospital 49 days.
16	48	M.	Severe concussion of spine.	Struck by a locomotive.	Rest, etc.	Cured.	In hospital 2 months, 2 days.
17	70	M.	Severe concussion, with linear fracture.	From fall down a cellar.	Palliative.	Died.	Was admitted to the medical ward. No history obtained from himself. Died in a few hours, and condition revealed on post-mortem.
18	14	F.	Slight concussion, with scalp wound.	From blow on head.	Rest, etc.	Cured.	In hospital 3 days.
19	42	M.	Severe concussion. (Brain and cord.)	From scaffolding giving way.	Palliative.	Died.	There was partial paralysis; convulsions followed, and death took place 5 hours after admission.
20	19	M.	Severe concussion of spine.	From large piece of coal falling on him while mining.	Rest and general treatment.	Improved.	There was paralysis, also fracture of neck of thigh-bone. In hospital 110 days.
21	26	M.	Severe, of brain.	From fall, striking on curbstone.	Carbonate ammonia, brandy, etc.	Cured.	Patient was pulseless at wrist for a time. In hospital 4 days.
22	27	M.	Severe, of brain.	From fall, 46 feet.	Restoratives, etc.	Died.	In hospital 5 days. There was fracture of inferior maxilla. Possibly fracture of base, but all symptoms were those of concussion. No post mortem allowed.

23	22	M.	Severe, of brain.	From fall through a coal-hole.	Restoratives, caps, etc.	Died.	There was fracture of left clavicle. Fracture of base suspected. No post-mortem reported. In hospital 2 days.
24	10	M.	Slight, of brain, with scalp wound.	Fell from freight-train while in motion.	Rest, ol. tigli, injections, etc.	Cured.	In hospital 6 days.
25	25	M.	Slight, of brain, with scalp wound.	Struck by falling derrick.	Rest, cold, etc.	Cured.	In hospital 6 days.
26	...	M.	Slight, of brain.	Fell down hold of vessel.	Rest, etc.	Cured. Eloped.	In hospital 2 days.
27	13	M.	Slight, of brain.	Fell from a height.	Rest, cold, etc.	Cured.	In hospital 16 days.
28	23	M.	Slight, of brain, also scalp wounds.	From a blow on head.	Rest, Rochelle salts, etc.	Cured.	In hospital 3 days.
29	22	M.	Of brain.	From blow.	Rest, laxatives, and potass. bromid.	Improved.	In hospital 5 days.
30	19	M.	Of brain.	Fell down hatchway.	Rest, laxatives, and bromide.	Cured.	In hospital 7 days.
31	22	M.	Profound, of brain.	Struck with a block while on board ship. There was wound of scalp. No fracture.	Ol. tigli, enemata, potass. bromid.	Died.	In hospital 6 days.
32	18	M.	Of brain.	Fell from height.	Turpentine enema, rest, bromide, etc.	Cured.	In hospital 25 days.
33	12	M.	Of brain.	Struck by wagon.	Purgative, bromide.	Cured.	In hospital 2 days.
34	17	M.	Profound, of brain.	Fell down hatchway.	Restoratives and palliatives.	Died.	In hospital 1 day. Post-mortem showed injection of vessels, but no laceration of brain or fracture of skull.
35	23	M.	Slight, of brain.	From fall.	Potass. bromid. and rest.	Cured.	In hospital 12 days.
36	34	M.	Of brain.	Blow from belaying-pin, five weeks ago, while at sea.	Potass. bromid., etc.	Improved.	In hospital 29 days.
37	20	M.	Slight, of brain.	Struck on head with brick.	Potass. bromid., etc.	Cured.	In hospital 12 days.
38	24	M.	Of brain.	From blow on head.	Purge, blister, potass. bromid.	Cured.	In hospital 5 days.
39	17	M.	Profound, of brain and cord.	Struck back of head against bridge while on top of train in motion.	Calomel, potass. bromid., etc.	Died.	In hospital 2 days. No autopsy.
40	39	M.	Profound, of spine.	Fell 10 feet.	Rest, purge, catheter, etc.	Died.	In hospital 51 days. Post-mortem,—Collapse of left lung, effusion in cavity of pleura. Brain and cord unusually hard; gray matter light in color.
41	9	M.	Of brain.	Caught in belting of machinery.	Potass. bromid. and rest.	Cured.	In hospital 8 days.
42	30	M.	Of brain.	From fall down stairs.	Palliative.	Died.	Died same day.
43	24	M.	Of brain.	Fell down hatchway.	Restoratives, cupping, etc.	Died.	In hospital 1 day. Post-mortem showed clots pressing on brain. No fracture.
44	19	M.	Of brain.	Fell from roof of house, 25 feet.	Enemas, rest, potass. bromid., etc.	Cured.	In hospital 10 days.
45	18	M.	Of brain.	From blow with brick.	Purging, rest, potass. bromid., etc.	Cured.	In hospital 8 days.
46	28	M.	Of brain.	From fall into basement.	Croton oil, potass. bromid., etc.	Died.	In hospital 2 days.
47	45	M.	Of brain.	From being beaten.	Purging, bromides, beef-tea, etc.	Died.	In hospital 4 days.

WILLIAM HUNT.

CANCER.

Of the 71 tabulated cases under the general head of cancer, 18 were of the breast, 8 of the lower lip and chin, 5 of the tongue, 4 each of the face, rectum, and penis, 3 of the nose, 2 each of the neck, lower jaw, wrist, hand, and leg, and 1 each of the mouth, hip, testicle, inner canthus, right orbit, antrum, scrotum, foot, axilla, anus, groin, vulva, shoulder and ribs, forehead, and thigh.

There were 43 males and 28 females. The average age on admission was 48.86 years. 34 of the cases are set down as epithelioma, 11 scirrhus, 23 cancer or carcinoma, and 3 encephaloid. Of the 28 female cases, 18 had cancer of the breast.

48 were subjected to surgical operations, mostly by the knife. Of these 5 died. It is scarcely necessary to state that the term "cured" simply refers to the fact that the patient recovers from the local lesion. There is no doubt that with the great majority of them the disease returns, and eventually proves fatal. Most of the patients are admitted for operation, and are generally benefitted for a time at least. With some the disease never returns, but so sad is the general history that the writer is disposed to think the exceptions are those of mistaken diagnosis. These remarks may not perhaps apply to the early stages of some epitheliomas, as of the lip, for example, but even here the prognosis should be guarded.

In general, however, the proximate effects of operations are encouraging enough to justify their performance. The immediate mortality, it will be noticed, is very small, and the patient is often relieved of agonizing pain, of disgusting and fetid growths, of horrible deformities, and of mental anguish.

In a few cases, caustics were used, but there is no encouragement whatever to advocate the replacement of the knife by these agents. Since the introduction of anæsthetics there can be no comparison as to the amount of pain caused by the two methods. That by the knife is soon over, that by caustic must be an enduring agony. Again, after the knife we mostly have a clean wound, with a good skin covering to treat, while the caustic is too apt to destroy the good with the bad, and to leave an open granulating surface, which very often does not heal at all.

A wise discrimination is generally aimed at in the selection of cases for operation. Where the general condition is very bad and there is a marked glandular involvement, nothing but palliative treatment is advised.

After operation the case is to be treated on general surgical principles, with strict attention to nutrition and hygiene. Specifics are absolutely useless, both generally and locally. Placebos may be given for moral effect, and pain must be assuaged, but the freer the stomach and wound are kept from irritants of all kinds the better.

Cancer.

No.	Age.	Sex.	Nature.	Cause.	Treatment.	Result.	Remarks.
1	49	F.	Scirrhus of breast.	....	Removed by oval incisions; dry earth dressing.	Cured.	In hospital 18 days.
2	37	M.	Epithelioma of rectum.	....	Had been operated on ten weeks before admission, with some relief.	Discharged, unrelieved.	On examination disease had extended four inches up the rectum. In hospital 3 days.
3	49	F.	Cancer of face. Old, of eleven years' standing.	....	Palliative.	Incurable.	Had been treated with dry earth dressing in this hospital five years before present admission, with no benefit. In hospital 10 days.
4	57	M.	Epithelioma of lower lip.	....	Operation.	Cured.	
5	60	F.	Scirrhus of breast.	Supposed from having struck breast on a chair, against which she fell, twenty years ago. From that time a lump came, which lately grew very fast.	Operation.	Cured.	In hospital 22 days.
6	53	M.	Epithelioma of penis.	....	Amputation with écraseur.	Cured.	Had a slight after-bleeding, which was easily checked. In hospital 17 days.
7	54	M.	Epithelioma of mouth.	....	Gargles for cleansing.	Discharged by request.	In hospital 14 days.
8	44	M.	Cancer of lip.	....	Palliative.	Unrelieved.	In hospital 28 days.
9	42	F.	Cancer of right breast and axillary glands.	Supposed to be from a fall, and striking breast against step of a car, a year previous to admission.	Operation.	Cured.	In hospital 35 days.
10	60	F.	Epithelioma of cheek, with glandular enlargements in neck.	....	Operation, removing both the growth on cheek and glands.	Cured.	In hospital 23 days.
11	50	M.	Cancer of tongue. Dorsum and under surface involved.	....	Removed one-half of tongue by écraseur. One vessel ligated.	Wound entirely healed. Relieved.	The disease was of two years' standing. In hospital 23 days.
12	55	M.	Epithelioma of upper lip, and also of nose.	....	Removed diseased parts and slid covering flaps from cheeks.	Relieved.	Disease had existed four years. To out department. In hospital 2 months.
13	50	M.	Epithelioma of lower lip.	....	Removed by V-incision.	Cured.	In hospital 10 days.
14	27	M.	Epithelioma of inner canthus of right eye.	....	Earth dressing.	Improved.	Discharged for misconduct. Disease was of five years' standing.
15	37	M.	Cancer of testicle. Of two months' standing when admitted.	....	Removed by operation. Vessels controlled by pins and ligatures.	Died.	Disease extended and involved the abdomen, especially the mesenteric glands. In hospital 4 months, 18 days.

Cancer.—(Continued.)

No.	Age.	Sex.	Nature.	Cause.	Treatment.	Result.	Remarks.
16	70	F.	Extensive epithelioma of neck.	....	Palliative. Operation not advised.	Unimproved.	Had been operated upon nine months previously, and sought relief for return of disease. In hospital 8 days.
17	55	M.	Cancer of lower jaw.	....	Palliative. Operation not advised.	Unimproved.	Disease advanced very rapidly. Patient declared it to be of only six weeks' standing when admitted. In hospital 8 days.
18	50	M.	Carcinoma of lower jaw.	....	Diseased portion removed by operation.	Cured.	In hospital 25 days.
19	52	F.	Scirrhus of breast.	....	Removed by operation.	Cured.	In hospital 18 days.
20	45	M.	Cancer of right orbit.	....	Eye and diseased parts enucleated.	Improved.	There was a family history of cancer. In hospital 66 days.
21	14	F.	Encephaloid of legs.	....	Topical dressings; exploratory opening. Amputation above knee.	Died.	Disease appeared in head, lungs, and spine. The advance was rapid at first. Mother died of cancer of breast. In hospital 6 months, 1 week.
22	42	M.	Epithelioma of penis.	....	Amputation.	Cured.	In hospital 23 days.
23	62	M.	Epithelioma of lower lip.	....	Operation.	Cured.	In hospital 24 days.
24	53	M.	Epithelioma of whole of lower lip.	....	Disease removed by knife, and a sliding plastic operation performed.	Cured.	In hospital 66 days.
25	32	M.	Epithelioma of nose.	....	Chloride of zinc and chloral dressing.	Improved.	In hospital 58 days.
26	18	M.	Of wrist, clavicle, etc.	....	Tonics, etc.	Improved.	In hospital 41 days.
27	60	M.	Of face. (Lympho sarcoma.)	....	Tonics, etc.	Improved.	Patient had a severe hemorrhage. In hospital 37 days.
28	24	M.	Of antrum.	....	Disease removed by operation, the superior maxilla being taken away, except orbital plate. Tonics, arsenic, etc.	Stationary.	In hospital 46 days.
29	38	M.	Epithelioma of tongue.	....	Tongue removed by écraseur.	Improved.	In hospital 3 months, 1 week.
30	25	M.	Epithelioma of scrotum, left side.	....	Removed by knife.	Improved.	In hospital 16 days.
31	84	M.	Epithelioma of hand.	Followed cutting a wart (a year since) at base of left thumb.	Removed by knife; dressed with carbolized oil.	Improved.	In hospital 11 days.
32	61	F.	Scirrhus of breast.	....	Breast removed, and wound dressed with carbolized oil.	Died.	In hospital 9 days.
33	45	F.	Scirrhus of right breast.	....	Breast removed; dressed with carbolized oil.	Cured.	In hospital 28 days.
34	40	F.	Scirrhus of breast.	....	Breast removed; carbolized oil dressing.	Died.	In hospital 21 days. Autopsy showed cancerous nodules in liver and general peritonitis.

35	66	F.	Scirrhus of breast.	....	Breast removed, and carbolized oil dressing used.	Improved.	In hospital 44 days. A case of three years' standing. Had been treated with caustics, etc.
36	28	M.	Carcinoma of axilla.	....	Palliative.	Eloped.	In hospital 10 days.
37	53	M.	Carcinoma of rectum.	....	Tumor ligated at base and excised.	Relieved.	In hospital 11 days.
38	90	F.	Epithelioma of foot.	....	First ligated then excised.	Cured.	In hospital 29 days.
39	50	F.	Carcinoma of breast.	....	Removed by knife.	Cured.	In hospital 26 days.
40	62	M.	Epithelioma of nose.	....	Chromic acid, cod-liver oil, etc.	Improved.	In hospital 13 days. Patient was re-admitted 8 days after discharge, and improved greatly. In hospital 29 days.
41	45	F.	Cicatricial nodules remaining after second operation on breast.	....	Nodules removed by knife; third operation.	Cured.	In hospital 21 days.
42	50	F.	Cancer of nates.	....	Palliative.	Sent to Philadelphia Hospital.	In hospital 4 days.
43	60	M.	Extensive cancer of tongue, pharynx, and palate.	....	Palliative.	Sent to Philadelphia Hospital.	In hospital 2 days.
44	23	M.	Epithelioma of wrist (left).	....	Iodoform, chloral, etc.	Improved.	In hospital 4 months and 12 days.
45	47	M.	Epithelioma of nose.	Says was struck by piece of stone on the spot where the disease came five years afterwards.	Operation; ulcer excised, and parts covered in by skin-flap from forehead.	Cured.	In hospital 2 months.
46	50	F.	Of old cicatrix following removal of breast.	....	Palliatives.	Unimproved.	In hospital 12 days. There were indurated axillary glands and skin nodules besides the cicatrix.
47	53	F.	Scirrhus of breast.	....	Operation declined.	....	In hospital 1 day.
48	25	M.	Epithelioma of left hand (palm).	....	Excision of part.	Improved.	In hospital 21 days.
49	57	M.	Cancer of right leg over [illegible].	Thinks a slight bruise 4 years ago started the trouble.	Exploratory excision; Fowler's solution and tonics, etc.	Improved.	In hospital 4 months.
50	60	M.	Epithelioma of penis.	....	Amputation by écraseur; carbolized oil dressing.	Cured.	In hospital 15 days. Disease began by what was thought to be a small wart, near root of penis. Began 1½ years ago.
51	60	M.	Cancer of rectum (epithelioma).	....	Excision of three inches of rectum. The rectum was encircled by deep incisions starting from base of scrotum. Then by laceration with fingers and nicking with scissors Dr. Levis, the operator, was enabled to free and draw down the gut until a healthy portion was reached. The vessels were tied as they were cut, and the bowel was stitched to the skin after the removal of diseased mass. Patient had no difficulty as to retaining fæces, and at times had to be relieved by enemata.	Cured.	In hospital 73 days. Disease appeared a year ago, and occupies anterior part of rectal wall.

8

Cancer.—(Continued.)

No.	Age.	Sex.	Nature.	Cause.	Treatment.	Result.	Remarks.
52	38	F.	Cancer of breast.		Removed; second operation.	Cured.	In hospital 2 months.
53	62	M.	Epithelioma of lip.		Removed, pins and twisted (fig. 8) suture.	Cured.	In hospital 42 days.
54	46	M.	Epithelioma of lip.		Removed, pins and ligature.	Cured.	In hospital 20 days.
55	46	M.	Epithelioma of left cheek.		Mass removed by operation.	Relieved.	In hospital 50 days.
56	60	F.	Scirrhus of glands of groin.		Palliative.	Stationary.	In hospital 12 days.
57	38	F.	Epithelioma of tongue.		Palliative.	Stationary. Discharged by request.	In hospital 9 days.
58	32	M.	Epithelioma of rectum.		Rectum excised, as in Case 54.	Death.	In hospital 7 days. There was general peritonitis.
59	40	F.	Scirrhus of breast.		Removed by knife.	Cured.	In hospital 29 days.
60	48	F.	Scirrhus of breast.		Removed by knife.	Cured.	In hospital 50 days.
61	39	M.	Epithelioma on thigh, very extensive.		Palliative. Operation not advised.	Stationary.	In hospital 8 days.
62	40	M.	Carcinoma of neck, extensive.		Palliative. Operation not advised.	Stationary.	In hospital 5 days.
63	42	M.	Epithelioma of penis.		Declined operation.	Stationary.	In hospital 2 days.
64	36	F.	Epithelioma of vulva.		Palliative. Operation not advised.	Stationary.	In hospital 1 month.
65	52	M.	Encephaloid, left shoulder and ribs.		No operation advised.		In hospital 3 days.
66	58	F.	Scirrhus of breast (right). Three years' duration.		Breast removed.	Cured.	In hospital 36 days.
67	63	M.	Epithelial cancer of tongue.		Palliative.	Unimproved.	In hospital 2½ months. Disease began 10 months before admission. Patient was discharged by request. Died about one month after left hospital.
68	41	M.	Epithelioma of whole of lower lip and chin.		Operation declined. Patient to the out department.		In hospital 1 day.
69	63	F.	Epithelioma, forehead. Two years' duration.		Removed, and wound closed with sliding flaps.	Cured.	In hospital 49 days.
70	38	F.	Scirrhus of breast (right).		Removed by operation.	Cured.	In hospital 75 days.
71	65	F.	Encephaloid of breast (right), with cauliflower excrescence.		Removed by incisions around mass, and forcibly tearing this away.	Cured.	The mass was seven inches long and five wide, and horribly offensive.

WILLIAM HUNT.

PAINFUL AFFECTION OF THE FOOT.

In January, 1876, I published an account of "A painful affection of the foot, involving particularly the fourth metatarso-phalangeal articulation,"* and then stated that this malady had, I believed, never been described. Since then I have observed quite a number of cases, and have received so many inquiries from medical friends and others all over the country on the subject, that I am convinced that the disease is quite common. The affection may be either of a mild or of a severe type, and the former, it would seem but rarely, degenerates into the latter, but the more severe forms are more or less so from the very commencement, or at least becoming so quite early after the first neuralgic symptom. The milder form of the disease may continue, with perhaps an occasional paroxysm of greater pain, often through a period of years, with but little change. In my account of this affection the joint of the fourth toe was referred to as being always involved, but since then I have seen cases where the third joint was also affected. Dr. Erskine Mason, of New York, published an account of a case in which the disease involved the joint of the third toe,† but these instances are quite the exception.

I have treated one case of this malady in the Pennsylvania Hospital, the notes of which were sent by the patient too late for publication in my former paper. Under date April 22, 1876, Mr. Wm. K. M., of Rome, Noble County, Indiana, sent me the following account:

"I am a carpenter, and have been a hard laboring man all my life. In March, 1866, was seized suddenly with pain in the joint of my fourth toe of the right foot. At times was obliged to pull off my boot and sock, which gave me relief. Many times I would wear my boot without my sock. At times the toe would puff up and become numb. During the summer and fall the pain was somewhat better. In the spring of 1867 it was worse. I wore a larger boot, but was often very lame. In 1870 I had to use a cane. Kept at work until August, 1872, when I was unfit for all business, for the pain was nearly constant. The disease was called rheumatism, gout, etc., and I used a great variety of liniments, sometimes thinking I was better, but in a few days the malady was as bad as ever. Rest and quiet proved the best remedy for the time. At this time the neuralgia extended to the back of my heel, and was an aching, heavy, dead pain, so that I could not even bear the weight of the bedclothes."

Mr. M.'s sufferings so continued until the close of 1872, when he came on from the West, and was admitted into the hospital, December 20 of that year. On examination I found that there had never been any injury of the foot, at

* *Amer. Jour. Med. Sciences.* † *Ibid.*, Oct. 1877, p. 445.

least so far as the patient was aware of, nor had he any idea of the origin of the disease. Pressure on the joint of the fourth metatarso-phalangeal articulation gave severe pain. The foot was quite natural in appearance, and there was no swelling nor redness. The patient stated that he had met with but one other instance of the same trouble. As there was some slight tenderness of the joint of the third toe, and as he was unable to remain a long time or return, I excised not only the joint of the fourth, with a portion of the shafts of the metatarsal and phalangeal bones, but also removed the articulating surfaces of the third. A very rapid recovery followed the operation. He was discharged, cured, January 27. Four months afterwards he wrote me:

"My foot was tender for some time after I came home, but at the present time I can say that it is well. I have no pain, and am almost free from lameness. The foot tires out a little sooner than the other. I have been to work a month. I think I can safely say that it is a permanent cure, and feel satisfied that the operation has saved my life, which I am very grateful for."

In May, 1876, Mr. M. wrote me that he was perfectly well, without the least lameness; and that he had worked steadily since the spring of 1873; and sent me the accompanying picture of the foot, which shows the shortening of the third and the fourth toes.

Under date of May 12, 1876, Dr. Thomas, of Savannah, Georgia, wrote me as follows:

"Dear Sir,—I have read with much interest your report of cases affecting the fourth metatarso-phalangeal articulation of the foot, published in the January number of the *American Journal of Medical Sciences*.

"I am doubtless the more interested because I think your article throws some light on a case which has been under my advice for several years, and I have looked in vain through some of the surgical authorities for guidance. The patient is a young lady, with every comfort around her. For the past three or four years she has suffered with a severe pain in the fourth metatarso-phalangeal articulation of the left foot, sometimes brought on by long walks, and at other times without any immediate provoking cause. She does not know that she has ever hurt the part, though possibly has sprained it, but not enough to have noticed it. There has never been any swelling or redness perceptible, but always more or less sensitive to pressure or the touch. The joint when handled appears to be too loose, otherwise no abnormal indication save the pain.

"I have tried many things in the way of treatment, but all without proper success. She has applied tincture of iodine, belladonna plaster, capsicum plaster, etc. There are times when she thinks that a comfortable-fitting shoe feels better than a looser one. My impression is that this neuralgia, or whatever it may be called, is more common than we would at first think, and the profession may be grateful to you for the article mentioned."

The following report is interesting and valuable because it is from the pen of a medical gentleman who has himself long been a sufferer from the more severe form of this affection:

"PHILADELPHIA, January 1, 1877.

"MY DEAR DOCTOR,—I have read with pleasure and profit your paper on a 'peculiar and painful affection of the foot.' Thinking it might be interesting to you I send you an account of my own case.

"I have suffered intensely at intervals from this affection for many years, and in all this time have never found medical man or layman who understood what I meant when I complained of it or alluded to it. It has been pronounced by surgeons, who have examined my foot, to be a subluxation or a malformation of the articular surface of the first phalanx of the fourth toe, where it articulates with the fourth metatarsal bone, the concavity not being sufficiently concave. This I have long been convinced is an error. By other eminent authorities I have been conveniently 'pooh-poohed.' By persons out of the medical profession I have been in my agony comforted by an 'Oh, tight boots.' I had at last despaired of making anybody understand my ailment, summoned all my patience, and suffered in silence.

"My own sensations have convinced me that the pain is caused by pressure upon a nerve, but what pressed upon the nerve I was unable to tell. The immediate necessity of removing the boot, and the relief afforded by manipulating my foot in a manner learned by experience, pointed to a dislocation; but the reduction of the displacement was never sufficiently sudden and marked to confirm the belief that there had been a dislocation.

"Now, after living for more than half a century, practised my profession for over thirty years, and suffered half my life with an affection not understood and ranked with a disease so trifling as a corn, I find myself enlightened and the mystery cleared up by your valuable paper on the subject.

"The first paroxysm occurred in my boyhood, and was produced by tight lacing of skate-straps. On unbuckling the straps the cramp, as I called it, was at first soon relieved and thought nothing of, but a continuance of this system of squeezing by tight straps and tight boots, and riding for hours on horseback with the flexors of the leg and foot in violent action and the toes turned in, the attacks became more frequent, more painful, and the abnormal condition of the parts became chronic. These were in my case undoubtedly the causes predisposing. The causes determining the accession of a paroxysm are the wearing of a badly-fitting boot, especially if the sole were narrow, a long and fatiguing walk, particularly on a hot day, over a hot pavement, a long ride on horseback, a wet boot sticking to the sock, a wet sock sticking to the toes, long-continued flexion of the knee-joint, as in a railroad car, carriage, or lecture-room, treading on an uneven surface, as a cobblestone pavement, and should the nervous system be depressed from any cause these exciting causes will act more powerfully.

"The symptoms of an attack in my case are most intense pain, 'cutting to the heart,' sickening, a feeling that it is unendurable, faintness, cold sweat, total incapacity for the time of directing the mind or will to any other subject, a horrible increase of torture on the use of the boot-jack, and all this with no redness, no swelling, no abrasion of the skin, no callosity, no visible displacement of bones, at least after removal of the boot.

"The suddenness of the attack is noteworthy. I have been obliged to drop everything and remove my boot, sometimes in company, sometimes in my wagon. I have even been obliged to sit down on the curbstone and remove the boot. I have dismounted from my horse, and sent home for slippers before I could proceed. I have tied my horse to a tree and lain on the ground, unable to ride farther.

"I have spoken of a tight boot, and of removing the boot, but I have had tight boots which were great favorites, because they would not 'let my toe out of joint.'

"The remedies from which I have obtained relief are removal of the boot, and then manipulating the toes, straightening them out. When inconvenient to take off the boot, I have found grasping the foot tightly round the metatarsal region to answer, and have sometimes worn a circlet of india-rubber band binding the foot round the instep. Putting on a dry boot and dry stocking is of great benefit, and the boot should be well sprinkled inside with powdered soapstone before putting on. Frequently an attack has been relieved completely, without other means, by rest and a cup of strong tea."

The following cases occurred in two members of the same family:

Mr. R., aged 46, from Ohio, consulted me September 29, 1876. He attributes his disease to the following injury: On entering a meeting-room on a winter night, some twelve years since, he stamped violently to remove some snow which covered his boots; the latter were new and fitted tightly; he experienced sudden intense pain in his left foot, and has never been free from it since. Even in bed, occasionally, the pressure of the clothes is sufficient to induce an attack. On examination the region of pain was that of the joint of the fourth toe.

Mrs. R. W., daughter of the above, aged 22, has been suffering more or less for six years. She attributes her malady to a pair of tight, narrow-toed shoes, with a high heel. The neuralgia has been confined to the joint of the fourth toe of the right foot. In the house Mrs. W. always wears slippers; constantly is obliged to remove the shoe when walking, and generally avoids buttoning the shoe. No redness or swelling has ever been noticed. Now and then the pressure of a stocking or slipper is sufficient to bring on an attack. In both of the above cases the temperament was markedly nervous.

Mrs. M. L. M., of Germantown, wrote me:

"Case No. 13, described in your paper, is just like mine. If I wear shoes such as you recommend and bathe the limb freely perhaps I can be cured. The reason I think so is that merely by accident I had commenced wearing a pair of shoes laced up in front just about the time I called upon you, and I have not suffered so much since. Rubbing with a coarse towel or bathing with arnica relieves my pain."

Under date of August 16, 1879, Dr. T. M. Woodson, of Gallatin, Tennessee, wrote me:

"DEAR SIR,—I have read with much interest in the *American Journal of Medical Sciences* for January, 1876, your article describing a peculiar and painful affection of the fourth metatarsal phalangeal articulation. Since then I have met with a case which in every particular coincides with your description; it certainly can be nothing else, for there is no local evidence of any disease. The patient, a well-to-do and very intelligent farmer, aged 36, rather small in stature, weight 130 pounds, has suffered from the pain in the locality described for eleven years. He does not recollect or know of any strain or injury; suffers most on active exercise, so much so as to rob him of all rest night and day. There are occasionally times when the pain is slight, especially on rest and in wearing loose shoes, but the affection incapacitates him from attending to his farm duties.

"I showed him your article and explained his trouble, and he now insists on my performing the operation described by you. What is your observation since the publication of your paper? Have you any new suggestions to offer? Do you still advise an operation in suitable cases? Would it not be better to remove the toe of the fourth metatarsal? Would this weaken the foot?

"The gentleman mentioned is exceedingly anxious for relief by the operation, and as my information is solely obtained from your article, I take the liberty of addressing you on the subject."

Portions of the foot excised have been carefully examined. In some instances nothing abnormal could be found. In one case an abrasion or indentation was found on the outer part of the head of the fourth metatarsal, which was so slight, however, that it did not seem at all likely to be the cause of the pain, but was unquestionably the result of the pressure of the fifth metatarsal. With this exception the parts were normal. A vertical section through the bones showed no evidence of any disease. The occurrence of neuralgia in this locality may be explained by the anatomy of the parts. The metatarso-phalangeal joints of the first, second, and third toes are found on a line with each other; the head of the fourth metatarsal is from one-eighth to one-fourth of an inch behind the head of the third; and the head of the fifth is from three-eighths to half an inch behind the head of the fourth. The joint of the third is slightly in advance of the joint of the fourth, and the joint of the fifth is considerably behind the joint of the fourth.

Fig. 1.

The joint of the fifth metatarsal is so much posterior to that of the fourth that the base of the first phalanx of the little toe is brought on a line with the head and neck of the fourth metatarsal, while the head of the fifth is opposite the neck of the fourth (Fig. 1).

On account of the peculiar tarsal articulation there is very slight lateral motion in the first three metatarsal bones. The fourth and the fifth have greater mobility, the fifth more than the fourth, and in this respect it resembles the fifth metacarpal. Lateral pressure brings the head of the fifth metatarsal and the phalanx of the little toe into direct contact with the *head and neck of the fourth metatarsal*, and to some extent the extremity of the fifth metatarsal rolls above and under the fourth metatarsal.

From the external plantar nerve there are superficial and deep muscular branches. The superficial separates into two digital nerves, which supply the outer and inner side of the fifth toe and the outer side of the fourth. Small branches are distributed freely between the fourth and fifth toes, about the metatarso-phalangeal joints (Fig. 2).

The neuralgia may be ascribed to the position which the fourth metatarso-phalangeal articulation bears to that of the fifth, the great mobility of the latter, and, lastly, to the proximity of the digital branches of the external plantar nerve, which under certain circumstances may be bruised by the contact of the fourth and the fifth metatarsals.

At one time I was inclined to believe that females were more often troubled with this affection than males, but from further examination of a large number of cases it would seem that the cases are about equally divided between the sexes, and that the left foot is not more often affected than the right. The disease seems to be essentially a neuralgia, without any structural disease, or change in the joint or toe. Most of the cases observed have resulted from injury, as a sudden twist in walking, especially in mountainous regions or uneven roads, and occasionally from dancing or jumping. In fact, any exercise which may suddenly displace the toes, when confined in the shoe, may produce the trouble, and several of the cases have had their origin in Swiss walking tours. The dissections of several of the specimens which were removed by operation presented simply negative results; and in no case has there been found any evidence of inflammation or redness, and seldom any swelling, simply intense pain localized in the joints of the fourth or third metatarsal phalangeal articulations has been the most prominent symptom. From an examination of the articulated foot, and especially the peculiar position of the metatarso-phalangeal articulation of the toes, it would seem that the third and fourth joints, or the nerves adjacent, are liable to suffer from pressure or injury, from the fact that the head of the fourth metatarsal lies in contact with the base of the phalanx of the little toe and the top of the shaft of the third toe. Again, the liability of the toes to be squeezed together, while at the same time the great mobility of the outer part of the foot allows the fourth toe to be rolled upon by the fifth. The nerves also supplying the outer part of the foot lie so closely upon these movable bones, and are so deeply lodged, that the opportunity for sudden pressure or pinching is quite possible.

Fig. 2.

Plantar nerves, with the digital branches o the external plantar to the fourth and fifth metatarso-phalangeal articulations, with the deep branches.—From Henlé.

Treatment.—When this form of neuralgia has been induced by injury the treatment should be local,—blood-letting, anodyne applications, with rest, and the foot should be elevated until all sensitiveness has disappeared. Then, as in these less severe cases which do not require operative treatment, a thick, broad-soled, somewhat square-toed, low-heeled shoe should be worn. Thin soles should never be used; the shoe or boot ought to lace up on the front; and in some cases great comfort is derived by the two sides of the shoe opening on the front to a point beyond the irritable joint. Care should be taken that the toes are not subjected to any lateral pressure, and the shoe should be sufficiently broad, so that the toes can slightly spread in walking.

Great comfort is also afforded by the application of a narrow flannel bandage applied around the foot, so as to firmly but moderately give support to the toes, and thus prevent any rolling or friction of the joints of the toes.

In the severe cases no treatment except excision of the metatarso-phalangeal articulation, or amputation of the toe, with the removal of the joint in question, will likely be of any service.

The cases which have been subjected to operation have all done well, and in no instance has there ever been the slightest return of the neuralgia.

THOMAS G. MORTON.

ELEPHANTIASIS.

No record appears of the treatment of any case of this interesting disease previous to 1869. Since then, however, and within the past five years, five cases of this malady have been admitted into the surgical wards. All of these were males, four between 25 and 35, and the remaining one 55 years of age. One of the cases, a negro, was twice received, there being an interval of four years between the times of his admission. In 1869 an interesting example of elephantiasis of the scrotum was received. The patient was a white man, and was unable to assign any cause for the appearance of the disease. After deep, but free incisions, made in a linear manner, I exposed and ligatured the spermatic arteries with signal benefit. The wounds suppurated very freely, after which, by strapping and rest, an almost normal condition resulted. No subsequent trace of the patient has ever been obtained, so that whether the improvement has been permanent or not is unknown.

Fig. 1.

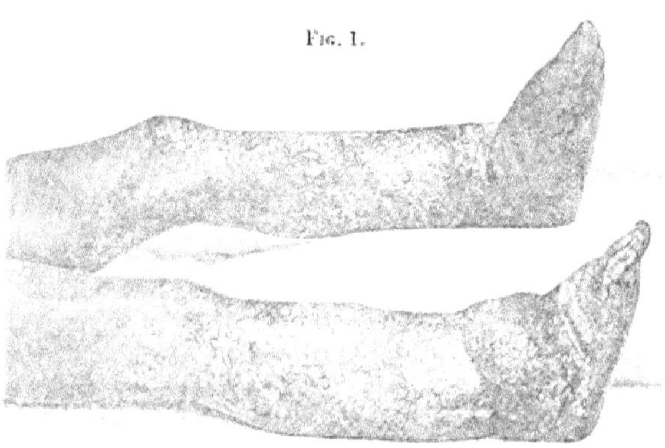

In 1873, a negro, 30 years of age (Case No. 35 of Ligations of Arteries), was admitted with marked elephantiasis (Fig. 1) of the right lower extremity; the disease had been growing fourteen years. The case seemed an excellent one for ligation of the main artery of the limb. This operation was performed on December 12, 1873. Compression by elastic bandage was also made as a regular portion of the treatment. Three months afterwards the patient was discharged, improved. But during this period the limb was kept elevated and at rest, with continuous pressure by a bandage, so that it could not be said that the subsidence of the enlargement was due to the ligation of the femoral artery. That the operation had little or no effect, at least of a permanent character, upon the disease was soon demonstrated, for almost at once after leaving the hospital the limb increased to its usual size, being almost as large as it was before the operation. Shortly afterwards the patient applied and was received into the Philadelphia Hospital, West Philadelphia. The disease, which at the time of his admission into the Pennsylvania Hospital had been almost entirely confined to the right lower extremity, and had barely attacked the left limb (there being only a trifling swelling of the left ankle and some thickening of the skin), soon increased, and became a source of considerable discomfort; moreover, the original disorder returned, so that he was unable to walk except with great difficulty. This condition increased from month to month until April, 1878, when he again sought admission to the Pennsylvania Hospital. His condition was so distressing that he was anxious for an amputation of the limb. The case now seemed out of the reach of any surgical procedure. Thinking over the fact, prominently brought to mind by several cases of wounds of nerves where atrophy had followed nerve section, it occurred to me that a division or section of the trunk of the sciatic nerve might be followed by such change in nutrition and subsequent diminution in size as to give the patient in this case some relief; his condition being so as to warrant such an experiment, even if the limb were to a great extent palsied, it would be better than the risk of such a serious operation, without considering the deformity from amputation, which, as the disease existed at least as far as the middle of the thigh, would require to be made very high up on the thigh.

I also recalled the fact that a section of the sciatic would not necessarily involve a permanent loss of usefulness of the limb, for the nerves would, doubtless, in time reunite, and, even if the progress of the disease were not permanently arrested, much would be gained.

The probability of limited gangrene occurring in a limb altered in its nutrition by division of so large a nerve-trunk was discussed, but this remote possibility was not considered such a sufficient objection as to warrant us in withholding the possible advantages of the operation.

The man had a phthisical history, and his general condition of health was not very favorable, yet his digestion was fair, the urine was normal, and after a residency in the ward of ten days I concluded (with the consent of my colleagues) to prepare him for the operation. (Case 2, p. 117.)

On November 17 I cut down, making a longitudinal incision in the median line, and exposed the sciatic without dividing any vessel; I then excised an inch and a quarter of the trunk of the nerve.

After the effects of the anæsthesia had passed off, it was found that sensation existed to about the middle of the leg, but a part of the limb below this point was still sensitive.

The temperature two hours after the operation of the operated limb at the middle of the calf was 97½°, two hours later 99°.

On the 19th it was found that a marked diminution in the size of the limb had taken place. From twenty-three inches (which the calf measured prior to the nerve-section) it was now reduced to seventeen inches,—a gain of fully five inches in the circumference. On the 2d of December, two weeks after the operation, the limb measured fifteen inches in circumference, and on January 6, seven weeks after the operation, the calf measured only eleven inches, which was less than half its original size. Most noteworthy and curious changes occurred in the condition of the skin itself. The dense thick rugose skin gradually peeled off in large masses of epithelial flakes, at first more especially about the ankles; and this continued until the skin of the limb as far as the groin became as soft as the skin of an infant. Had it not been for a severe attack of pleuro-pneumonia which the patient was seized with he would been able to have been about the ward in less than a fortnight.

This severe attack of illness set in suddenly without assignable cause, and at once threatened the patient's life. During this period of chest difficulty it was possible to give but slight attention to the operated limb. It was allowed to remain in the same position most of the time, for the patient could not be moved without the greatest difficulty, the limb being supported by pillows, but considerable pressure was necessarily made upon the heel, so that when more carefully examined some days after the chest inflammation it was found that a spot about the size of a silver dollar had become the seat of a slough over the calcaneum, but the edges of the ulcer were quite healthy.

Had the patient not been colored this condition could hardly have remained undetected, but as there was no difference during this time in the color of the heel-tissues and the adjacent part of the foot, it escaped observation. However, granulations soon filled up the cavity, and cicatrization rapidly occurred. The right limb was in such an excellent condition that the patient would soon have been about on crutches. On the twenty-fifth day he was actually up in the ward wheel-chair. On December 22 he complained of pain in the *left* thigh, the temperature rising to 101° in the evening. On January 27, the thigh having enlarged very much, with exquisite pain, an exploratory incision was made, but no pus could be found. On February 28 an incision demonstrated a large, deeply-situated abscess, which was at once evacuated. During the month of March the cough, which had continued more or less constantly since the acute attack, now increased. There was some chronic pulmonary consolidation, and the temperature record indicated pneumonia phthisis. The drain from the thigh abscess greatly prostrated the patient, and it was now very evident that the lung was seriously involved. Although the right limb was giving rise to no source of irritation, the left was the centre of great drain, which, with the pulmonary complication, steadily undermined the patient's vital powers, and it became evident that, in spite of large amounts of food and stimulants, there was a daily decline of strength.

The change was very gradual, but finally the patient sank, and died April 4, nearly five months after the operation.

It cannot be said that the nerve-section was the cause of death, although it may be thought to have been so indirectly, for the limb upon which the nerve-section was performed quite recovered its functions, and was in fact apparently healthy. The occurrence of pleuro-pneumonia, and subsequent catarrhal pneumonia, readily opened the way for the development of phthisis, while the abscess in the left thigh, from the broken-down condition of health, was readily accounted for.

The excision of a portion of the sciatic accomplished exactly that which was expected; but the test for sensation by the æsthesiometer, which was made as early as could be after the operation, showed an anæsthetized territory far less than I had anticipated. The extent of the loss of sensation will be seen by an examination of Figs. 2, 3, 4, taken by Dr. Hand, the resident physician, to whom I am indebted for the notes of the case, and for his untiring devotion to the patient throughout his illness. In another instance I should be inclined at first to expose and stretch the sciatic, and, if this was not followed by a sufficiently marked atrophy, to simply divide the nerve.

In connection with the formation of abscess in the left thigh in this case, we observe that in Case 3 of this series, a large abscess formed spontaneously, which caused the death of the patient by exhaustion.

CASE 1. *Elephantiasis of the scrotum: deep incisions; ligation of spermatic arteries; great improvement.*—Patrick S., aged 27, was admitted August 20, 1869, with an immense enlargement of the scrotum. The tissues were hard, and presented the usual elephantine appearance. The disease had been growing steadily for a number of years. Patient had never been in the East Indies (?).

September 1.—Spermatic arteries were tied, and very deep incisions were made in a number of places in the scrotal tissues. After this the whole part was covered with a poultice, and free suppuration ensued. A rapid diminution of the size of the part followed, and the patient was so far well that he was discharged October 23, 1869.

CASE 2. *Elephantiasis Arabum of the right lower extremity: ligation of the femoral artery.*—John P. T., colored, aged 30, was admitted December 1, 1873. The patient, a farm laborer, had been a slave in Virginia, where he was born and always lived until after the war, when he came to Philadelphia. His father and mother were healthy. He and a younger brother were the only ones out of fifteen children who suffered from enlargement of the limbs. Fourteen years ago he first noticed his right leg increasing in size. Had some pain at first, but none of late years. He had great inconvenience from the weight of the limb, and from the serous oozing, which kept his foot wet and cold. The disease mainly was confined to the right inferior extremity, the skin being greatly hypertrophied and hanging in large folds over the instep.

December 2.—Dr. Morton tied the femoral artery. Subsequently the limb was enveloped in a flaxseed poultice to remove the old epidermis. The ligature came away on the twenty-first day. Compression was used as a regular portion of the treatment, and the limb steadily decreased in size.

March 21.—He was discharged, improved.

There was a marked improvement after the separation of the ligature, but, as all along, compression was kept. With the limb in a horizontal position, it was not easy to determine how much was due to the operation. A year after this I had an opportunity for examining the limb, and found it about the same as prior to the operation, although the patient considered himself as improved, and thought the ligation had arrested the growth of the disease. (See Case 5.)

CASE 3. *Elephantiasis; abscess of thigh; death from exhaustion.*—Frederic Y., aged 55, was admitted April 22, 1876. Had good health until 1863. Wounded at the battle of the Wilderness in right arm. He fell and hurt his left hip at same time. Eighteen months ago first noticed great toe of left foot swell, with a burning sensation. After this some pus was discharged from the toe. Has now severe pain in left hip; standing or lying equally painful. For one year has had to sleep in a chair. Patient much emaciated, except the left limb, which is greatly enlarged. Foot around the instep measures 13 inches; ankle to heel, 16 inches; calf, 16¾. Patient was placed on cod-liver oil; chloral dressing to limb; large doses of quinine given, and blister to relieve hip pain; twenty grains of chloral at night. After a time a lotion containing lead-water and laudanum was applied to the leg. An abscess afterwards subsequently developed in the gluteal region, which was aspirated.

June 29.—Losing ground. Measurements: instep, 13 inches; ankle, 16½; calf, 14½.

July 10.—Died. No operation had been performed for the relief of the original disease.

CASE 4. *Elephantiasis of left lower extremity.*—Patrick O'D., aged 34, married, Ireland, was admitted March 16, 1877. There was no evidence of any hereditary tendency. At the age of twenty-two he had a chancre, followed by suppurative buboes, but there were no constitutional symptoms. Eight years ago he noticed that the ankle began to swell, accompanied by slight pain in the leg; afterwards pain in the thigh. Then the limb gradually increased progressively in size. On the front of the leg was a patch about the size of an ear. The cuticle was horny and thick, except the general induration. The skin was otherwise normal, and the urine was healthy. Pressure, with various ointments, was applied, and the patient was kept at rest in the recumbent position. When the patient was discharged, June 5, 1877, the limb was somewhat smaller and the tissues softer. The measurements on admission were as follows:

Around the right instep,	13½	inches;	left, 15	inches.
" " " ankle,	10	"	" 13½	"
" " " calf,	13½	"	" 19	"
Above the right knee,	14	"	" 19	"
Middle of the right thigh,	19	"	" 22½	"

CASE 5. *Elephantiasis; ligation of femoral artery; temporary improvement; gradual increase of the disease; exsection of sciatic nerve; death five months afterwards from phthisis pulmonalis.*—John P. T., aged 34, was admitted November 9, 1877. Patient states that his present trouble began fourteen years before. His right leg was first the seat of the enlargement, and later the left. Two years ago he was admitted into this hospital, when Dr. Morton ligated the right femoral artery (see Case 35, Aneurisms), with marked temporary advantage, reducing the size of the limb greatly. (See *American Journal of Medical Sciences*, April, 1876.)

On re-admission both legs are enlarged, the right to a greater extent than the left. There is a band of hypertrophied integument encircling the right ankle about four inches wide, which projects midway to the toes. The skin is rough and fissured, and from it there exudes a very offensive discharge. Measurement of legs: *right*, at calf, 21 inches; at knee, 19; at ankle, 20. *Left*, at calf, 16½; at knee, 15. Patient ordered to bed. Quinine, gr. viij, and best diet given: whiskey, ℥iij. Examination of urine negative.

November 17.—Under ether Dr. Morton removed one and a half inches of the sciatic nerve at middle of right thigh, making an incision of about seven inches long; nerve apparently was somewhat enlarged. The edges of incision were brought together by nine silver sutures, carbolized lint dressing was applied to wound, the leg wrapped in cotton batting, and a roller bandage applied. Two hours after operation it was found that there was a loss of sensation in the toes and foot, extending up to middle of leg. Temperature of right leg after the operation, 97⅗°; two hours later, 99⅘°; evening, 99°. Has no pain; leg placed on pillow.

November 18.—Passed a comfortable night; no further loss of sensation (for temperature and measurement of leg, see chart). Whiskey, ℥vj, during the twenty-four hours; ℥ij sol. morphiæ given at night. Pulse slow and full.

November 19.—He is able to lift the leg from the pillow; loss of sensation remains about the same; appetite and general condition good.

November 20.—Bandage removed; slight discharge from the skin, which can be peeled off in some places; limb has greatly diminished in size; doing well.

November 21.—Limb examined; skin not so rough; zinc ointment used as dressing to leg and foot; has no pain; is able to raise his leg and bend knee; pulse full but frequent. Ordered quinine, gr. vj, pil. cath. co., No. ij, as bowels have not been open for two days.

November 22.—Bowels open freely during the night; wound healing.

Fig. 2. Fig. 3. Fig. 1.

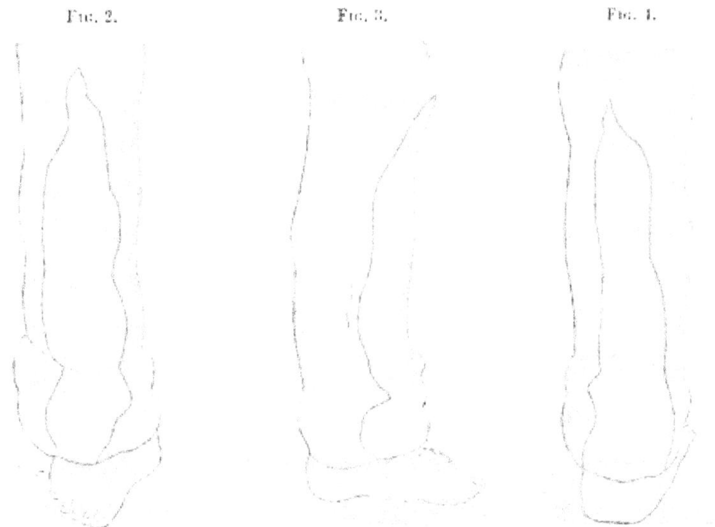

The black line shows the region of total anæsthesia after the operation; the dotted line the alteration and increase of region of sensation three months after the operation.

November 24.—Size of leg has diminished to fifteen inches at greatest point; skin peeling off; appetite continues good.

November 26.—Cosmoline used as dressing to leg; roughness of skin has almost disappeared; new skin, smooth and soft; doing very well. The loss of sensation has extended from middle of leg up to the upper third, and on the outer side of calf. He was able to lift leg from pillow.

November 29.—Skin has almost entirely peeled off; new skin, soft and smooth; cosmoline used as dressing.

December 2.—Chill this P.M. Ordered quiniæ; pot. cit., gr. x, every three hours; whiskey, ℥iij, during the night; sodii sulphit., gr. x to ℥j, to leg.

December 4.—Complains of some pain in legs and groin. Ordered tr. ferri chlor., gtt. xx, every three hours; quinine, gr. xij, daily. Has some pain in right chest. Epistaxis this P.M.; stop sodii sulphit.

December 5.—Worse this A.M. Ordered amm. carb., gr. v, brandy, ℥ss, every two hours. Complains of severe pain in right side; respiration short and hurried; pulse frequent and small. Left lung dull posteriorly, with feeble inspiratory murmur; crepitant râles can be heard throughout the remainder of the chest; has some cough, with slight expectoration. Beef-tea, ℥ij, given every three hours; milk, ℥iij, during day and night. P.M., condition unchanged; ♏v Magendie's sol. morphia given hypodermically; respiration not so frequent.

December 6.—Pulse frequent and small; does not complain of so much pain; cough easier; ♏v Magendie's sol. morphia given hypodermically.

December 7.—Carb. amm., gr. v; brandy, ℥ss, every three hours; not so much pain; complains of some pain in left thigh.

December 9.—Had a fair night; not so much pain in chest; respiration not so frequent; pulse weak and accelerated; *left thigh* swollen and painful; flax-seed poultice applied; carb. amm. and brandy, ℥ss, every four hours; whiskey, ℥j, every hour.

December 10.—Slough over heel; poultice applied; cough not so troublesome; expectoration slightly increased; poultice continued to left thigh.

December 12.—Got out of bed, and was in wheel-chair contrary to orders. Not so much pain in chest; some difficulty in breathing; carb. amm., gr. v, and brandy, ℥ss, continued; chest not so dull; portion of slough over right heel removed; packed with lime-water and charpie; swelling of left thigh subsiding; not so much pain; poultice.

December 13.—Right leg washed; bandage applied; poultice to heel; improving slowly; whiskey reduced to ℥j every two hours; skin of right leg cracking, and on removing portions of it a granulating surface is exposed. Temperature up this P.M.; slight chill; quinine, gr. iv.

December 15.—Some pain in chest; cough increasing; expectoration of a rusty brown hue; swelling of left leg has subsided. Stop carb. amm. and give milk-punch; whiskey, ℥viij, daily.

December 19.—Not so much pain; same dressing to leg; slough at heel granulating.

December 21.—Measurements of feet: *right*, 2 inches from toes, 11½ inches; 3 inches from same point, 11 inches. *Left*, 2 inches from toes, 12½ inches; 3 inches from same point, 13 inches.

December 22.—More pain in left leg this P.M.; leg (left) bandaged; ♏v Magendie's sol. given hypodermically. Lung not so dull; still coughing; line of incision almost healed.

January 3, 1878.—Line of incision healed; leg dressed only every other day; able to lift leg from the bed; loss of sensation remains the same.

January 5.—Sat up in chair for two hours; no bad symptoms; complains of pain in *left* thigh.

January 13.—No change in general health, which is poor. Has been up in chair for past few days for an hour or two. Diet unrestricted.

January 20.—Appetite very poor. Ordered mist. gent. co., ℥ss, t. d. Some swelling and pain in left thigh; no fluctuation can be detected; condition of right leg remains the same.

January 25.—Swelling and pain increasing in left thigh; poulticed.

January 27.—An exploratory incision into left thigh at upper third or anterior surface; no pus found.

February 3.—Has considerable cough; swelling of left thigh has not diminished; complains of pain.

February 28.—Opened an abscess of left thigh at middle or anterior surface; a large amount of very fetid pus discharged.

March 3.—Left leg discharging freely.

March 6.—Complains of severe pain in left thigh. ♏v Magendie's sol. given hypodermically; t. d. whiskey, ℥iv, during the night.

March 17.—Discharge diminishing.

March 21.—Discharge from left leg diminishing, as is the pain. No improvement in general condition; appetite very poor.

March 23.—Placed on air-bed.

April 2.—Cough very troublesome. Morphia given hypodermically, one-sixth of a grain, four times daily. Gradually sinking.

April 9.—Very little discharge from left leg.

April 12.—Slight delirium.

April 13.—Death at 7 P.M.

Measurements of Right Leg of J. P. T.

	Upper Third.	Middle Third.	Lower Third.		Upper Third.	Middle Third.	Lower Third.
1877.				1877.			
November 11....	19	23	20	December 21....	13	14	14
" 22....	17	16	15	" 24....	13	13¾	13¼
" 23....	14½	16½	15	1878.			
" 25....	15	15	15	January 6.......	12	12½	12½
" 26....	16	16	17	" 13.......	12	12	12
" 30....	14	14½	14	" 20.......	11¾	12	12
December 2....	13¾	15	15½	" 24.......	11½	12	12
" 13....	13½	15	16	February 10.......	11¾	11¾	12
" 16....	14	14¼	16				

Chart of the Temperature of Legs of J. P. T.

	Right.		Left.			Right.		Left.	
	M.	E.	M.	E.		M.	E.	M.	E.
1877.*					1877.				
November 18.	96⅗°	97°	96°	95⅗°	December 2.	100⅗°	102⅗°	99⅗°	98°
" 19.	97½	98	98½	98⅗	" 3.	101⅗	101⅗	99⅗	101⅗
" 20.	100	102⅗	102	101½	" 4.	101⅗	100⅗	100⅗	102¼
" 21.	100	101	98	100	" 5.	103⅗	100⅗	103½	103⅗
" 22.	99⅗	101	98⅗	101	" 6.	103	100½	105	106½
" 23.	99⅗	100	98	99	" 7.	101	105	101	105
" 24.	96½	98	96	97	" 8.	100⅗	100	100	100
" 25.	98	98⅗	98	98⅗	" 9.	99⅗	102⅗	99⅗	102¾
" 26.	98⅗	99½	98⅗	99⅗	" 10.	99⅗	102	99½	100
" 27.	98⅗	100	98⅗	99⅗	" 11.	101⅗	103	101	103
" 28.	99⅗	99	99⅗	99	" 12.	102	102⅗	102	103
" 29.	99½	99	99⅗	98½	" 13.	100	100½	100	100
" 30.	99	100⅗	98⅗	98	" 14.	99½	101	99½	100⅗
December 1.	98	99	98	99	" 15.	99	102	99	102

* *November* 17.—Right leg: temperature two hours after the operation was 97⅗°; two hours later, 99⅗°; evening, 99°.

Post-mortem of J. P. T.—Emaciation extreme; adipose tissue in very small amount in the subcutaneous connective tissue. In some parts, especially the upper portions of the thigh, there was a considerable amount of serum in the connective tissue.

The abdominal cavity was normal, except the presence of a considerable amount of clear serum.

Thorax.—Some adhesions (old) of both pleural membranes, especially on right side; considerable clear serum.

Pericardium normal. The heart cavities, valves, and orifices were normal. The heart muscle was very flabby, evidently undergoing degeneration.

Lungs were much congested and somewhat œdematous. Their substance was filled with numerous small shot-like nodules, scattered pretty generally and evenly over all parts. A number of larger yellowish-white, cheesy masses were found, especially in their upper portions; none of very large size; some firm, others with central softening. In the upper and posterior parts of both lungs were a few cavities of small size, or rather some of them became cavities after section was made, through the emptying out of their thickly fluid contents.

The abdominal organs presented no especial abnormal conditions.

An incision was made in the tissues of the posterior part of the right thigh, and the sciatic nerve removed from its point of exit from the pelvis to the popliteal space. At the seat of operation the tissues around the nerve were dense and fibrous, giving an appearance of a localized swelling of the nerve. The portions of the nerve above and below the point of excision appeared of normal size and appearance.

A careful dissection of the nerve was made by laying open its sheath above and below the point of excision, and by tracing its ends through the thickened tissue to their junction. In the first place, there was found in the tissues of the thickened sheath a large artery having a calibre the size of a crow's quill. From this main trunk, whose open, cut end presented itself on the outer surface of the sheath, were given off four or five branches, sufficiently large to admit of passing the bulb of a silver probe from a pocket-case, which were distributed up and down and around the nerve running in its sheath. Their walls were very thin for such large vessels, but they were abundantly supported by the surrounding fibrous tissues. The walls were, however, distinctly arterial in structure.

Near the place of reunion the sheath was very tightly adherent to the nerve, which at one end (distal) appeared rounded and smaller in size than the portion of nerve immediately above it, and at the other end (proximal) appeared flattened, and to be placed, in the newly-formed tissue of the sheath, behind and out of the direct continuous line of the nerve.

The dissections showed that the nerve divided higher up than the place of excision, but that the two branches continued together within the same sheath until reaching the usual point of separation, viz., the lower third of the thigh.

The examination showed that undoubtedly reunion had taken place between the ends of the divided nerve. The union, however, was apparently only complete between the portions of the smaller branch of the nerve, viz., the part that finally became the external popliteal. In this portion the bond of union had a size much smaller than the normal-sized trunks which it reunited. The newly-formed nerve-tissue was found surrounded by very dense fibrous tissue, and the cross-section of the nervous tissue found in it was triangular in shape rather than rounded, as were the nerve-trunks which it reunited.

The main portion of the sciatic nerve, viz., the portion which lower down was given off as the internal popliteal branch, was certainly not reunited. The proximal end, as already stated, was pressed backwards and flattened against the tissues of the thickened nerve-sheath, and ended abruptly in the fibrous tissue filling the gap between the two ends of the nerve, whilst the distal end was rounded, and showed an extension into the newly-formed connection in the gap of what was apparently newly-developed nervous structure.

In making a dissection of the tissues on the anterior aspect of the left thigh to discover the condition of the femoral artery, which had been ligatured years previously, it was noticed that the branches of the anterior crural nerve, found on removing the skin, seemed unusually large and conspicuous. Several of these nerves, which ordinarily are quite small at the middle and lower portion of the thigh, were seen to be large rounded trunks, quite equalling in size the ulnar nerve at the wrist, for example. This part of the examination was very hastily made, and was done, as already mentioned, merely with a view of removing the femoral artery, so that later it was impossible to carefully examine the extent of the enlargement of these nerves. The left femoral artery was found obliterated for only a short distance above and below the point of ligature, perhaps for two inches, in which part it was converted into a fibrous cord. Above and below the calibre was patulous, but distinctly reduced in diameter.

Portions of the heart and lungs and the enlarged branches of the left anterior crural nerve, as well as the right sciatic nerve, were examined microscopically. Several large pieces of skin were also removed; one from the dorsum of the foot, and others from the affected portion of the skin of the face and forehead.

The heart muscle showed fatty atrophy. The changes in the lung-tissue were those of catarrhal pneumonia, and they presented various stages of the process.

The enlargement of the cutaneous nerve branches was found to be due to a great increase of connective tissue, placed not only around the bundles as a whole, but also within the nerve, separating the nerve-bundles from each other with great distinctness. In this connective tissue, as well between the bundles as around the whole nerves, was found adipose tissue, which, from microscopical appearance, contributed largely to the increase of size. The nervous tissue appeared normal.

The condition of the resected sciatic nerve in relation to reproduction of nerve-tissue, as already described, was confirmed microscopically.

Note on temperature of J. P. T.—After the operation the patient's temperature gradually rose until the evening of the 21st, when it reached 103½°. The patient being at the same time constipated, a purgative was given, and the next morning it fell to 99°, but reached 102° the day following, after which it did not exceed, and indeed only once reached, 101°, until the evening of December 2, when it rose to 105°, and the patient had a chill. On the next evening the temperature rose to 106°, and chest symptoms appeared. While on the two following nights 106½° was reached. The patient's temperature now gradually fell, although there was a rise to 105° on the 8th, with some pain in thigh, and on the mornings of the 9th and 10th marked 100°. On the 13th it marked 105°, and there was a slight chill, but from this date it declined to 99½° on the 18th. On the evening of the 22d it rose from 101° to 103½°, and there was increased pain in the thigh, but this point was not again reached until the 6th of January, the rise as usual being attended with increased pain. From this time the temperature fluctuated much as before, its extreme limits being 100° and 104°, and, as had always been the case, with a sharp evening rise, until February 28, when an abscess of the left thigh was opened without much apparent effect, but on the whole the temperature showed a tendency to decline, and frequently marked 99° in the morning.

CASE 6. *Elephantiasis of both lower extremities.*—William R., aged 25, single, a farm hand, born in Berks County, Pennsylvania, was admitted April 9, 1878. Three years ago, having been quite strong and well, he began to have pain and swelling in his right ankle, finally the leg and thigh became involved. Some six months after this the left leg and thigh became stiff and swollen in the same manner. The patient's muscular power was good; cutaneous sensibility somewhat impaired. He would have been able to work, except his limbs had grown, lately, so unwieldy, and he had great fatigue after walking. During the past three years he had increased from 145 pounds to 190 in weight, the gain being principally due to the increase in his lower extremities. The skin of the lower limbs was thick and rugose and quite characteristic, being thrown into folds and with deep furrows. The patient declined any operative treatment.

Resumé of Cases of Elephantiasis.

No.	Sex.	Age.	PART INVOLVED.	TREATMENT.	RESULT.	REMARKS.
1	M.	27	Scrotum.	Spermatic arteries ligated and deep incisions.	Improved.	The subsidence was so marked that the disease may be considered as cured.
2	M.	30	Lower extremities.	Ligation, femoral artery. (See Case 5 of this table.)	Improved.	Some months after leaving the hospital limb as large as before.
3	M.	55	Lower extremity.	Rest, pressure, tonics.	Died.	An abscess found in gluteal region.
4	M.	34	Lower extremity.	Rest, pressure, ointments.	Improved.	
5	M.	34	Lower extremities.	Excision of sciatic nerve.	Improved.	Afterwards died of catarrhal pneumonia and scrofulous abscess.
6	M.	25	Lower extremities.	Rest, alteratives, etc.	No improvement.	Declined treatment.

THOMAS G. MORTON.

NEURALGIA, AND NERVE-SECTIONS.

By the involvement of nerve-trunks in contracting cicatrices, or the development of painful neuromata after amputations, patients sometimes suffer from intractable neuralgias that are only amenable to surgical operation. Such cases are not unfrequent in hospital practice; but, in addition to these, it is found that occasionally a case of ordinary neuralgia resists all the usual resources, and then applies to the surgeon for relief by excising, dividing, or stretching the nerve. Eighteen cases of neuralgia are recorded as having been admitted into the surgical wards. Four of these were instances of stump pain, following leg, ankle, foot, and thigh amputation, for the relief of which excision of the posterior tibial and popliteal nerves was performed, the first three cases resulting in recovery. In the re-amputation of the thigh, however, only an improvement is recorded. There were also four cases of facial neuralgia, two of which were treated by excision of the infra orbital nerve with excellent results. In one of the remaining cases the neuralgia was caused by a blow; but with the administration of quinine and arsenic great improvement followed. One case, after failure of the customary treatment, declined surgical interference.

In one case of traumatic neuralgia of the brow, involving the supra-orbital nerve distribution, an excision of a part of its trunk at its point of emergence from the skull at the supra-orbital foramen was made, with entire relief of the pain.

Neuralgia of the testicle, following specific inflammation in one case, and in another after an operation for varicocele, yielded in each instance to rest, sedatives, quinine, and iron.

Two cases of persistent hip neuralgia after an injury, one case a blow from a shell, were improved by hypodermic injections of morphia and atropia, with the use internally of large doses of iodide of potassium. In a case of blepharospasm, with involvement of most of the muscles of the face, both supra- and infra-orbital nerves were excised. The anæsthesia having been subsequently overcome and replaced by the return of normal sensation, with renewed muscular twitchings, an excision of a portion of tissue about the point where the nerve emerges on the brow again was followed by anæsthesia, showing that minute filaments had brought about a reunion of separated nerve-trunk. In a case of tic douloureux, with almost constant facial twitchings, the infra-orbital was excised, with such amelioration of the symptoms that after a year's interval the patient presented himself perfectly cured, but, having the same affection on the left side, urged a repetition of the operation on the left infra-orbital.

The excision was performed, and the patient made a speedy recovery, and was able to be about the wards, but, just previous to the time when he intended leaving the hospital, he was seized with erysipelas of the head and died.

In Case 12, an instance of neuralgia of the leg and foot, following injury from an iron drill, the pain was only temporarily, and in a measure, relieved by hypodermic injections of morphia and general treatment. Excision of the lacerated popliteal nerve or amputation were suggested, but any operative treatment was declined by the patient. The limb was quite useless, palsied, and more or less atrophied, but, with the application of an apparatus, the patient was able to walk much better, but the pain was only relieved by a constant resort to anodynes.

There was also a case of fatal injury, caused by a bar of iron, which was forced through the body of a workman at one of our iron tube manufactories. The bar entered the hypogastric region, passed through the bladder and intestines, and then, on emerging in the back, fractured the vertebral column, lacerating the spinal cord.

The section of the sciatic for the relief of elephantiasis represents the only operation of the kind recorded in modern surgery, it being the first instance where an attempt has been made to overcome the hypertrophied or enlarged condition of the extremity by attacking the nutrition through the nerve-supply.

It is generally conceded that in the treatment of elephantiasis Arabum, ligations of the main artery supplying the parts—an operation which had once been performed in this hospital—have been productive of only temporary benefit, if any at all; being generally very unsatisfactory, as in fact all other forms of treatment have heretofore proved to be in this very obstinate affection. Should a similar case present again for treatment, I should, in all probability, advise that at first the main nerve should be stretched. This would for a time at least diminish the nutrition of the limb, and perhaps result in more or less atrophy. Should this not exert a sufficiently decided influence, I should again stretch the nerve, and then, if this should not be sufficient, either simply divide the trunk or excise a portion of it. The history of the case herein presented has conclusively demonstrated the value and importance of this operation. There was an enormous decrease in the disease, and, in truth, with much less loss of power than had been expected. It is highly probable that stretching would temporarily so alter the nerve supply that sufficient time would elapse to favor and encourage the disappearance of the disease.

Case 1. *Neuromata of stump, with spasmodic neuralgia; bulbous nerves: re-amputation; cure.*—J. W. C., aged 24, was admitted October 12, 1865. His right foot was shattered by a shell at the siege of Richmond, September 29.

1864. Amputation of the leg was performed on the battle-field a few hours after the injury. On the twenty-first day the stump sloughed, causing considerable loss of skin. Suppuration continued for months, and during this period pain became a prominent symptom, with marked twitching of the muscles. The cicatrix was irregular, with a general transverse direction, and the integument was firmly attached to the ends of the bones. Small painful nodules were found at the extremity of the stump upon each side: under pressure they were exquisitely sensitive. Re-amputation of the leg was made, followed by entire relief from the symptoms.

CASE 2. *Excision of the trunk of the infra-orbital.*—E. L., aged 60, was admitted May 18, 1870, for intense neuralgia of the second branch of the fifth pair. Had just returned from Florida. He was enfeebled, suffered from a bronchial catarrh, and was in wretched health. Used opium and stimulants constantly, which only gave partial and transient relief. The pain was located in the branches of the infra-orbital. The least pressure at the place of emergence on the face brought on an attack of painful tic.

On May 24th the ordinary incisions were made, the antrum was trephined, and an inch of the nerve was removed from a point as far back as possible. Total anaesthesia of the side of the face followed, with relief from pain. Considerable suppuration ensued. The patient was much reduced by his temporary confinement. The bronchial catarrh assumed a very severe form, and a large pulmonary abscess formed, and the symptoms became alarming. With the excessive expectoration great wasting occurred. Finally he improved, and remained quite free from neuralgia. He left for his home in the East on the 6th of June, having entirely recovered.

CASE 3.—*Stump neuralgia, after Pirogoff amputation; excision of posterior tibial and external popliteal nerves.*—J. W., a laborer, aged 52, was admitted August 19, 1870, with a compound fracture of the right foot. Pirogoff amputation was performed. An excellent recovery followed without any undue tension of the cicatrix.

Previous to the patient's discharge stump neuralgia appeared, the pain at times being quite severe. Although the patient was able to walk, and even to bear his entire weight on the stump, each step was accompanied by pain, referred principally to the great toe, with a sensation as if the toe were being puckered up or tied in a knot. Pain was also referred to the little toe, and slightly in each of the others.

This neuralgic condition prevented the man from engaging in any work, and in the month of August, 1871, he sought admission again for the purpose of a re-amputation. The tissues around the cicatrix in front of ankle were found very tender, the heel being in a normal condition. Apparently no bulbous condition of the nerves existed. Since the stump was an exceedingly well-formed one, amputation of the leg was not deemed advisable, but excision of the posterior tibial nerve was determined upon. An inch of the nerve was removed. It was found enlarged and thickened. The pain was materially relieved; the cicatrix remained sensitive.

In December following the patient re-entered the hospital. Dr. Morton made an excision of the external popliteal nerve at the edge of the biceps tendon. The patient was shortly afterwards discharged quite well, with a useful stump, long enough to walk without any artificial appliance, and able to bear his entire weight upon it.

CASE 4. *Excision of the trunk of the infra-orbital branch of the trigeminus for neuralgia of thirty years' duration.*—S. S., aged 61, a native of England, was first attacked with neuralgia in 1834. The pain then was looked upon as an ordinary facial neuralgia of severe type, which continued off and on for three months. She experienced occasional spells of pain during the next two years. In 1836 she had eight attacks of intense pain, which, however, were relieved by mild opiates and anodyne applications. Yearly attacks were experienced until 1840, when she had a terrific spell of facial neuralgia, and during 1841 and 1842 she was confined to bed for eighteen months, experiencing but little benefit from powerful opiates. In 1853 the patient went to England, having suffered for fifteen years, and some benefit was experienced from the voyage. During a period of six months there was a comparative freedom from pain. In 1854 she was again in bed for six months, having returned to her home in Delaware. She now had all the teeth of the upper jaw extracted, but without any good result. The neuralgia continuing, another voyage to England was made in 1864, with some benefit, which lasted for several months. From 1866 to the time the patient came under my care there had been a condition of almost continuous, intense, unremitting suffering.

January, 1871.—On admission into the hospital it was found that the pain was most intense, and apparently originated on the left side of the face, about the infra-orbital foramen, and was always located at this point, and thence radiated over the face and head. The nerve at this place was exquisitely sensitive, the least pressure inducing an attack. Either eating, drinking, talking, or swallowing was quite sufficient to provoke a paroxysm.

January 17.—Excision of the trunk of the infra-orbital nerve was performed. After the infra-orbital foramen was exposed the cutting forceps were applied, and that portion of the edge of the orbit above the foramen was removed until the canal containing the nerve-trunk was reached; passing the hook under

Nerve hook devised by Dr. Morton for isolating the trunk of the infra-orbital nerve in the canal.

the diverging filaments, the instrument was pressed backwards, and the roof of the nerve canal broken up; the nerve was then excised, about an inch and a quarter being removed.

August 12, 1873.—The patient after the operation, now for more than two and a half years, has been *absolutely* free from *all* pain. There is also a complete restoration of sensation on the left side, where the nerve was excised. About one year after the operation a feeling of numbness took the place of anæsthesia; this, in turn, was followed by a very peculiar tingling, more apparent when the skin was rubbed; following this an itching, increased by friction, preceded the return of normal sensitiveness, which latter condition has been observed for about six months, but the painful tic has not reappeared.

CASE 7. *Stump neuralgia after Chopart amputation; excision of posterior tibial and external popliteal nerves.*—W. F., aged 40, was admitted September 27, 1873, with the left foot considerably crushed; sloughing followed, and secondary amputation through the tarsus was performed (October 15). Very slow union, and a wretched condition of general health followed, and neuralgia became early a prominent symptom.

April, 1874.—Stump unhealed; neuralgia continues. During the succeeding three months quite a number of small superficial abscesses formed in the foot, ankle, arm, and other parts of the body, which caused considerable exhaustion. The painful state of the stump was not relieved by any treatment.

September, 1874.—Nearly a year has elapsed since the amputation, and the neuralgia has assumed a more severe form; deep-seated pain extending up the leg in the course of the posterior tibial nerve; pressure in the cicatrix, which is very tight, produces intense suffering. The stump has closed up, and no evidence of a bulbous enlargement of the nerve is present.

September 5.—Dr. Morton excised an inch and a half of the posterior tibial nerve just above the ankle. Entire relief followed this operation in the lower part of the stump and on the sole of the foot. As pain continued in the upper part of the tarsus and anterior part of the foot, supplied by the anterior tibial, on October 3, in the presence of Prof. Erichsen, of London, and others, Dr. Morton excised the external popliteal on the edge of the biceps tendon. Complete relief followed.

It will be seen that the division of the posterior tibial only involved loss of sensation in the sole of the foot following the excision of the external popliteal. Loss of sensation extended over the front of the tarsus, which extended up the anterior face of the leg for several inches, but sensation in the internal and external malleolar regions was not in the least involved by the excisions. Within the region of loss of sensation a needle can be well introduced without producing pain.

October 17.—The patient is walking about with entire comfort, without even the aid of a cane, and is free from pain. Movements of extension and flexion are readily performed. Discharged December 7, 1874.

January 10, 1878.—As some sensitiveness was found in the distribution of the external popliteal, W. F. was readmitted to the ward. Dr. Morton opened the cicatrix above the knee, and searched for some connecting filament, but none could be discovered.

January, 1879.—W. F. is and has been for two years employed in the hospital. He is able to do as much as any other attendant, walks without difficulty, no cane or crutch being required.

CASE 9. *Traumatic neuralgia of brow; excision of supra-orbital nerve.*—In July, 1875, the patient, a male, aged 42, was admitted with a lacerated wound of the eyeball, upper lid, and brow. The brow wound in cicatrizing involved in its contraction fibres of the supra-orbital nerve. Pain became constant, in spite of ointments of atropia and veratria and general treatment.

August 20.—Excised the trunk of the supra-orbital nerve.

August 26.—The usual space of anæsthesia is found. The wound closed up by first intention, and patient discharged entirely cured.

CASE 10. *Infra-orbital neuralgia, non-traumatic.*—E. B., male, aged 76, was admitted December 16, 1875, for intense pain, from which he has been suffering for nearly twelve years. Some three years ago Prof. Pancoast excised the nerve at its exit from the cranium. As nothing could be promised, the patient declined treatment.

CASE 12. *Traumatic neuralgia from an iron drill, which was driven through the popliteal region, severing the artery, vein, and nerve.*—J. B., aged 27, was admitted, with following account from Dr. Geo. S. Wentz:

Section of the iron drill, actual size.

"EBERVALE, LUZERNE COUNTY, PA.,
January 24, 1876.

"J. B., aged 30 years, received a wound of the left thigh five weeks ago, while engaged in mining coal. A large mass, falling from a height overhead, struck an iron drill (see figure) which he was using. The butt or dull end of it, being about an inch and a half in diameter, was driven through his thigh, passing from the inner to the outer side, grazing the posterior surface of the femur in its course, and completely lacerating not only the skin and muscles but the main artery, vein, and nerves, causing profuse hemorrhage, great prostration, and severe shock, from which it was doubtful for several hours whether he would rally. I saw him shortly after the accident. Arterial bleeding had then stopped. I cleansed the wound of pieces of coal and clothing, placed him on a suitable bed, with limb elevated, and ordered him to be kept absolutely quiet. Stimulants were given, with warm applications to the foot (which was cold and devoid of sensation), and anodynes to procure rest.

"The thigh wound is nearly well, as far as the healing of the wound is concerned, though he still suffers more or less pain in the foot from the injury sustained by the nerves.

"You will notice some small patches of skin on the foot and leg, which have sloughed. These may have been caused in part by the applications having been made too warm, as well as deficient nerve and blood supply. In my opinion, considering the severity of the wound, the loss of blood, the great shock to the system, and the irritability of the patient, the progress of the case has been satisfactory, though he thinks he should now be able to walk, inasmuch as '*no bone was broken.*' I might remark that the course of treatment pursued obviated the necessity of ligating the ends of the divided artery, as there was no secondary hemorrhage."

On admission the patient was very anæmic and weak. The wound of entrance and exit of the iron bar is seen about four inches above the knee, although almost healed. On the ball of the great toe there is a deep ulcer.

Measurements.—Right leg above knee, 12 inches; left, 12¼ inches.
" " at knee, 10½ " " 11¼ "
" " at calf, 10¾ " " 11¼ "

Temperature.—Right popliteal space, 97°; left, 97½°.
" toes, 94°; " 94½°.

Loss of sensation and motion was almost complete. The atrophy is seen by reference to the above table, and the temperature also was less on the right side. The pain consequent upon the injury of the nerve was at times very severe, and required the use of hypodermic injections of morphia. The limb was ordered to be regularly rubbed by an attendant; and, in order that he should go about, an apparatus was made similar to the united fracture brace for the femur. With this he was able to walk fairly. As no operation was deemed advisable at the time, he was discharged, February 19, improved.

On September 25, 1877, B. was again sent to the hospital, on account of the continued pain in the foot. Atrophy of the limb was great, and it became a question of amputation, or stretching, or excision of the nerve at the seat of injury; but the patient, not desiring at the time to submit to an operation, left unrelieved.

CASE 17. *Blepharospasm; excision of both supra-orbital nerves.*—W. P., aged 64, from Hazleton, Pennsylvania, had been annoyed for more than three years with twitching of the eyelids. He was able, however, to continue his work until some three or four months prior to this time, when the spasms became much more intense, the involuntary closure of the lids being incessant, unaccompanied, however, by the slightest pain. The trouble had become so aggravated that he required assistance even in walking about. Light seemed to be the exciting cause, as during the evening and at night the twitching was very much less. The trouble subsequently extended, so that most of the muscles of the face within the past four or five months have been involved. The incessant action of the orbicularis has induced a feeling of tension of the brows, with great weariness.

The patient's general health has been uniformly good. Firm pressure on the supra-orbital nerves exercises a very marked control over the twitching, but does not stop it entirely.

Prior to his admission into this hospital an excision of the right and left nerve had been made by Dr. Goodman, who reported that the operation was quite difficult from the great depth of the nerve, consequent upon a very thick brow and an unusual amount of adipose tissue. After an incision along and under the eyebrows, the notch was sought, and with the handle of the knife the nerve was exposed, and traced up on the brow. A blunt hook was carried around the trunk at its emergence, and divided just under the arch of the brow. The diverging filaments were then collected and divided, the main nerve being well drawn out, acting as a guide, and half an inch was then removed. The other supra-orbital nerve was next excised in a like manner.

The wounds united by first intention. Total anæsthesia was found to exist over the entire forehead, the line of loss of sensation extending from the extreme outer portion of the eyebrow on either side, directly upward, outward, and backward, until on a line with the upper surface of the occipital bone, then across, until meeting a similar line from the opposite side.

The relief of the blepharospasm, after the healing of the wound, was complete, the strongest light being borne without the least flinching, and the result thus far had been of the most satisfactory character. The union of the wounds by first intention was probably due in the first place to the general good condition of the patient, and also to the fact that immediately after the operation the lids were firmly closed, and considerable compression maintained by several turns of bandage, which was not removed for four days. This not only excluded all light, but insured absolute rest of the muscles.

After this Mr. P. returned home; was able to continue work. Gradually the old trouble reappeared, and became so intolerable that he sought admission into Pennsylvania Hospital in September, 1878, four years after the first operation. On examination the anæsthesia which followed the supra-orbital nerve-sections had disappeared, sensation having returned to the parts supplied by these nerves: under the impression that union of the sectioned trunks or filaments had taken place. He was put upon quinia, eight grains daily.

September 28.—Dr. Morton made an incision over the right brow, and dissected down to the supra-orbital foramen, but no nerve-trunk could be found. It was then deemed best to remove all that portion of tissue included in an area of half an inch square to the bone; by this means any uniting nerve-filaments would be excised.

October 1.—There is again complete anæsthesia, representing the area of supply by the supra-orbital on the right side. As yet there is little or no diminution in the muscular spasms.

October 12.—Excision of the infra-orbital branch on the right side was made. The fibres of the nerve were collected from the point of exit from the foramen, the trunk being drawn out, half an inch was excised.

October 14.—There is a great deal of redness and swelling of the lower lid and upper portion of the cheek. Applied a flaxseed poultice.

October 27.—Patient succeeded this afternoon in opening his right eye to a limited extent, which he has not succeeded in doing for a long time.

October 29.—Able to open both eyes.

November 2.—From date of last note there has been a steady decrease in the amount of the spasm. Discharged, very much improved.

In September, 1879, this patient reported that he had been greatly benefited by the excision of the infra-orbital on the right side, having been able to attend to his pastoral duties for the past year. He at this time again sought admission, in order that the infra-orbital of the left side should be treated by excision, so much relief having been experienced from the former operation. He was now about 70 years of age, and in less favorable condition than before. After a residency in the hospital of a week, to prepare him for operation, the left nerve was excised in the usual manner, and the wound did well, uniting mainly by first intention. At the end of a week he was so well that he attended the public religious services in the library of the hospital with the other patients. After this he had a chill, erysipelas supervened, and on the eighth day he suffered from suppression of urine and suddenly died on October 16. No post-mortem was obtained.

The following case of injury to the spinal cord is introduced in this place, on account of its resemblance to Case 12, and the rare form of accident it illustrates.

CASE 19. *Fracture of the spinal column, with laceration of the cord, from an iron bar; death in fifty-two hours.*—A. R., male, aged 44, was admitted suffering intensely from the effects of an injury received an hour previous to his being brought to the hospital. The man was employed in one of our large iron tube works. His duties were to seize the hot pipe as it is sent out of the furnace upon an iron rod, and then at a certain moment to direct the rod as it is forced from the tube. By some carelessness he remained in front of the iron bar, when it suddenly struck him with great force upon the front of the abdomen to the right of the median line, passed through the bladder, intestines, and then fractured the vertebra, and in its way through lacerated the spinal cord. Several feet of the bar went entirely through his body, and it was only with great effort and with terrible suffering that it was drawn back through his body. When admitted he was in great shock, and required large doses of morphia to give even a measure of relief. He lived fifty-two hours.

Section of the iron rod, actual size.

The following interesting case, illustrating neuralgia from a gunshot wound which involved the sciatic, is thought to be quite worthy of record:

CASE 20. *Deep pistol-wound of the buttock; ball not found; subsequent localized neuralgia in the calf of the leg, with œdema of the limb; examination of the sciatic nerve, and a small mass of clothing found imbedded in the nerve; extraction; recovery.*—John T. D., aged 24, a brass-founder, was admitted November 16, 1879. This morning a bullet from a small pistol entered his right buttock, about an inch and a half above the ilio-femoral fold, and directly over the course of the sciatic nerve. A careful examination showed that the bullet had penetrated deeply, but it could not be found.

November 25.—There is some œdema of the leg and foot, and pain localized in the centre of the calf of the leg, in a space corresponding to about three inches in diameter. Position in bed allows the œdema to mostly disappear.

December 12.—The pain increased. A careful examination reveals a small hard nodule, apparently in or overlying the nerve,—pressure increases the pain.

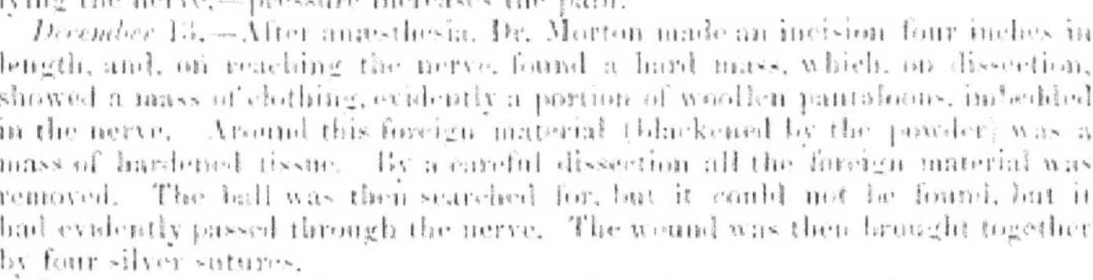

December 13.—After anæsthesia, Dr. Morton made an incision four inches in length, and, on reaching the nerve, found a hard mass, which, on dissection, showed a mass of clothing, evidently a portion of woollen pantaloons, imbedded in the nerve. Around this foreign material (blackened by the powder) was a mass of hardened tissue. By a careful dissection all the foreign material was removed. The ball was then searched for, but it could not be found, but it had evidently passed through the nerve. The wound was then brought together by four silver sutures.

December 28.—Pain has been entirely relieved; patient convalescent.

Neuralgia; Nerve-sections; and Injuries of Nerve-trunks.

No.	Sex.	Age.	Nature and Part involved.	Operation and Treatment.	Result.
1	M.	24	Neuralgia in stump, after amputation of leg.	Excision of anterior and posterior tibial nerves.	Cured.
2	M.	60	Neuralgia, facial.	Excision of infra-orbital nerve.	Improved.
3	M.	52	Neuralgia, after Pirogoff amputation.	Excision of posterior tibial and external popliteal nerve.	Cured.
4	F.	61	Neuralgia, facial.	Excision of infra-orbital nerve.	Cured.
5	M.	24	Neuralgia of testicle, after gonorrhœa.	Rest, sedatives, etc.	Cured.
6	M.	19	Neuralgia of testicle, after varicocele operation.	Rest, sedatives, quinine.	Cured.
7	M.	40	Neuralgia in stump, after Chopart amputation.	Excision of posterior tibial and external popliteal nerve.	Cured.
8	M.	20	Neuralgia in hip (traumatic).	Iodide of potassium, rest, morphia.	Improved.
9	M.	42	Neuralgia in traumatic brow; lacerated wound.	Excision of supra-orbital.	Cured.
10	M.	76	Neuralgia, facial.	Declined treatment. Three years before, excision had been performed, with complete relief for over two years.	
11	F.	38	Neuralgia, facial (from a blow).	General treatment, quinine, arsenic.	Improved.
12	M.	27	Neuralgia in leg (from injury of popliteal nerve).	Apparatus to assist walking.	Improved.
13	M.	34	Neuralgia in hip (struck by shell during late war).	Atropia, morphia, iodide of potassium.	Improved.
14	F.	18	Neuralgia of finger and arm (after felon).	Iodide of potassium and chloral dressing.	Improved.
15	M.	28	Neuralgia of leg and foot. (See Case 12.)		
16	F.	24	Neuralgia in stump, after thigh amputation. (See Case 71, Amputations.)	Re-amputation and division of nerves high up.	Improved.
17	M.	64	Blepharospasm.	Excision of supra- and infra-orbital nerves.	Subsequently died of erysipelas, after successful operation.
18	M.	33	Elephantiasis. (See Case 5, Elephantiasis.)	Excision of sciatic nerve. (See Elephantiasis.)	Subsequently died of catarrhal pneumonia, after successful operation.
19	M.	44	Laceration of spinal cord.	....	Died from shock.
20	M.	24	Neuralgia, gunshot injury.	Removal of portion of clothing from the body of the nerve.	Cured.

Thomas G. Morton.

OSTEITIS AND PERIOSTITIS.

It will be noticed that the parts involved, in all of the 15 cases reported under this head, were the lower extremities. Of these the femur was the seat of disease in 3 and the tibia in 12. The left tibia gave us 7, the right one 3, and in 2 the side is not reported. 7 cases resulted from injuries, 6 from disease, 1 of these from exposure, and of 2 we have no history.

The good effect of operations in appropriate cases of osteitis and periostitis is well marked. Where there is manifest enlargement, particularly if acute, free longitudinal incisions will almost always relieve the intense pain which accompanies the disease. If there is reason to believe the shaft of the bone itself is involved, the trephine or gouge may be used to open the bone into the cancellated structure or into the medullary cavity. Often the seat of hidden suppuration is opened, and thus not only is relief obtained, but a positive cure is accomplished of what would otherwise lead to most formidable destruction. As stated under the head of necrosis, there is every reason to believe that much of it might be avoided if prompt operative measures were resorted to when the opportunity occurs. Eight of the cases in the table were cured, and there is not one that was not relieved or improved.

Osteitis and Periostitis.

No.	Age.	Sex.	Nature.	Cause.	Treatment.	Result.	Remarks.
1	17	M.	Enlargement, upper part left tibia.	Supposed to have come from a severe fall, while skating, eight years since.	1st. Potass. iodid. and hyd. bichlorid. 2d. Cod-liver oil. 3d. Operation. Head of tibia was trephined, and abscess was found. A fluidrachm of pus was discharged; dressed with carbolic solution.	Cured.	In hospital 62 days.
2	24	M.	Of left tibia, which is larger in diameter and one inch longer than right. Much pain. No sinuses.	No history of injury, or syphilis.	Trephined head of tibia, and found pus and some dead bone; dressed with tent; washed out daily with sol. potass. permang.	Cured.	In hospital 70 days.
3	46	M.	Periostitis of right tibia.	Followed fall and blow five weeks before admission.	Rest, elevation, poultice, etc.	Cured.	In hospital 16 days.
4	33	M.	Of left tibia.	Syphilis.	Potass. iod. and hyd. bichlorid.	Relieved.	In hospital 7 days.
5	24	M.	Of right tibia.	Syphilis.	Potass. iodid. and hyd. bichlorid.	Relieved.	In hospital 9 days.

Osteitis and Periostitis.—(Continued.)

No.	Age.	Sex.	Nature.	Cause.	Treatment.	Result.	Remarks.
6	15	F.	Of left tibia.	Accompanying malarial fever.	Iodine paint; incision; some suppuration.	Cured.	
7	32	M.	Of tibia.	Syphilis.	Potass. iodid. and blister.	Cured.	In hospital 18 days.
8	27	M.	Of tibia.	Followed typhoid fever.	Incisions, evacuating pus, poultices, potass. iodid., etc.	Improved.	Discharged for misconduct. In hospital 3 months.
9	11	F.	Of left tibia.	Followed blow below knee.	Free incision; pus evacuated.	Improved.	In hospital 2½ months.
10	18	M.	Of femur (left).	Followed injury from fall.	Moss and salicylic acid; potass. iodid. internally.	Improved.	In hospital 96 days.
11	9	M.	Of femur (right).	No history of injury.	Incision; bone scraped; ol. morrhua, etc.	Improved.	In hospital 5 months
12	7	M.	Of right tibia.	Fell, striking right knee.	Trephining of bone and free incisions; Stromeyer splint.	Cured.	In hospital 84 days. There was anchylosis of the knee, which was also cured.
13	18	M.	Of femur.	Exposure.	Rest, blister, potass. iodid.; small abscess formed and was opened.	Improved.	In hospital 3 months, 17 days.
14	28	M.	Of left tibia. Periostitis and osteitis.	From fall when child.	Incisions and gouge.	Cured.	In hospital 80 days.
15	42	M.	Of left tibia.	From fall 6 months ago.	Incision and gouge.	Cured.	In hospital 66 days.

WILLIAM HUNT.

OPERATIONS FOR THE REMOVAL OF CALCULI AND FOREIGN BODIES FROM THE BLADDER.

1.—LITHOTOMY.

THE first operation for the removal of a stone from the bladder by lithotomy recorded as having been done in this hospital, was performed by Dr. Redman, of the surgical staff, on October 22, 1756. From that time until October, 1878, 133 lithotomy operations were recorded, with 114 successful results. In the same period lithotrity was performed 19 times, with 2 deaths.

The early lithotomy operations, or those performed prior to the present century, appear to have been uniformly successful. Twelve cases in succession were relieved in this manner, without a single death. The records of these cases are not as complete as could be desired, and they fail to furnish the data that would be needed to establish a rigid comparison between the different methods of operating in order to estimate their relative mortality. As the rule, however, the lateral operation of Cheselden has been practised. In some exceptional cases the bilateral has been resorted to. Upon four occasions the median operation was performed. From 1832 to 1852 a series of successful results were also obtained. During this period 35 lithotomy operations were attended by only 1 death; and during the ten years, from 1868 to 1878, out of 20 such operations, there was likewise only 1 fatal case.

The total mortality of the 133 cases of lithotomy was 19, or 14.28 per cent. Post-mortem examinations were held upon these fatal cases, with the result of demonstrating that there coexisted in all of them more or less advanced organic disease of the kidneys or bladder, or both.

In one instance where lithotomy was performed no stone existed in the bladder, although the rational signs were present. The patient subsequently recovered, with a sinus in the perineum, but his cystitis was greatly improved by the operation.

Anæsthetics have been used in all the later operations, ether being preferred. No accidents of a serious nature occurred during the performance of any of these operations.

In 6 cases, foreign bodies, some of them more or less encrusted with urinary deposits, have been removed from the bladder (4 being males, 2 females). These objects, consisting of a piece of a lead-pencil, a steel bar from a watch-chain, a large hair-pin, a glass thermometer-tube, and a gum catheter were severally removed by lithotomy or dilatation of the urethra. In one instance a bone "piercer" was extracted from the bladder of a young girl by the latter method.

The usual method of performing lateral lithotomy is adopted; but preparatory treatment, by rest and good food, with tonics when necessary, is considered of primary importance.

The after-treatment adopted is simple. Immediately after the operation, hemorrhage having been controlled, a full anodyne is given by the mouth or hypodermically in an adult, or in the form of suppository in the case of a child. A soft towel is placed between the knees, and another between the ankles. A roller bandage is now applied to the knees, so as to keep the limbs at rest. The patient is allowed to lie either upon the back or side, a draw-sheet having been placed under the hips to catch the discharges. This cloth, being readily renewed, enables the patient to be kept dry and comfortable. No catheter is introduced into the wound, but the urine is allowed to find its way freely through either channel.

The diet is regulated with care for the first few days. The bowels having been fully evacuated on the morning of the operation (by castor oil and an enema), they are not allowed to be disturbed, but are locked up by opiates for four or five days. If, at the expiration of this time, they should not be spontaneously moved, the lower bowel is stimulated by an enema.

The knife used in performing most of the later operations is seen in the figure, which is the exact size.

The following table furnishes a statistical account of the cases of lithotomy from 1756 to the end of 1878. It is followed by notes of interesting cases, and short histories of several cases in which foreign bodies were removed from the bladder:

Table of Lithotomy Operations at the Pennsylvania Hospital from 1756 to 1879

No.	Name.	Age.	Admission.	Discharge.	Result.	Operator.	Remarks.
1	Ann Fost	...	Oct. 27, 1756	Jan. 12, 1757	Cured	Redman	Stone of unusual form and size.
2	James Miller	Child	Aug. 27, 1759	Oct. 10, 1759	Cured	Shippen	Stone very large.
3	James Clark	4	April 14, 1763	Sept. 5, 1765	Cured	Bond.	
4	James Child, Jr.	...	May 4, 1765	June 19, 1765	Cured	Bond	Weight, 2 ounces, 2½ drachms. Size of a hen's egg.
5	John Harper	...	Aug. 25, 1765	Oct. 16, 1765	Cured	Bond.	
6	Jeremiah Tracy	Child.	Oct. 2, 1765	April 2, 1766	Cured	Bond.	
7	Dominick Harken	...	March 5, 1777	May 9, 1777, by request.	Relieved.		
8	Ann Coyle	...	June 27, 1782	July 10, 1782	Cured.		
9	Michael Fisher	...	Nov. 10, 1786	Dec. 12, 1786	Cured.		
10	Jacob Felkner	Youth	Nov. 29, 1788	April 30, 1789	Cured	Jones.	
11	James Bennet	...	Sept. 1, 1792	Nov. 28, 1792	Cured	Foulke.	
12	James Fox	2 years, 8 months.	March 20, 1798	April 7, 1798	Cured	Physick	3½ by 2¾ inches in circumference; weight, 2½ drachms.
13	Abner Lamb	7	Jan. 1, 1800	Jan. 21, 1800	Died	Physick	5¾ by 4 inches; weight, 1 ounce, 2 drachms, 49 grains.
14	James Shaw	...	Dec. 13, 1801	July 31, 1802	Died.		
15	John Carrigan	7	May 13, 1802	June 16, 1802	Cured	Wistar.	
16	Isaac Vanderwalker	4 years, 2 months.	June 6, 1804	Aug. 22, 1804	Cured	Wistar.	
17	John Bruhaken	...	May 9, 1809	June 29, 1809	Died.		
18	Joseph Bently	13	Nov. 6, 1809	June 15, 1810	Cured	Physick	Weight, 5 drachms.
19	Thomas McDowell	...	March 27, 1810	April 23, 1810	Cured.		
20	George Wall	...	Nov. 7, 1810	Feb. 7, 1811	Cured.		
21	John Brown	...	Nov. 13, 1811	Jan. 25, 1812	Cured	Physick.	
22	Francis Welsh	10	Feb. 22, 1812	April 25, 1812	Cured	Physick	Weight, 164 grains.
23	Thomas Allen	...	April 25, 1812	April 14, 1813	Cured.		
24	Wm. P. Price	13	April 29, 1813	May 15, 1813	Died 3 days after operation.	Physick	Weight, 194 grains.
25	Needham Bryan		May 12, 1813	Aug. 4, 1813	Cured.		
26	Nathan Cattell		Nov. 1, 1813	Jan. 22, 1814	Cured.		
27	Gideon Goodwyn		Sept. 21, 1814	Dec. 29, 1814	Cured	Dorsey	Weight, 1¼ ounces, 30 grains.
28	Shadrach Mears	...	June 17, 1815	Aug. 21, 1815	Cured	Dorsey.	
29	Jacob Tussey	8	Jan. 15, 1816	March 30, 1816	Cured	Dorsey	Weight, 6 drachms.
30	James Parker	...	July 6, 1816	Aug. 1, 1816	Died	Dorsey.	
31	Robert Davis	...	May 1, 1816	Aug. 10, 1816	Cured	Dorsey	Weight, [illegible] ounces, 2 drachms.
32	Jane Maise	...	Sept. 1, 1816	Nov. 15, 1816	Cured.		
33	Clatterin Loughkin	...	June 21, 1817	Aug. 1, 1817	Cured	Dorsey	2 stones.
34	Thomas Allen	22	June 22, 1817	Aug. 30, 1817	Cured	Dorsey	Weight, 1 ounce, [illegible] grains.
35	Isabella Berry	9	June 23, 1818	Aug. 5, 1818	Cured	Parrish	Weight, 3 drachms, 11 grains.

Table of Lithotomy Operations at the Pennsylvania Hospital from 1756 to 1879.—(Continued.)

No.	Name.	Age.	Admission.	Discharge.	Result.	Operator.	Remarks.
36	James Moses	...	Oct. 14, 1818	Jan. 26, 1819	Cured.		
37	Richard Harris	...	Dec. 5, 1818	April 15, 1819	Relieved.		
38	Robert S. Hooper	38	July 1, 1819	Oct. 3, 1819	Cured	Parrish	Weight, 1½ ounces.
39	Penrose Fuhr	...	Dec. 7, 1819	Feb. 5, 1820	Cured.		
40	Samuel J. Herron	...	Dec. 8, 1819	Jan. 31, 1820	Cured.		
41	Isaiah Baptiste	12	Sept. 9, 1823	Oct. 25, 1823	Cured	Barton.	
42	Edward Roebuck	6	Sept. 21, 1823	Jan. 7, 1824	Cured	Barton	Weight, 86 grains.
43	Samuel Hustin	...	Feb. 17, 1824	April 5, 1824	Died.		
44	Osmin Henvis	26	Dec. 30, 1824	July 6, 1825	Cured	Barton	Rough; circumference, 4 inches; weight, 6½ drachms.
45	Abraham Margerum	...	Nov. 14, 1825	Jan. 2, 1826	Cured.		
46	James Barber	...	April 14, 1826	June 9, 1826	Cured.		
47	John Chandler	...	July 21, 1828	Nov. 15, 1828	Cured	Barton.	
48	Samuel Suter	...	Dec. 20, 1828	Dec. 29, 1828	Died.		
49	Charles Lex	...	Jan. 21, 1829	March 28, 1829	Cured.		
50	Samuel McDonald	...	April 2, 1829	July 6, 1829	Cured.		
51	Michael Engles	...	Oct. 24, 1829	Nov. 28, 1829	Died	Barton.	
52	Robert Fry	...	Sept. 25, 1829	April 3, 1830	Cured	Barton.	
53	William Eastwood	6	April 28, 1832	June 2, 1832	Cured	Barton.	
54	Houston Sigman	4	Sept. 10, 1832	Nov. 1, 1832	Cured	Barton.	
55	Joseph Purlz	3	Oct. 1, 1832	Oct. 28, 1832	Died	Barton.	
56	Grayson Nelson	14	April 30, 1833	July 3, 1833	Cured	Barton.	
57	Patrick H. Thorp	19	Nov. 7, 1833	Jan. 1, 1834	Cured	Barton.	
58	Peter Spyers	17	Feb. 8, 1834	July 19, 1834	Cured	Hewson	Weight, 92 grains.
59	Daniel Gillan	6	May 31, 1835	Aug. 5, 1835	Cured.		
60	James Driver	16	June 4, 1835	July 29, 1835	Cured.		
61	Stephen Black	6	Dec. 22, 1835	Feb. 9, 1836	Cured	Barton.	
62	Ellen Clincy	10	Feb. 29, 1836	May 11, 1836	Cured.		
63	Robert Thomson	24	March 17, 1836	June 16, 1836	Cured.		
64	William McElroy	12	Jan. 14, 1836	March 4, 1837	Cured.		
65	Samuel Darby	16	April 3, 1836	July 18, 1836	Cured	Norris.	
66	Anthony Stresler	43	Oct. 25, 1837	Jan. 25, 1838	Cured	Randolph.	
67	Bernard McKenna	7	Dec. 15, 1838	Feb. 22, 1839	Cured.		
68	John Hughes	4	Jan. 19, 1839	Feb. 6, 1839	Cured.		
69	John Ransley	2	March 12, 1839	June 6, 1839	Cured.		
70	John Hughes, 2d	5	Nov. 11, 1839	Dec. 4, 1839	Cured.		
71	John Ramsey	3	Jan. 8, 1840	Feb. 22, 1840	Cured.		
72	W. R. Patterson	3	May 5, 1840	June 5, 1840	Cured.		
73	William Bradley	8	Nov. 11, 1840	May 20, 1841	Cured	Norris.	
74	Thomas Carlin	4	Dec. 30, 1840	Feb. 6, 1841	Cured	Norris.	
75	James Wharton	4	Jan. 10, 1842	April 14, 1842	Cured.		
76	Morgan Morgan	5	Aug. 27, 1842	Oct. 24, 1842	Cured	Norris.	
77	John McIntyre	3	Sept. 21, 1842	March 11, 1843	Cured.		
78	John McConnell	3	May 17, 1843	July 1, 1843	Cured	Norris.	
79	William Houston	3	May 2, 1843	June 21, 1843	Cured	Norris.	
80	Henry Huey	5	Sept. 19, 1843	Oct. 11, 1843	Cured.		
81	C. C. Goldsborough	21	Oct. 31, 1843	Jan. 15, 1844	Cured	Norris.	

82	Esther McDermott	...	Nov. 30, 1843	April 1, 1844	Relieved.		
83	Thomas Hildert	5	Oct. 22, 1844	Dec. 2, 1844	Cured	Norris.	
84	Samuel Jarvis	4	Dec. 2, 1844	Dec. 13, 1844	Died	Norris.	
85	John O'Neill	11	Dec. 17, 1844	Feb. 5, 1845	Cured	Norris.	
86	Charles Ross	11	Aug. 18, 1845	Oct. 22, 1845	Cured	Peace.	
87	James Farpday	4	Sept. 17, 1845	Nov. 8, 1845	Cured	Peace.	
88	David Williams	...	June 29, 1846	Oct. 9, 1846	Relieved.		
89	John Sharkey	14	Sept. 18, 1847	Oct. 29, 1847	Cured.		
90	John Beck	10	Jan. 26, 1848	May 5, 1848	Cured	Norris.	
91	Alexander Russell	19	May 18, 1849	July 7, 1849	Cured	Peace.	
92	Thomas Seal	11	Nov. 29, 1849	Feb. 15, 1850	Cured	Norris.	
93	John Turfee	$2\frac{1}{2}$	Dec. 11, 1850	Jan. 12, 1851	Died.		
94	Daniel McMenomy	19	Jan. 9, 1851	June 25, 1851	Cured	Norris.	
95	William Briery	18	Nov. 20, 1851	Dec. 11, 1851	Cured	Norris.	
96	Patrick O'Donnell	25	April 17, 1852	July 21, 1852	Cured.		
97	William Anguin	...	Oct. 9, 1852	Dec. 8, 1852	Relieved.		
98	Bennett Lynch	3	April 27, 1853	June 12, 1853	Died.		
99	William Griffiths	26	June 6, 1854	June 21, 1854	Died	Norris.	
100	William McCreight	6	Nov. 22, 1854	Dec. 15, 1854	Died	Norris.	
101	William Passwater	5	July 18, 1855	Sept. 17, 1855	Cured.		
102	Michael Reynolds	28	Sept. 19, 1856	Nov. 27, 1856	Cured.		
103	John Slown	29	Dec. 19, 1857	Jan. 15, 1858	Cured	Norris.	
104	Henry Grater *	2	May 5, 1858	May 27, 1858	Cured	Norris.	
105	Mary Ann Poutz	18	June 11, 1860	Aug. 16, 1860	Cured	Pancoast.	
106	James Flanagan	58	Feb. 6, 1861	April 8, 1861	Died	Hartshorne.	
107	Barney McIlvaine	46	July 12, 1861	Aug. 19, 1861	Cured	Hewson.	
108	Jacob Tanner	52	April 11, 1862	April 18, 1862	Died	Pancoast.	
109	William S. Kirk	67	April 25, 1862	July 12, 1862	Relieved	Pancoast.	
110	John Mack	[illegible]	April 15, 1864	April 27, 1864	Died	Hartshorne.	
111	John Fisher	7	Oct. 5, 1863	Nov. 10, 1863	Cured	Hartshorne.	
112	John C. Roberts	8	May 23, 1864	June 14, 1864	Cured	Morton.	
113	William F. Williams	21	Dec. 4, 1865	Dec. 15, 1865	Died	Morton.	
114	John Malseed	20	Feb. 5, 1867	May 27, 1867	Cured	Agnew.	
115	Ed. Armstrong	25	April 25, 1867	June 6, 1867	Cured	Hewson.	
116	Michael Rogers	23	Dec. 30, 1867	Feb. 15, 1868	Cured	Morton.	
117	Fred. K. Bentz	2	June 6, 1868	July 11, 1868	Cured	Agnew.	
118	William Davidge	17	July 14, 1868	Aug. 19, 1868	Cured	Hewson.	
119	Edward Mount	8	Nov. 18, 1868	Dec. 30, 1868	Cured	Hunt.	
120	Edward Arlenspach †	40	Jan. 27, 1869	July 9, 1869	Cured	Hewson	Median operation.
121	Joseph F. Corte	49	May 18, 1869	June 12, 1869	Cured	Agnew	Large stone.
122	Chas. T. Sweeney	4	Nov. 13, 1869	Feb. 21, 1870	Cured	Hunt	Weight, 230 grains.
123	Nicholas Malloy	47	March 25, 1872	May 11, 1872	Cured	Hewson	Weight, 400 grains.
124	John McRea	51	June 21, 1872	Sept. 17, 1872	Cured	Morton	Median operation.
125	Barney Riley	9	May 29, 1873	July 17, 1873	Cured	Hunt	Weight, 143 grains.
126	James Sweeney	2	July 28, 1873	Aug. 4, 1873	Cured	Morton	Weight, 160 grains.
127	Chas. P. McCullough	30	May 6, 1874	June 1, 1874	Cured	Morton	Weight, [illegible] grains, large.
128	Evan W. Jervis	5	Aug. 17, 1875	Sept. 15, 1875	Cured	Andrews	Weight, 21 grains.
129	Benj. B. Potts	65	Aug. 23, 1875	Dec. 7, 1875	Died	Lewis.	
130	James G. Thompson	7	Oct. 21, 1875	Nov. 19, 1875	Cured	Lewis	Weight, 102 grains.
131	James Lakey	8	May 18, 1876	Dec. 9, 1876	Cured	Hewson	[illegible]
132	Rudolph Ritzous	3	June 20, 1878	July 3, 1878	Cured	Hunt	Weight, 81 grains.
133	Charles Haward, colored	5	Oct. 11, 1878	Dec. 11, 1878	Cured	Morton	Weight, [illegible] grains.

* Case of congenital urinary calculus, weight, 1[illegible] grains. N. Y. Med. Jour. [illegible], July, 1858, p. 7[illegible].
† No stone found.

Results of 133 Cases of Lithotomy.

Cured	108 } or 85.72 per cent.
Relieved	6 }
Died	19, or 14.28 per cent.
Males	126
Females	7
Total	133

Of 120 cases in which the age was recorded, there were—

	Operations.	Deaths.		Operations.	Deaths.
Under 5 years	36	4	From 30 to 40 years	9	2
From 5 to 10 years	32	1	" 40 " 50 "	7	1
" 10 " 20 "	22	1	" 50 " 60 "	4	2
" 20 " 30 "	8	1	" 60 " 70 "	2	1

Notes on Operations since 1860.

CASE 105.—Mary A. P., aged 48. Stone large. Lithontripsy was attempted, but found impracticable; lithotomy was then performed. The stone weighed 2 ounces, and was removed through an incision in the median line of the vagina (coppocystotomy). The wound was at once closed with silver sutures. Recovery.

CASE 106.—James F., aged 58. Stone was held behind the pubis, on the right side, by a sacculated bladder; was dislodged by the forceps. Death.

Post-mortem.—A sac was found in the bladder, in front where the stone (which weighed 14 drachms) had been lodged.

CASE 111.—John F., aged 7. Stone was retained in the fundus of the bladder, to which it adhered and nearly escaped detection.

CASE 113.—William F. W., aged 21. For seventeen years had severe vesical irritation. Calculus weighed 318 grains. Death from uræmia occurred on the eleventh day.

Post-mortem.—Left kidney enlarged, the right almost absorbed; evidences of old abscesses in all directions.

CASE 120.—Edward A., aged 40, a German baker. Had pain for a long time, with severe vesical symptoms and considerable discharge from the bladder, which continued over a year. When admitted into Dr. Morton's ward a careful examination of the bladder failed to detect any stone. The case passed into the care of Dr. Hewson, February 1, 1869, who pronounced the presence of a stone in the bladder.

February 24.—Before the clinical class Dr. H. again thought he felt a stone, and subjected his patient to lithotomy. No stone could be found. Convalescence was slow, and the patient when discharged had a small urinary fistula, but the cystitis was much relieved.

CASE 124.—John McR., aged 54. Had been a subject of stricture for years. Of late the symptoms were much more marked, with signs of calculus. Dr. Morton dilated the urethra and found a stone. The usual median operation was performed. The calculus was found at the neck of the bladder. A part projected into and dilated the urethra. Stone weighed 112 grains; was phosphatic. It measured one and a half inches in length.

CASE 125.—Barney R., aged 9. Had suffered for two years. Some hemorrhage followed the operation, which was controlled by alum water.

CASE 126.—James S., aged 2. Symptoms of stone for six months. Calculus weighed 160 grains.

CASE 127.—Charles P. McC., aged 36. Bladder and urethra extremely irritable. On ninth day had a severe chill, but no further indisposition. Discharged, well, on twenty-first day.

CASE 128.—Evan W. J., aged 5. Suffered for two years. The day after the operation the pulse was 140; temperature, 101°. On the third day the temperature was 107°; pulse 165, and delirium. Quinine and stimulants were freely given. Gradually convalescence occurred.

CASE 129.—Benjamin B. P., aged 65. Thirty years previously was operated upon for stricture. In 1855 he thought he had stone. Ten years ago, at San Francisco, had a false passage made, and for this he underwent an operation, on account of urinary extravasation. Eight years ago a calculus was detected and crushed, and had been free of trouble until a few months ago. An examination on admission determined the presence of a stone, and lithontripsy was determined upon.

September 4.—A No. 12 bougie was easily passed.

September 20.—Stone was crushed.

October 29.—Operation repeated, but the main part of the stone was found to lie so firmly behind the prostate that it could not be grasped.

December 1.—Lithotomy; stone removed. After the operation reaction was imperfect, and exhaustion and death followed.

Post-mortem.—Heart: hypertrophy, particularly of left side; dilatation of right ventricle. Kidneys: surfaces uneven; nodulation marked; few spots of granulation; section shows them to be pale and firm, with cortex cysts, some as large as a walnut. Bowels showed local peritonitis: several coils of intestine were bound down to the fundus of the bladder. Bladder: walls much thickened; connective tissue surrounding much thickened, and infiltrated with pus. Mucous membrane discolored, and covered with lymphoid points. Rectum thickened: walls black.

CASE 131.—"James L., from Mahanoy City, Schuylkill County, Pennsylvania, aged 8, was admitted to St. Joseph's Hospital in the latter part of October, 1874. He was then suffering from severe vesical symptoms; there was frequent micturition, accompanied by straining, and by marked protrusion of the mucous membrane of the rectum. On sounding, a calculus of good size was easily detected. His mother stated that three years previously he had been cut for stone by some one near his home, and that a calculus of the size of an almond had been taken away.

"November 4 I performed the operation of left lateral lithotomy through the old cicatrix, and removed a calculus weighing 2 drachms and 35 grains. On subsequent section this proved to be phosphatic in its character on a uric acid nucleus. At the time of the operation an old fistulous track was discovered leading from the rectum to the bladder. The rectal orifice of the fistula was partially occluded by a small pendulous polypoid growth, which moved up and down, acting apparently as a sort of stop-valve. I was subsequently informed by Dr. Bissell, who treated the child after the performance of the first lithotomy by another practitioner, that this fistula made its appearance shortly after the first operation. He stated that he and others had on several occasions felt the second calculus with a probe passed up through the rectal fistula into the bladder, and that the boy was in the habit of passing much of his urine by the rectum. For the first few days after my operation, on November 4, the child did not do very well. His stomach seemed unusually irritable, and he had difficulty in retaining his food.

"November 21 he vomited a large lumbricoid worm, and afterwards he passed one or two by the rectum. He then became very much better, the wound granulated, and the fistula seemed to be contracting, and gave scarcely any trouble. The urine and fæces passed by the natural passages, and the patient appeared to be in a fair way to recovery.

"The 28th of November his mother was forced to go to her home, and insisted on taking the child with her, despite of earnest advice to leave him in the hospital. He thus passed from my observation."*

May 18, 1876, this lad was admitted into the Pennsylvania Hospital with marked symptoms of stone. On examination there was found on the left of the median line a double linear cicatrix; a sound carried from the meatus towards the bladder was prevented entering by a foreign body; in the perineum there was a decided enlargement.

May 20, urethra was dilated as far as the foreign body; a forceps carried in was made to grasp the stone, and it was then brought to the meatus; an incision was then made, and it was extracted; a staff was then introduced into the bladder and a grooved director into the fistulous opening in the rectum, and the director was then carried forwards and upwards until it came in contact with

* Note from Dr. John H. Brinton.

the staff; an incision was then made in the perineal centre dividing the tissues to the grooved director, in fact, a lithotomy operation. A stone was found in the bladder about the size of a date-stone: the rectal walls were approximated by silver sutures; an S-shaped catheter was allowed to remain in the bladder. In December the patient was discharged with a small recto-vesical fistule.

CASE 132.—Rudolph R., aged 3. Stone symptoms very marked; had been examined at one of the city hospitals, and as no stone could be found the symptoms were supposed to be due to phymosis, and circumcision was performed, but without relief to the vesical irritation. After etherization a stone was found by Dr. Hunt.

CASE 133.—Charles H., colored, aged 5. The stone weighed 33 grains, was large, rough, and hard. Recovery was rapid, but a fistule remained some weeks. The urine when examined was found to contain albumen in a large amount. Was discharged to go to the country. Two months afterwards the child was well, the fistule had closed, and the albuminuria had disappeared.

Table of Foreign Bodies removed from the Urinary Bladder.

Case 1. Lead-pencil. Case 2. Bone piercer. Case 3. Hair-pin. Case 4. Glass thermometer-tube. Case 5. Portion of a gum catheter. Case 6. Steel bar from a watch-chain.

Notes of Cases.

CASE 1. *Lead-pencil introduced into the male urethra: removed from the bladder by median operation: recovery.*—This patient had introduced into his urethra, some weeks prior to his admission, a portion of a carpenter's lead-pencil, which found its way into the bladder soon after, and at once excited irritation. An unsuccessful attempt was made to seize the piece and to withdraw it by the urethra. A fragment, however, came away some time after with the urine, but the organ was so irritated by the remaining portion that it was removed by Dr. Morton by the median operation. The remnant measured two inches in length, and was incrusted with calculus.

CASE 2. *Bone piercer removed from the female bladder after dilatation.*—The patient, a sewing-girl, aged 27, and unmarried, stated that four months previously a wooden seat had broken under her, and that a splinter had penetrated her person through the vagina. She was unable to sit or stand upright without great suffering; constantly had to maintain a stooping posture; had great pain over the bladder, with all the symptoms of stone. A sound at once detected a foreign body. The urethra, after anæsthesia, was rapidly dilated by Dr. Morton, who introduced the forefinger, and the extraction of the foreign body was readily effected by the dressing forceps. It proved to be an old-fashioned bone piercer, two and three-quarters inches in length. One end was very sharp, the other blunt and partly broken. The shaft was incrusted with stone. The patient afterwards admitted having introduced the instrument blunt end first into the urethra, and finding difficulty in withdrawing it, let go her hold upon it, when it slipped back into the bladder, which was then distended with urine. Within twenty-four hours she was able to retain her urine, was free from pain, and was discharged well, five days after the operation.

CASE 3. *Hair-pin introduced into the female bladder: removed after dilatation.*—Millie L., colored, aged 12, was admitted May 24, 1876, with vesical irritation. On examination a foreign body was discovered in the bladder. After etherization and dilatation, Dr. Hunt extracted a hair-pin (see figure) two and one-half inches in length, which the child afterwards confessed to have introduced into her vagina some

days before. Instead of placing it in the vagina, as she supposed, the urethra was the part into which the curved end of the pin was introduced. It was only removed after dilatation, and the pin was turned so that the curved end was first delivered (see figure).

CASE 4. *Glass thermometer-tube removed by lithotomy from the male bladder.*—A lad was admitted into the hospital who confessed to have been in the habit of introducing into his urethra the stem of a glass thermometer. On the last occasion it broke, and the lower fragment passed into the bladder. There it at once produced grave vesical irritation, but it was tolerated for several years. The lad was taken out of the hospital to a private house close by, and Dr. Hewson, assisted by Dr. Hunt, operated by the lateral section, removing the tube, which measured about four and one-half inches in length. The lad made an excellent recovery, without an unpleasant symptom.

CASE 5. *Portion of a French gum catheter removed from the male bladder.*—In 1875 a young man came from Kansas. In using an ordinary flexible catheter it became fractured about an inch and a half from the end, and, on attempting to withdraw it, the end separated, and remained in the bladder. Various attempts were made at the time to remove it, which were unsuccessful. He then came East, and on admission the foreign body was detected, and removed by aid of the lithotrite, without difficulty.

CASE 6. *Steel bar from watch-chain, with piece of string attached, voided from the bladder.*—Dominick McD., aged 17, was admitted October 28, 1870, with symptoms of vesical irritation. He stated that he had been in the habit of passing a steel bar (about an inch and a half long) into his urethra; and had fastened a string to the middle of the bar, so that he was able to push in the bar any distance along the canal, and then at pleasure to draw it out. On the last occasion the bar slipped into the bladder, and, becoming fixed crosswise, he was unable to get it out. When admitted the string was hanging out at the end of the urethra.

October 31.—On making slight traction the already weakened string gave way, and the bar remained in the bladder. It was readily felt by a sound. Several attempts were then made to extract it, without success. The urethra was then well dilated, with a view to its passage, should it present favorably at any time. The patient was directed to allow the bladder to become well filled with urine, and then to assume a position on his knees, bending forward. At the first effort the bar was passed. On examination, the string to which it was attached was found had been broken about three inches from the bar. Patient was discharged November 7, 1870.

Two Cases of Stone, one lodged in the Male Urethra, the other of Considerable Magnitude, removed after Dilatation.

CASE 1. *Stone lodged in the urethra.*—James B., aged 28, was admitted April 6, 1870, with a calculus in the urethra, where it had been for two months, gradually coming forward. The stone was found about one inch from the meatus. An incision was made and the stone readily removed. Was discharged April 7, 1870.

CASE 2. *Calculus removed after dilatation.*—Margaret G., aged 9, was admitted December 12, 1868, having suffered with severe symptoms of stone for three or four years. Was greatly emaciated, and much worn by long irritation. The stone was readily discovered, and under ether the urethra was dilated, December 16, 1868, but the stone was found to be large and hard. A grooved director was introduced, and a small incision was made in the urethra, when the stone was removed. Almost immediately after the operation vomiting and abdominal tenderness ensued, and peritonitis developed at once. Death occurred on the second day. The calculus weighed 162 grains, and measured $3\frac{1}{2}$ by $2\frac{1}{4}$ inches.

2.—LITHOTRITY.

This operation was introduced into the hospital by Dr. J. Randolph in 1836, who performed it in a number of cases with marked success. 19 cases are recorded in which crushing was practised, 2 of which died. The cause of death in 1 case was not stated, in the other an encysted stone was found in a sacculated bladder. In 15 cases the patients were completely cured by the operation.

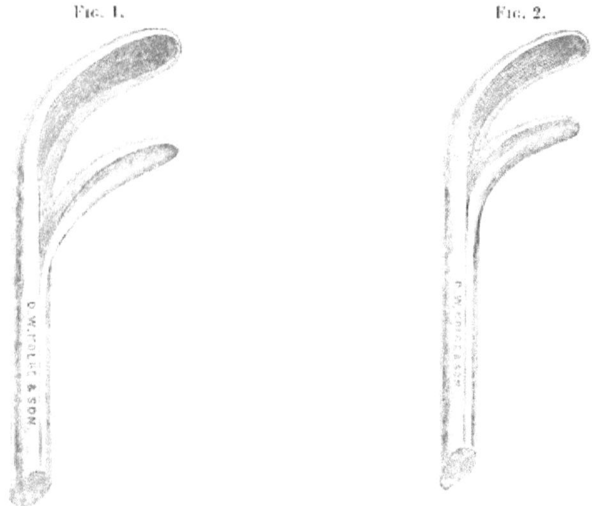

Fig. 1. Fig. 2.

The crushers used in the hospital are of two sizes and are not fenestrated (Figs. 1 and 2); they have been drawn their exact size.

Table of Lithotrity Operations performed from 1836 *to* 1878.

No.	Name.	Age.	Admitted.	Discharged.	Result.	Operator.
1	William Askine	32	Oct. 26, 1836	Nov. 16, 1836	Cured	Randolph.
2	Silas Tompkins, M.D.	36	Dec. 15, 1836	May 17, 1837	Cured	Randolph.
3	Margaret Beestall	4	Jan. 8, 1838	March 19, 1838	Cured	Randolph.
4	George House	15	March 30, 1838	May 26, 1838	Cured	Randolph.
5	Chester Cavender	23	Aug. 22, 1838	Sept. 27, 1838	Died	Norris.
6	Daniel McMenomy	10	March 9, 1839	Aug. 13, 1839	Cured	Randolph.
7	I. Edwards	19	Jan. 30, 1840	March 1, 1840	Cured	Randolph.
8	William Brackerly	8	Nov. 11, 1840	March 15, 1841	Cured	Norris.
9	John B. Stewart	25	July 1, 1841	Aug. 20, 1841	Died	Peace.
10	Francis McFadden	7	Sept. 2, 1846	Jan. 6, 1847	Cured	Randolph.
11	Lucian A. Harris	18	Nov. 4, 1847	Dec. 18, 1848	Cured	Randolph.
12	Serina Grass, negress	49	June 15, 1859	Aug. 13, 1859	Cured	Neill.
13	John Fritz	17	Oct. 21, 1856	April 10, 1858	Cured	Pancoast.
14	John A. Hunter	28	April 5, 1868	Sept. 5, 1868	Cured	Agnew.
15	Joseph F. Corbe	19	Dec. 1, 1869	Jan. 7, 1870	Cured	Morton.
16	Benj. B. Potts	65	Aug. 23, 1875	(See Lithotomy.)	Improved	Levis.
17	Benj. Tarleton	69	Sept. 22, 1874	Oct. 16, 1874	Relieved	Levis.
18	Charles P. McCullough	38	June 13, 1876	June 27, 1876	Cured	Morton.
19	James Daly	25	Nov. 8, 1877	Transferred to medical ward.	Cured	Morton.

Remarks on the Lithotrity Operations.

Case 4.—This patient had been cut by Dr. Randolph eleven years previously for stone.

Case 6.—A fragment of calculus became entangled in the urethra, and was removed by an external incision five weeks after his leaving the hospital.

Case 11.—A large fragment was afterwards cut from the urethra by Dr. Norris.

Case 13.—Was admitted for paralysis and calculous disease (a fall from a tree caused fracture of the spine). After several operations the stone was completely crushed and the bladder entirely relieved.

Case 15.—This patient had long-standing syphilitic disease. The calculus, which was soft and quite large, after several sittings was successfully crushed.

Case 16.—This patient was cut for stone in the hospital (Case 121, Lithotomy), May 22, 1869, and at time of discharge was free of all symptoms. Applied for admission December same year for vesical irritation. On examination a calculus was found.

December 8.—Lithotrite was introduced and the stone was crushed; the fragments weighed 62 grains.

January 10, 1870.—Discharged, quite well.

Case 17.—Was admitted September 22, 1873. About ten years since first noticed irritation of the bladder, a slight scratching of the urethra, when passing urine; there was also a deposit in the urine after standing. For eight years has had frequent micturition, which has of late much increased.

October 1.—Bladder was injected, and Dr. Levis succeeded in crushing the stone across the smaller diameter.

October 2.—Chill followed the operation.

October 8.—Stone again crushed; large amount of stone passed.

October 16.—Patient was well enough to be up, and was discharged at his own request, improved.

Case 18.—See history, Case 127, Lithotomy. Two years previously was operated on by lithotomy, and was quite well for eighteen months, then symptoms of stone reappeared. The calculus removed by crushing weighed 63 grains.

Case 19.—Was admitted with vesical irritation. A small calculus was found, and crushed at one sitting. The patient was afterwards admitted into the medical ward. Died June 7, 1878. The post-mortem showed a calculus in the kidney; bladder healthy.

Results of 19 *Lithotrity Operations.*

Cured	15
Relieved	2
Died	2

Arranged according to the Age.

Under 5 years	1 operation.	
5 to 10 "	3 operations.	
10 " 20 "	5 "	
20 " 30 "	4 "	with 2 deaths.
30 " 40 "	3 "	
40 " 50 "	1 operation.	
60 " 70 "	2 operations.	
	19 "	

Mortality . . . 10.52 per cent.

Thomas G. Morton.

REFRACTURE OF BADLY-UNITED FRACTURES OF THE EXTREMITIES.

REFRACTURE, or rupture of the callus, is only one of the methods of dealing with deformity following mal-union after fracture, but it combines the advantages of availability and freedom from risk. It is the method most frequently adopted in the hospital, and, as a number of cases have occurred within a few years, it is possible to draw some valuable conclusions from the study of their notes, as recorded at the time of treatment.

Gradual extension and pressure at the seat of deformity has been spoken of by some writers as a method of dealing with deformed union, but, as is easily seen, this will be of service only at the earliest stages of solidification, and in truth should constitute the treatment of the fracture from the very start, if there is tendency to angular displacement or overlapping. When cases present themselves for the relief of distorted limbs the union is too firm to be affected by any such method as this, which can only mould soft or semi-solid callus.

In addition to pressure and extension, tenotomy may not unfrequently be of service in the early weeks of fracture by weakening the power of displacing muscles upon the broken fragments. It may also be needed as an assistant to refracture, when the surgeon chooses that method of managing the troublesome bones; but of itself it is, of course, valueless in meeting any deformity dependent on deposition of callus.

By the term refracture I mean causing a separation of the fragments at the original site, which may be done by the application of external force alone, or by introducing, as a preparatory step, a perforator to weaken the union due to the large mass of callus usually deposited about fractures when the apposition is not good. The latter method makes, in many instances, a compound fracture, but still comes under the appellation of a subcutaneous operation. I see no reason why a narrow saw, or a chain-saw, as has been suggested, might not be introduced in the same way, and subcutaneous osteotomy done, as in cases of anchylosis; but I do not know that this method has been used to any extent in separating badly-consolidated fractures.

The more serious operations, such as excision of the ends of the displaced fragments and amputations of the limb, which may occasionally be adopted in instances of very great deformity and inability to use the extremity, must be very seldom justifiable, and should only be employed when other means have failed. With these procedures the present paper has nothing to do, but will be restricted to the consideration of refracture, as described above, which has been the method employed in most of the cases treated during the last few years.

The first question in regard to this operation that must be discussed is, What are proper cases to be subjected to refracture? This query can only be answered by investigating the amount and character of deformity that demands operative treatment, the locality of the original fracture, and the length of time after the primary injury that refracture of the mal-union can be accomplished.

By far the most frequent reason for having recourse to this operation is angular shortening of the lower limbs, preventing perfect locomotion. There is, of course, no algebraic formula to tell us how much shortening shall be considered equivalent to an operation; and it is well known that a considerable amount of shortening, at times nearly an inch, exists in many persons with uninjured legs without inconvenience. The most favorable cases for operation are those where, on account of imperfect treatment, the fragments have united at their extremities, but not in the same line, and thus cause angular deformity and consequent shortening. The union in such instances is more readily ruptured, because the surfaces united are somewhat limited in extent, and the angularity of their position gives a good opportunity for disruption by applying pressure at the salient angle. In addition to being more amenable to separation, it is evident that the prevention of a recurrence of the deformity is more easily met by after-treatment than if the distortion was dependent upon overriding of oblique smooth surfaces.

Such cases, then, are eminently proper ones for treatment by refracture. Shortening, resulting from lateral apposition of the cylinders of bone, caused by want of attention to overcoming muscular displacement during the original treatment of the fracture, appertains to many cases, and such cases have by some been considered unsuitable for operative interference. The deformity in these instances is most probably due to overriding of very oblique surfaces of fracture, and can be treated by refracture almost as readily as other forms. In attempting to dissever the fragments the force should be applied in a manner to bring transverse strain upon the uniting fibro-cartilaginous bond. If the bones are apposed with anterior surface to posterior surface, the power should be exerted laterally; if side to side, the surgeon must use his force in an antero-posterior direction.

The general rule to govern the selection of cases for refracture may be formulated somewhat in this wise,—when the deformity is attributable to causes incidental to the first fracture, but avoidable in the second, and is of sufficient degree to interfere with perfect motion, or to cause persistent pain or great unsightliness, refracture is to be employed as a means of relief. Unfortunately, cases presenting themselves for relief are usually those that have not been under observation during the treatment of the fracture. It is, therefore, difficult to appreciate the causes of the deformity, whether it be due to great obliquity and smoothness of the fractured surfaces, to great muscular irritability, to the unwillingness of the patient to submit to restraint, or to the ignorance of the original attendant. Consequently it is at times a question whether the displacing causes can be avoided in the after-treatment, and the surgeon has to operate without knowing exactly what agents are to give him most trouble. The mere matter of great muscular development certainly should not deter us as much as formerly, for with the improved methods of extension and counter-extension, it is possible to tire out any muscular masses that tend to produce displacement, and in certain instances tenotomy may be called upon as an adjuvant to extension.

The locality of the fracture has some bearing on the propriety of the operation and the exact manner of performing it. A slight amount of shortening, or even a decided crook in the humerus or clavicle, is of little detriment to a man, while the same distortion occurring in a female, or in the lower extremity of either sex, would perhaps be sufficient cause for undertaking an operation. Again, a degree of displacement of little moment in the shaft of a bone may be a serious impediment to the free use of a ginglymoid articulation, or may interfere greatly with the tendons as they lie upon the expanded extremity of the bone. This matter of location is also of importance respecting the feasibility of refracture, for it is easily seen that it is exceedingly difficult to apply the rupturing power when the deformity is close to the hip- or shoulder-joint, for the upper fragment is not readily steadied during the manipulative processes.

Another desideratum is the determination of the time after the original fracture that it is possible to tear asunder the united fragments. This necessarily must be variable, as solidification in fractures depends on sex, age, character and situation of lesion, and a host of other contingencies. The operation, however, has often been done many weeks after the original fracture, and in some instances as late as twelve or thirteen months subsequently. In one of the cases recently treated in the hospital a period of about six months had elapsed, and yet a good cure was effected.

In truth, it may be said that absolute time has nothing to do with the question, for the surgeon must be guided by the practicability of refracturing the callus uniting the displaced ends, and this cannot be determined until trial has been made on each individual case. The only limit would seem to be the injury of the soft parts that might result from applying great power. This is fortunately giving us a wide range, and it is astonishing how much rude handling the surrounding tissues will tolerate without resenting it. Looking at the cases reported, we find that severe inflammatory reaction occurred in none of them. If we recollect that high inflammation subsequent to ordinary fractures, which are not compound, is comparatively rare, and that, when found as a sequel, it is in great part due to the sharp spiculated fragments forced into the soft parts at the time of accident or during transportation, we can understand its conspicuous absence after refracture, where the ends are more or less smooth and rounded, and the bones are placed at once in proper retaining apparatus.

Having discussed the subject of selection of cases, I must turn now to consider the various procedures and appliances employed in effecting the desired rupture of the bond of union. The most simple method, when it is possible, is to bend the distorted limb into position with the hands, as is done when it is desired to straighten partial fractures occurring in childhood. This is only practicable when the union is of recent date or the bone small. Although the callus in its early stages is fibrous and capable of being moulded into shape by slow laceration, yet, in the majority of cases presented to the surgeon, the bond of union is firm, and, when ruptured, gives evidence of its solidity by a sharp snapping sound. When this is once heard the problem is solved, and restitution can usually be effected. Hence it is that by a sudden force the uniting bond can be ruptured easier than by a gradual strain brought to bear upon it, just as a boy breaks by a sudden bend a stout stick that he could not tear apart by continuous pressure.

When the fracture is too firmly consolidated to allow replacement by mere bending and rotation with the hands, it becomes necessary to use some method that gives a better opportunity of utilizing the strength of the operator. Thus, he can make a fulcrum by bringing the limb across the end of a table, or by placing a hard block or pad under the convexity of the angular displacement. Again, he can bind a straight splint along the limb from the seat of fracture to or beyond the distal extremity, and thus control the joints and gain leverage; and, at the same time, by placing a similar support upon the limb above the point to be fractured, can render the occurrence of fracture at a new situation impossible. This, however, is probably never necessary. Another device of a mechanical kind is a screw-press or clamp, by which great pressure can be exerted to break down the arch formed by the angular union.

These methods are especially adapted to cases where the fragments have united at an angle. When the deformity depends upon lateral application of the two bony cylinders, rupture must be attempted by flexion across the bond of union, combined with rotation in the axis of the limb and strong extension and counter-extension. As has been stated, these cases are less appropriate for treatment by refracture than the former variety. Indeed, in some cases of this kind, as well as in others where the shortening is great, it may be well to keep the patient etherized and continue the excessive extension for an hour or so after rupture, in order to gain as much length as possible before dressing the fracture. In one of the cases here reported in which the time after the injury was six months, and where callus existed in large quantity, the two methods were conjoined; and while powerful extension by pulleys was made, strong sudden force was applied to the convexity of the deformed thigh (see Case 9).

In order to obtain a firm hold upon the limb when making extension by the compound pulleys, and yet to distribute the pressure so as not to bruise the tissues, Dr. Levis has recently devised an effective appliance by which to

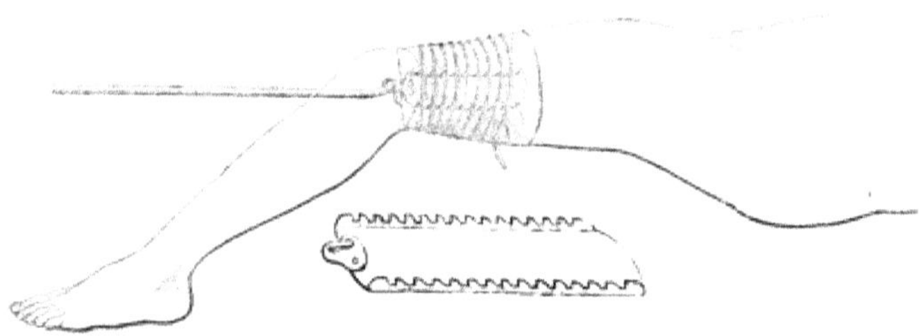

attach the pulley rope. It consists of a strong iron hook, the upper part of which expands into a flat plate about seven inches long and two wide. This plate is slightly curved transversely to apply itself to the limb, and is roughened. On the upper surface of this plate, at each side, there extends a longitudinal ridge serrated by deep notches cut into it. The appliance is adjusted in the following manner: After a bandage has been applied to the limb at the point selected, the plate and hook are laid upon it, then a strong cord is carried around the limb circularly and over the plate, catching in the saw-like teeth on the ridges above. This binds the hook and plate firmly to the limb just as sailors splice two spars together by a rope. Finally the compound pulleys are applied and extension made as required without danger of the skin and muscles being lacerated by the rope, since the pressure is distributed over a large surface.

It will now be appropriate to give a synopsis of the cases treated by re-fracture during the five years from January, 1873, to January, 1878, in which the operation was done without any preparatory drilling of the callus.

	Operator.	Date.	Sex and Age.	Time after original Fracture.	Bones refractured, and situation.	Deformity before the operation.	Result.	Time of Treatment.	Remarks.
				Weeks.				Weeks.	
1.	R. J. Levis.	1876.	M., 12	10	R. and U., at middle.	Angularity.	Improved.	4	No inflammatory symptoms.
2.	A. Hewson. Same patient as No. 7.	1874.	M., 15	3½	H., at middle.	Much deformity.	Some overlapping.	6½	No mention of inflammatory symptoms. Refractured femur at same time.
3.	T. G. Morton.	1875.	M., 26	6	T. and F., upper third.	Angularity.	Good position.	6½	No mention of inflammatory symptoms. Had both humeri broken when leg was first fractured.
4.	R. J. Levis.	1874.	F., 13	5	T. and F., at ankle.	Foot displaced outwards.	Scarcely any deformity.	5	No inflammatory symptoms.
5.	T. G. Morton.	1877.	M., 37	15	T. and F., lower third.	Foot displaced inwards.	No appreciable deformity.	5½	No inflammatory symptoms.
6.	R. J. Levis.	1873.	M., 17	6	F., upper third.	Angularity, shortening, 3½ inches.	Shortening, ½ inch.	6½	No mention of inflammatory symptoms. Had double fracture of one forearm when femur was first fractured.
7.	A. Hewson. Same patient as No. 2.	1874.	M., 15	3½	F., middle.	Shortening, 1½ inches.	Shortening, ½ inch.	6½	No mention of inflammatory symptoms. Refractured humerus at same time.
8.	Wm. Hunt.	1875.	M., 17	6 (?)	F., above condyles.	Shortening, 1½ inches.	No deformity, no shortening.	9	No inflammatory symptoms.
9.	R. J. Levis.	1876.	F., 19	21	F., middle third.	Angularity, shortening, 2¾ inches.	Shortening, ¼ inch.	10	No inflammatory symptoms.

In this connection it is proper to record a remarkable case treated by Dr. Hunt some time previous to the dates given above. As drilling and tenotomy were employed as adjuvants to the refracturing process, the case is not put in the table, which includes only the cases where simple refracture was done. Instances of excision of the deformed union and of drilling have occurred at intervals in the hospital, but are not tabulated in this paper, since the conditions are in my estimation quite different. The liability to suppuration and erysipelas is certainly increased by a resort to preparatory perforation with the drill. The following case resulted in such perfect cure that it merits insertion at this place:

Refracture of the bones of the leg, after perforation with the drill, and tenotomy for the cure of faulty union.—The case was reported by Dr. Hunt several years ago, but the greatness of the deformity and the success of the complicated operation render it one of special interest in this connection. A sailor, 37 years old, was admitted, February 19, 1872, with such deformity of the left leg that he could scarcely walk. In the summer of 1871 he had fallen from the rigging and fractured both bones of the leg near the middle. He remained on shipboard nearly three weeks, when he was transferred to a hospital at Liverpool, where he remained for several months. Examination showed remarkably firm bony union, with great angular deformity, and the foot in the position of such marked equino-varus that the sole presented towards the opposite side. There was also a cicatrix over the hip-joint of the same side, which was somewhat anchylosed, from an injury received at the time his leg was broken. After he had been in the hospital some little time a large drill was introduced through a small incision, and the bone bored completely through in three or four different directions, though there was but a single external wound. The surgeon then applied sufficient force to produce rupture at the old point of fracture, which produced a report like the discharge of a pistol. The next steps were the division of the tendo Achillis and the breaking up of the adhesions about the hip-joint. The leg was then put in good position, and showed very little shortening. The dressing was as follows: "A stout pasteboard splint, well fitted to the external aspect of the foot and leg, was applied. A moderate degree of extension by adhesive strips was maintained, and strips also were used to keep the foot in a position of slight eversion. The extension was abandoned after a short time as not requisite, and the limb, with the other dressing complete, was placed in a fracture-box." Slight suppuration followed, and several small pieces of bone were removed from the wound. Subsequently the patient was seized with an attack of what seemed to be hepatitis and pleurisy, and at the same time an ugly deep slough occurred at the wound over the fracture. At this time it seemed as though the man would die. There were grounds for suspecting a syphilitic taint, and he was therefore given iodide of potassium and the corrosive chloride of mercury. The result, however, was unexceptionally good, for when the slough separated it was found that the progress of union had not been interfered with. On August 1 the patient was able to walk, could bring the foot flat to the ground, and could perform all the natural movements of the limb. The union was solid, the tendo Achillis repaired, and the movements of the hip restored. On October 1 he was seen, and exhibited a slight limp in his gait, which was possibly due to the shortening, that amounted, however, only to one-quarter of an inch. This patient was seen in good health about two years ago.

A glance at the schedule given previously will show that during the time mentioned eight patients were treated in the hospital by refracture; the number of operations, however, was nine, because one patient had both an upper and a lower extremity operated on at the same time for deformity, resulting from mal-union. In all the cases except one (Case 1) there was very marked amelioration of deformity, and, as far as relief from lameness, etc., is concerned, we may consider the patients cured. This is very noticeable in the refractures of the femur, because we have the amount of shortening before and after treatment recorded. In the four cases it was reduced to ½ inch, ⅜ inch, none at all, and ¼ inch respectively; while before treatment there was 3½, 1¼, 1½, 2¾ inches. Another feature is the non-occurrence of marked inflammation after the violent manipulations necessary to effect rupture. In the hospital notes I find no mention of any such occurrence in any case of the whole number, and reference to the fact would hardly have been omitted in the running notes of the patients if it had occurred to any extent. The time of treatment after refracture will be found to be about the same as usually required for the uniting of ordinary fractures in similar situations. I have calculated the time by counting the number of weeks between the day of operation and the day on which the splints were discarded and the union found to be firm, though in cases of fractures of the lower extremity the patient was required to sustain a portion of his weight on crutches or canes for some time longer as a precautionary measure. This, of course, would vary much with different surgeons and attendants, and has therefore been left out of the calculation.

What are the objections that have been urged against refracture as a means of relieving deformity from mal-union, and what are the dangers that may be encountered? The first objection suggested to any one would be the possibility of the bone giving way at a point more or less distant from the original fracture. This, however, is to be considered almost impossible, unless the bone be diseased and softened. It is a very different condition of things from that in which strong efforts, employed for the rupturing of an anchylosed joint, are followed by fracture of the shaft of the bone. In the latter case we have two large expanded articular surfaces firmly bound together by adhesions, which have been deposited by a chronic inflammation, that has at the same time caused more or less disintegration of the neighboring bone. In the case under discussion there are two overlapping fragments much smaller, as a rule, than the articular ends, and generally united at an angle by fibro-osseous matter. This must be much weaker than the normal bone above or below, even if a large amount of irregular callus has been thrown out around the seat of fracture. In fact, to one who has not made the experiment, it will be surprising to find how much force is required to break even so small a bone as the radius across the knee, when the bone has been divested of all the tissues and dried. Much more strength is requisite to fracture one of the bones of the extremities when it is surrounded by periosteum and buried in the muscular masses surrounding it.

In order to test this matter I made some experiments in the hospital dead-house on a cadaver where there had been no fracture. I drew the leg of the subject over the end of a table, and while an assistant steadied the thigh and knee as well as he could, endeavored to fracture the tibia by the manipulations that would be employed in refracturing a deformed union of the leg. I repeatedly threw all my weight—over 150 pounds—with sudden force upon the leg, but only succeeded in fracturing the fragile fibula, which, being posterior and superficial, was brought in direct contact with the edge of the table without any muscular cushion. My assistant, who was stronger than I, failed also in his attempts to break the tibia. The only method that would avail to cause fracture of the tibia was to turn the subject on its face, in order to bring the subcutaneous surface of the tibia downwards, then, by suddenly striking this portion of the bone powerful blows on the sharp edge of the *iron* table, I succeeded in producing an oblique fracture at the lower part of the middle third.

These experiments show what a tremendous force would be required to fracture a sound bone by the manipulations used for refracturing deformed callus; for in that operation no one would ever think of selecting a sharp iron edge on which to strike the subcutaneous surface of the tibia. My efforts fully convinced me that there was no possible chance of my succeeding in fracturing the femur, buried as it is in a large mass of muscles. The radius and ulna were only broken by the same kind of sudden blow over the sharp table edge.

It is a long time before the permanent callus becomes as firm and compact as the original bone, as is shown by the fact that, after once sustaining a fracture, the patient, in falling, is exceedingly liable to refracture at the seat of original injury.

A rather remarkable case bearing on this subject was admitted a couple of years ago, but left the hospital before the final deformity was refractured, as was intended. The boy, who was 18 years old, had a distorted forearm, resulting from fracture about four and one-half inches below the point of the olecranon. His history was as follows: When eight years of age he fell out of a tree, and sustained a fracture of left forearm near the elbow. One year later he fell down-stairs, and broke it again in the same region, but says he thinks only one bone (?) was broken at that time. Three years subsequently, or when he was about twelve years of age, he fell into the hold of a boat, and sustained a third fracture, but does not know whether both bones were broken or not. After this he thinks the arm was as straight as the other, and straighter than it had been *previous to this third accident*, but there was some impairment of rotating power. Eight or nine weeks before these notes were taken, or nearly six years from the time of the third fracture, he fell from a coal-wagon, which, however, was not in motion, and sustained a fourth fracture, which from the present appearances must have involved both bones.

Here, then, is a case where refracture occurred in the same position, as near as may be, four successive times, at intervals of one, three, and six years, due evidently to the natural weakness of the bond of union, for the fractures resulted from injuries that would probably have been productive of other fractures if there had been any constitutional predisposition to fracture in the patient's skeleton; moreover, they occurred in a young person, when solidification is supposed to occur rapidly.

The objection to refracture, advanced by some, that erysipelas, abscess, necrosis, or pyæmia may result, can be dismissed in a few words. Every one knows the extreme unfrequency of these occurrences in non-complicated simple fractures. The original injury has, in nearly all these cases, been attended with more disturbance of the tissues than is caused by the operation of refracture; when the patient is in good condition, the bones have rounded ends and never lacerate, and the fracture is dressed without suffering the jarring and jolting of transportation. From the cases that I am cognizant of, I am inclined to believe that inflammation of a character sufficient to cause swelling and tension of the parts is exceedingly rare. There seems also to be no real danger of rupturing vessels that may possibly have become shortened by reason of the decreased length of the limb.

Is there any danger of non-union after refracturing a firmly-united bone? I believe this occurrence to be about as unlikely as non-union of soft parts after incision through an old cicatrix, provided the fracture receive proper treatment after the refracture has been accomplished.

Moreover, un-united fracture, unless there be constitutional reasons, is rare in any event except as the result of imperfect treatment, which has allowed mobility of the fragments, or of the presence of portions of tendon or muscle between the ends of the bone. These factors are eliminated in the case under discussion, where you have selected a favorable time as to the health of the patient, have dressed the limb with appropriate apparatus, and where there is little or no opportunity for tissue to become entangled between the ends of the bone.

It has been believed by some that these new fractures consolidate, as a rule, more readily than the primitive fractures. This may depend on the fact that there is required much less repair of muscular tissue and less absorption of blood, which processes are preparatory steps in the union of fractures of an ordinary variety.

As having some bearing on this part of the subject I may refer to Cases 2 and 7, where both the femur and the humerus were refractured at the same time, and union took place in six and a half weeks without difficulty. Again it will be observed that in Case 3 the patient had originally fractured both humeri as well as the leg, and yet, though the leg was rebroken after the lapse of six weeks, union occurred in six and a half weeks without any trouble. A similar instance is recorded in Case 6, where the boy had the radius and ulna of one arm broken in two places at the time of the fracture of the femur. Union of all the fractures occurred, and, after refracture of the femur, it united readily a second time in eight weeks. These facts are mentioned to show that there seems to be no narrow limit to the capability of the healthy system furnishing callus for the consolidation of multiple or repeated fractures.

The after-treatment of cases of refracture is sufficiently plain. In some instances it may be well to keep up great extension with the pulleys for an hour or so after refracture has been accomplished, in order to fully stretch the parts, and exert a certain paralyzing influence upon the muscles which tend to displace the fragments. The most successful treatment will probably be some form of extension combined with well-directed pressure of splints and pads over the situation of the angular displacement. The extension method is of course applicable in refracture for deformity of the upper as well as the lower extremity, though not so frequently required. The extending weight should be sufficient to do the work, and, if adhesive plaster slips, and will not bear the strain, it may be reinforced by the application of some apparatus made of straps and buckles, as in Case 7. If the fracture is low down near the ankle, it may be necessary to use a footboard, strapped to the sole of the foot, in order to get attachment for the extending weight. The necessity of employing splints of wood, pasteboard, or felt at the seat of fracture is obvious, since the extension merely obviates overriding, without keeping the fragments at rest or preventing angular or rotatory deviation.

In occasional instances tenotomy may be required to assist in preventing displacement, and this is not to be deprecated, for it does not add much to the severity of the operation, since the tendon cut is generally at a distance from the fracture, and hence the puncture does not make the fracture compound.

The lessons taught by an examination of these cases may be formulated as follows:

Refracture is the best method of correcting deformity after maltreated fractures, because it gives relief without suppuration, which is liable to follow drilling, and almost certain to result from the operation of excision.

It may be undertaken whenever there is a possibility of overcoming the causes that gave rise to the deformity after the original fracture; and is accomplished by bending, either alone or combined with powerful extension.

The time is only limited by the ability of the surgeon to rupture the bond of union.

If the bone be healthy, there is no danger of fracturing in any situation but that of the original lesion. The occurrence of erysipelas, abscess, necrosis, or pyæmia is too rare to be considered an objection to the operation. Non-union seldom, if ever, occurs.

The results as to correction of deformity and as to usefulness are uniformly satisfactory.

John B. Roberts.

NECROSIS.

THE parts affected in the 92 tabulated cases were as follows: tibia, 18; femur, 18; foot and ankle, 17; fingers and toes, 10; pelvis, 5; humerus, 4; lower jaw, 4; upper jaw, 3; elbow, 3; fibula, 2; sternum, 2; hand and wrist, 2; nose, 1; knee, 1; clavicle, 1; and patella, 1. In 52 the affection was directly traceable to injuries of various kinds, as reported in the table. 3 of these were due to frost-bite, and 2 to the fumes of phosphorus. In 20 the trouble arose from diseases, as syphilis, fever, etc., and in 20 the cause is not given. It would be fair to put down at least half of the latter as due to injuries.

Most cases admitted into hospital for necrosis present themselves at a late stage of the disease, and many not in the table necessarily arise in the house, as, for example, in the course of a bad compound fracture.

In acute periostitis, where necrosis is threatened, free incisions through the tissues and membrane down to the bone will no doubt frequently avert the death of the latter. The opportunity for doing this, however, both in hospital and private practice, is rare, as the conditions producing most cases of dead bone are slow and at first unlooked for where disease is the cause.

The bone having died, there is nothing else but for it to come away, and in most cases artificial assistance is required to accomplish this object. Where the amount involved is small, it may discharge itself in time through a sinus, and this will eventually heal without any operative interference, or without detriment to the general health of the patient.

The separation may be favored by the use of various stimulating washes and caustic applications. It is possible also for large masses of dead bone to be gradually broken down, and to be removed in the same way, but this slow disintegrating process is accompanied with great risk to the patient, who is in danger of succumbing to hectic and exhausting discharges. Unless, therefore, there should be some positive reason for non-interference, as position, for example, operations are generally advised and performed. It is wonderful to note the rapid change for the better that takes place in the patient, both as to general health and as to the local disease, after most operations for necrosis. The nicest point is to decide when to operate. In ordinary cases there is no doubt that we should wait until the necrosed portion has completely separated, and this is our general practice. Phosphorus necrosis appears to be an exception to this rule, as our experience has shown that the dead part had better be gouged out, or cut away at once, otherwise the whole bone will be speedily involved in destruction. Fortunately, this disease is rare, as the means for avoiding it are almost entirely under control, and should be well understood by those who are engaged in the business of match-making.

It is in operations for necrosis that we experience the full value of the Esmarch bandage. Whatever may be said against it in other cases does not apply here. Before its use these operations were, even in the best hands, ugly, mechanical, and blood-splashing efforts. They were also often abortive or incomplete, on account of the diseased parts not being seen. With its use the removal of dead bone may be done in such a way as to entitle the operation to a place among the "arts" in surgery.

Necrosis.

No.	Age.	Sex.	Nature.	Cause.	Treatment.	Result.	Remarks.
1	33	F.	Of malleolus of left tibia and joint.	Fall on ice a year ago. Some suspicion of syphilitic taint.	1st. Malleolus gouged. Carbolic dressings. No improvement, as disease was too extensive. 2d. Amputation.	Cured.	(See Amputations.)
2	55	F.	Extensive, of tibia; 20 years' standing.	Fell from carriage, wheel passing over legs, in 1853.	In 1862 had bone gouged; since discharged many pieces. Operation repeated September 3, 1873. Carbolic dressings.	Improved.	Discharged by request.
3	30	M.	Of ring finger of right hand.	Frost-bite, 9 months ago	Amputation of last phalanx of ring finger.	Cured.	In hospital 15 days.
4	17	M.	Of left foot and ankle.	Foot jammed in door when 2 years old. Repeated outbreaks after slight injuries.	A transverse incision was made. Astragalus was gouged out, and ends of tibia and fibula cut off.	Relieved.	Discharged. In hospital 6 months. Wound had healed, but there were still open sinuses at back of limb.
5	31	M.	Of left tibia.	Followed gunshot wound, received in April, 1863; 10 years before admission.	Removed fragments, gouged bone, and dressed with carbolized oil and charpie.	Cured.	In hospital 19 days.
6	26	M.	Of crest of left ilium.	Followed fracture of ribs and pelvis, a year previous.	Fragments removed. Poultice, blue-wash.	Cured.	In hospital 31 days.
7	17	M.	Of right femur.	Followed periostitis; a year before admission.	Examined under ether. Dead bone, but still very firm.	Discharged to wait separation.	In hospital 17 days.
8	23	M.	Of base of great toe.	Followed gunshot wound.	Loose pieces removed. Ung. zinci oxid.	Cured.	In hospital 17 days.
9	14	M.	Of bones of foot.	....	None.	Went to country.	In hospital 5 days.
10	19	M.	Of foot.	From fall, a year ago.	Tonics, quinine, cod-liver oil, etc. Amputation.	Cured.	Was in very bad general health, having ulcers, etc. (See Amputations.) In hospital 6 months.
11	9	M.	Of left ankle.	From bruise.	Poultice.	Improved.	Discharged for misconduct.
12	24	M.	Of metatarsal bone.	....	Opened, and dead parts removed.	Cured.	In hospital 27 days.
13	38	M.	Of pelvis.	....	Quinine, cod-liver oil, etc.	Improved.	In hospital 57 days.
14	22	M.	Of tibia.	Old abscess of bone.	Earth dressing, blue-wash, etc.	Improved.	In hospital 73 days.
15	27	M.	Of sternum.	Followed abscess.	Earth dressing. Operation; dead bone removed.	Improved.	In hospital 5 months, 9 days.
16	53	F.	Of ankle-bones.	Followed fracture.	Poultice.	Relieved.	To return for operation. In hospital 2 weeks. Returned, and had some small pieces removed. In hospital 1 month.
17	26	M.	Of right femur, near trochanter.	Followed gunshot wound, 3 years previously.	Operation, by cutting down on bone and removing the necrosed portions. Wet clay dressing.	Cured.	An abscess was also opened in bone, and discharged freely. In hospital 5 months.
18	35	M.	Of lower jaw. Venereal (?).	Says followed extraction of teeth.	Potass. iodid., etc. Declined operation.	Improved.	In hospital 2 weeks.
19	21	M.	Of ankle.	Followed compound fracture, 18 months ago.	Carbolized oakum, wet clay, etc. Parts discharged bone, etc., and healed under strapping.	Cured.	In hospital 2 months.
20	[illegible]	M.	Of left tibia.	Syphilis (?).	Large piece of bone removed by operation. Carbolic acid dressing.	Improved.	In hospital 28 days.
21	27	M.	Of second toe.	From being run over, 3 years since.	Poultice, etc.	Cured.	In hospital 11 days.
22	18	M.	Of tibia.	Gunshot wound, 7 months since.	Poultice, carbolized oakum. Operation; dead bone and ball removed.	Cured.	In hospital 21 days.

Necrosis.—(Continued.)

No.	Age.	Sex.	Nature.	Cause.	Treatment.	Result.	Remarks.
23	40	M.	Of femur.	No history of injury. Has existed from childhood.	Carbolized oil, laudanum and water, etc. Operation to remove bone, finally amputation.	Cured by amputation.	There was such great hemorrhage at the first operation that it was not completed, but after the bleeding was arrested the patient was put under treatment to prepare for amputation. (See Amputations.) In hospital 6 months.
24	25	M.	Of tibia.	....	Esmarch's bandage applied, bone scraped, and dead pieces removed. Dressed with carbolic oil, etc.	Improved. Eloped.	Patient had been operated on two years previously. In hospital 35 days.
25	67	M.	Of upper maxilla.	From fall, striking face against fence.	Poultice. Incisions and removal of débris; no large piece; repeated abscesses; opened.	Cured.	In hospital 95 days.
26	29	M.	Of upper maxilla.	Phosphorus fumes.	Dead bones removed by forceps and pliers. Very little cutting required. Cod-liver oil, etc.	Cured.	Patient had been in the match-making business since he was eleven years old, but a "dipper" only for the past three years. Had a tooth pulled, and incautiously went to work the next day. In hospital 10 days.
27	55	F.	Of tarsus.	Followed a sprain.	Incisions and gouge. Carbolic oil dressing.	Cured.	In hospital 3 months, 22 days.
28	26	M.	Of ilium.	....	Sinuses laid open. Dressed with carbolic oil.	Improved.	In hospital 3 months, 10 days.
29	35	M.	Of inferior maxilla.	Syphilis.	Dead bone removed. Potass. iodid., etc.	Improved.	In hospital 55 days.
30	29	F.	Of femur.	Followed injury from a needle.	Dead bone removed.	Improved.	In hospital 75 days.
31	6	M.	Of lower jaw.	Followed fracture.	Cod-liver oil and lacto-phos. lime.	Cured.	In hospital 53 days.
32	33	M.	Of bones of hand.	Followed fracture and abscess.	1st. Tonics, etc. Amputation.	Cured.	In hospital 33 days.
33	7	F.	Of knee.	Followed injury, when 2 years old.	Amputation.	Cured.	In hospital 35 days. (See Amputations.)
34	42	M.	Of tarsus.	Followed injury from kick of horse.	Amputation.	Cured.	In hospital 1 year, 2 months. (See Amputations.)
35	9	F.	Of tibia.	....	Dead bone removed; parts cleaned.	Cured.	In hospital 20 days.
36	10	M.	Of tibia.	Followed fall against a stone.	Bone removed.	Improved.	In hospital 60 days.
37	27	M.	Of right tarsus.	....	Sinuses opened and bone cleaned.	Improved.	In hospital 24 days.
38	22	F.	Of femur.	....	Bone removed. Salicylic acid and moss dressing.	Improved.	In hospital 78 days.
39	22	F.	Of inferior maxilla.	Denies specific history.	Declined operation.	Discharged by request.	
40	58	M.	Of bones of great toe and first metatarsal.	Followed frost-bite.	Amputation of diseased parts.	Cured.	In hospital 5 months, 11 days.
41	19	M.	Of right tibia.	Followed a gunshot wound with small pocket-pistol.	General health bad. Iron, quinine, etc. Abscesses opened, and small amount of bone discharged, with matter.	Relieved.	In hospital 23 days.
42	7	M.	Of fibula.	Followed dislocation of ankle (?).	General health attended to. Abscess opened; bone discharged.	Cured.	In hospital 3 months, 15 days.
43	42	M.	Of ankle.	Syphilitic.	General and local. Amputation declined.	Discharged by request.	In hospital 5 days.
44	45	M.	Of great toe.	Ran a shawl-pin into foot while walking across floor.	General, and amputation, including part of metatarsal bone.	Cured.	In hospital 4 months, 8 days.

45	20	F.	Of clavicle.	Suspicion of specific trouble, but denied.	General.	Improved.	Discharged to out department. In hospital 75 days.
46	20	M.	Of femur.	Followed injury from a fall.	General.	Stationary.	Patient had albuminuria. In hospital 16 days.
47	13	M.	Of left tibia.	Followed injury from fall.	Head of tibia trephined; pus discharged, and dead bone removed.	Cured.	In hospital 3 months, 13 days.
48	46	M.	Of elbow.	....	Excision.	Improved. Discharged by request.	In hospital 2 months, 5 days.
49	39	M.	Of bones of left thumb.	From bite by a man.	Amputation through middle of metacarpal bone.	Cured.	Patient had severe hemorrhages before operation. In hospital 50 days.
50	30	M.	Of right arm, bones at elbow.	....	Amputation; Teale method.	Cured.	In hospital 83 days.
51	22	M.	Of femur.	....	None.	....	Did not remain in house.
52	22	F.	Of femur.	....	Dead bone removed; an old case.	Improved.	In hospital 18 days.
53	22	M.	Of upper jaw.	Exposure to phosphorus fumes. Match-dipper.	Necrosed bone removed.	Eloped. Improved.	In hospital 17 days.
54	23	M.	Of crest of right ilium.	....	Some diseased bone had been removed before admission.	Stationary.	In hospital 55 days.
55	57	F.	Of left tibia.	Followed injury.	Chromic acid and poultice.	Improved.	In hospital 29 days. To out-patient department.
56	59	M.	Of metatarsal bones.	Followed fracture.	Carbolic dressing.	Stationary.	In hospital 28 days.
57	23	M.	Of left calcis.	Injury 4 years ago, by stepping on broken bottle.	Carbolic dressing, liquor villatte, etc.	Improved.	In hospital 114 days. To out-patient department.
58	54	M.	Of sternum.	Blow from a wagon-pole, 6 months ago.	Pure carbolic acid, salicylic acid, etc.	Stationary.	In hospital 29 days.
59	[illegible]	M.	Of right tibia.	Fracture, 1 year ago.	Operation; dead bone removed.	Improved.	In hospital 125 days.
60	16	M.	Of left foot.	Tramped on by horse, 3 years ago.	Salicylic acid, etc.	Improved.	In hospital 44 days.
61	43	M.	Of great toe.	Syphilitic.	Amputation of toe at metatarso-phalangeal joint.	Cured of this attack.	In hospital 107 days. This patient had repeated attacks of necrosis of bones of foot. Various operations were performed, and finally the foot was removed by Syme's operation.
62	18	M.	Of elbow.	Followed injury.	Salicylic acid, etc.	Improved.	In hospital 5 months.
63	44	M.	Of second toe.	Same case as 61.	Amputation of toe.	Cured.	In hospital 44 days.
64	40	F.	Of femur.	Followed fracture, 20 years ago.	Removed dead bone by operation.	Cured.	In hospital 5 months.
65	15	M.	Of right femur.	Followed injury, 8 years since.	Opened sinuses, drainage-tube, etc. Fragments of bone had come away from time to time.	Cured.	In hospital 2 months.
66	17	M.	Of left femur.	From fall and being caught in wheel of wagon, 22 months since.	1st. Rest, tonics, etc.; afterwards excision of head and neck of thigh-bone.	Cured.	In hospital 8 months, 21 days.
67	16	F.	Of left humerus.	Followed fracture, 2 years ago.	Tonics, etc. Shaft trephined and sequestrum removed.	Died.	In hospital 3 months. Patient improved after operation, but died ultimately of albuminuria.
68	24	M.	Of femur.	....	Spicula of bone removed.	Cured.	In hospital 87 days.
69	20	M.	Of ankle.	....	1st. Tonics, etc. Os calcis removed, and other diseased parts gouged. Carbolized charpie dressing.	Cured.	In hospital 7 months.
70	11	M.	Of tibia.	....	....	Removed same day by parents.	
71	26	M.	Of left tibia, acute.	Followed injury in mine, 3 weeks since.	1st. Tonics, etc. Tibia trephined and sequestrum removed.	Cured.	In hospital 7 months.

Necrosis.—(Continued.)

No.	Age.	Sex.	Nature.	Cause.	Treatment.	Result.	Remarks.
72	35	M.	Of right femur, involving hip-joint.	Followed a fall from height, 6 years since.	Palliative.	Discharged by request.	In hospital 6 days.
73	55	M.	Of nasal bones.	From wound with pitchfork.	Dead bone removed under ether. Carbolized oil dressing.	Unimproved.	In hospital 17 days. Patient had much discharge after the operation, with delirium, etc. Removed by request.
74	26	M.	Of head of humerus.	Followed attack of typhoid fever.	Refused treatment.	Discharged for misconduct.	In hospital 2 days.
75	13	F.	Of pelvis.	Followed old hip-disease.	Palliative.	Died from exhaustion.	In hospital 15 days.
76	16	M.	Of left humerus.	Followed abscess. No history of injury.	Sinuses opened and sequestrum removed.	Improved, to out-patient department.	In hospital 2 months.
77	45	M.	Of phalanges of toes.	Syphilitic.	Amputation.	Relieved.	In hospital 2½ months.
78	32	M.	Of tibia.	....	Dead bone removed. Carbolized oil packing.	Improved.	In hospital 51 days.
79	29	M.	Of left femur.	....	Opened, clasta enlarged, etc.	Improved.	In hospital 3 months, 10 days.
80	31	M.	Of ring finger of right hand.	From running a hook into finger.	Poultice, etc.	Improved.	In hospital 18 days.
81	16	F.	Of carpal bones.	Probably hereditary syphilis.	Bones removed; splint; hyd. bichlorid. and cod-liver oil.	Cured.	In hospital 3 months, 9 days. A remarkable case. Patient came in for amputation of arm, but under the view taken she recovered with a very useful limb.
82	36	M.	Of left tarsal bones.	....	Pieces of bones removed. Poultice, etc.	Improved.	In hospital 76 days.
83	35	M.	Of femur.	From strain.	Sequestrum removed.	Cured.	In hospital 2 months.
84	32	M.	Of phalanx of little finger.	Frost-bite.	Incision and poultice.	Cured.	In hospital 13 days.
85	13	M.	Of femur.	Followed coxalgia.	Abscess opened and small pieces of bone were removed; general health attended to.	Cured.	In hospital 2 months.
86	9	M.	Of head of right humerus.	From fall from wagon.	Excision of head of bone and two inches of shaft.	Cured.	In hospital 75 days. At first an enormous abscess was evacuated. The boy was in no condition for operation then. The excision was made a month after boy was admitted.
87	14	M.	Of left femur.	....	Etherized; incision and large pieces of bone removed.	Improved.	In hospital 21 days.
88	23	F.	Of right femur.	From fall on ice, 8 years ago.	Etherized; incision and dead bone gouged out.	Cured.	In hospital 56 days.
89	12	M.	Of tibia.	Followed typhoid fever.	Incision and dead bone gouged.	Cured.	In hospital 5 months.
90	15	M.	Of patella.	Followed fracture.	Sinus opened, and necrosed portion removed.	Cured.	In hospital 36 days.
91	13	F.	Of head of fibula.	....	Head of fibula removed.	Cured.	In hospital 3 months, 3 days.
92	9	M.	Of tibia.	....	Large amount of bone removed. Rest, bran and carbolic powder.	Cured.	In hospital 78 days.

WILLIAM HUNT.

ARTHRITIS AND SYNOVITIS.

There are 43 cases reported in the table. Of these the affection was the result of injury in 21, of disease in 6, and the cause is not given for the remaining 16. The knee was the seat of trouble in 31 of the cases. The others were of the wrist, 2; hip, 2; ankle, 6; toe, 1; and shoulder, 1. While the knee-joint furnishes by far the greater number of cases of arthritis and synovitis requiring hospital treatment, the proportion shown in our list between the lower and upper extremities as to liability to these affections is probably greater than the real truth, as many afflicted in the upper extremity would seek the out department for relief.

Of the 43, 20 were cured, 19 improved, 2 stationary, 1 died, and 1 was removed by friends.

It will be noticed that under the head of treatment nearly all of the various methods in use have been applied in the different cases.

No class of cases illustrates so well the benefit, in the acute stage, of rest, antiphlogistics, anodynes, and prompt operative interference for the removal of pus or excessive effusions. The aspirator is a grand agent for the latter purpose. In the later stages, tonics, full diet, fresh air, and movements, either passive or active, contribute greatly to recovery.

One of the nicest points for decision is as to when to begin this last treatment. The patient, through fear of pain, is rarely ready for it himself, in fact, almost always resists it. With females this may be said to be always the case, and it is for trying to make these patients do as they should that surgeons are often unjustly accused of cruelty.

The best test is the non-occurrence of acute inflammatory symptoms after the first few efforts to use the joint involved. If no such symptoms arise, the limb may be used with all the freedom that the patient can possibly give it, for his own sensations will be sure to keep him within bounds.

Arthritis and Synovitis.

No.	Age.	Sex.	Nature.	Cause.	Treatment.	Result.	Remarks.
1	13	M.	Synovial effusion of knee-joint; painful.	Wheel of a cart; fell and struck the knee; 17 days before admission.	Rest, blister, poultice, oxide of zinc ointment.	Improved knee, nearly normal.	In hospital 20 days. Discharged by request.
2	13	F.	Involves left knee.	Followed excision of right hip, probably from position.	Rest, etc.	Improved.	In hospital 9 months.
3	53	M.	Chronic, of right knee.	Sudden strain, 12 years since.	Extension and rest.	Improved.	In hospital 16 days.
4	36	F.	Of wrist-joint.	From strain while wringing clothes.	Extension on splint, iodine paint, etc.	Stationary.	Had symptoms of phthisis, for which took cod-liver oil. In hospital 39 days.
5	26	F.	Of right hip.	From fall, 3 years since.	Tonics, etc. No operation.	Stationary.	There were sinuses leading to diseased bone, above acetabulum.
6	23	M.	Of ankle.	....	Rest, wet clay.	Improved.	In hospital 9 days.
7	24	M.	Of knee-joint.	Followed fever.	Actual cautery, potass. iodid.	Improved.	In hospital 21 days.
8	64	M.	Chronic, of right knee.	....	Actual cautery and poultices.	Cured.	In hospital 27 days.
9	22	F.	Of left knee.	....	Tinct. iodine and soap plaster.	Cured.	In hospital 6 days.
10	41	F.	Of left toe-joints.	Followed sprain.	Ext. belladonna, and tinct. iodine painted twice daily.	Improved.	In hospital 7 days.
11	22	M.	Of right hip.	Probably strain while loading vessel.	Extension, rest, cod-liver oil, etc. Under ether adhesions broken up. Extension renewed, etc.	Improved.	In hospital 107 days.
12	19	F.	Of right knee.	From fall on ice.	Blisters and extension.	Cured.	In hospital 51 days.
13	12	M.	Of right knee.	From fall.	Blister and extension.	Cured.	In hospital 30 days.
14	35	F.	Of right knee.	....	Blister; potass. iodid., gr. x t. d.	Improved.	In hospital 3 months.
15	37	F.	Acute, of left ankle.	Struck against child's carriage.	Lead-water and laudanum; pressure by bandage.	Cured.	In hospital 11 days.
16	58	F.	Arthritis of wrist, chronic.	....	Incision and evacuation of pus. Splints.	Improved.	In hospital 2 months. There was erosion of cartilages.
17	18	F.	Synovitis, right knee.	....	Aspiration, and withdrawal of 6 ounces of fluid. Strapping with emp. de Vigo cum mercurio, and a bandage.	Cured.	In hospital 35 days.
18	34	F.	Of right knee.	From fall, 1 year since.	Iodine, emp. saponis, bandage, etc.	Improved.	In hospital 33 days.
19	27	M.	Of left knee.	From plank falling on it, 3 months ago.	Extension, actual cautery. Abscess formed; discharged freely.	Removed by friends.	In hospital 3 months.
20	23	M.	Of knee.	Exposure to cold, 9 months since.	Extension; aspirated twice; large evacuations of pus; fixed dressings; discharge continued.	Died.	In hospital 52 days. Died of hectic and exhaustion.
21	38	F.	Of knee.	From fall, 10 months since.	Iodine paint and posterior splint.	Cured.	In hospital 26 days.

22	22	M.	Of knee.	....	Extension, blister, incision, pus evacuated.	Improved.	In hospital 23 days.
23	14	M.	Of left ankle.	Strain, 1 year ago.	Emp. de Vigo, bandage, puncture, pus evacuated.	Cured.	In hospital 8 months, 9 days.
24	24	M.	Of ankle.	Sprain, 5 weeks since.	Iod. iron, wire splint, etc.	Cured.	In hospital 42 days.
25	13	M.	Of knee; 6 years' standing.	....	Wet clay, lactophosph. lime, and cod-liver oil.	Improved. Eloped.	In hospital 4 months.
26	28	F.	Of knee.	....	Chloral and salicylic acid.	Improved.	In hospital 2 months.
27	25	M.	Of right knee.	From fall on ice.	Rest, lead-water and laudanum.	Cured.	In hospital 4 days.
28	20	M.	Of left ankle.	No history of injury, rheumatic (?).	Iron, quinine, etc.; rest, poultice, etc.	Improved.	In hospital 2 months. Patient also had gonorrhœa.
29	18	M.	Chronic, of left knee.	Injured knee, 3 years since.	Potass. iodid. and hyd. bichlorid. internally, salicylic acid and bandage.	Cured.	In hospital 36 days. Abscess formed and was opened.
30	14	M.	Chronic, of knee.	....	Potass. iodid. and hyd. bichlorid. internally, ung. hydrarg., rest, etc.	Cured.	In hospital 36 days.
31	16	M.	Of right knee.	Injured by wagon.	Rest, lead-water and laudanum.	Cured.	In hospital 3 days.
32	50	M.	Chronic, of right knee.	Cold.	Rest, iodine paint, etc.	Improved.	In hospital 75 days.
33	2	F.	Of knee.	....	Rest, posterior splint.	Cured.	In hospital 39 days.
34	7	M.	Chronic, suppurative, of knee-joint.	....	Excision of joint.	Improved.	In hospital 1 year, 35 days. (See Excisions.)
35	16	F.	Of left knee.	....	Constitutional, passive motion, etc.	Improved.	In hospital 40 days.
36	26	F.	Of left knee.	Followed a sickness.	Splints, iodine, actual cautery, etc.	Improved.	In hospital 5 months.
37	49	M.	Of knee.	From blow with axe; 18 months before admission.	Actual cautery, rest, etc.	Cured.	In hospital 56 days.
38	42	M.	Of right shoulder-joint.	From rheumatism.	Actual cautery.	Cured.	In hospital 5 months.
39	21	F.	Of right knee.	From fall.	Actual cautery, resin cerate, etc.	Cured.	In hospital 5 months.
40	20	M.	Of left knee.	....	Rest, iodine, etc.	Cured.	In hospital 14 days.
41	48	F.	Of left knee.	....	Blister, iodine, silicate bandage, etc.	Improved.	In hospital 3 months, 18 days.
42	32	F.	Of left knee.	....	Rest, blister, iodine bandage.	Cured.	In hospital 21 days.
43	14	M.	Of left ankle.	From strain a year since.	Rest, emp. de Vigo, tight bandage, etc.; internally, potass. iodid. and hyd. bichlorid. Pus evacuated with exploring needle.	Cured.	In hospital 8 months.

WILLIAM HUNT.

ANCHYLOSIS.

Of the 26 cases of anchylosis reported, 16 originated from injuries of various kinds, including fractures and dislocations; 9 are put down as arising from diseased conditions, and in 1 the cause is not given. Nearly all are stated to be of the false variety, and it is nearly certain that there was no case of actual bony union of the opposing surfaces of the joints. Anaesthesia has demonstrated the rarity of this latter form of anchylosis. Great improvement follows treatment in nearly all cases, and often absolute cures are effected. In addition to the means used by the surgeon there is nothing so essential to success as co-operation on the part of the patient. After the adhesions have been broken up he can do a vast deal by systematically keeping up the movements of the part, and noting for his own encouragement the advances made from day to day. We think that the fear of disturbing fractures near joints, and the too firm dressing of such injuries, are responsible for much anchylosis. It is a question whether the risk of non-union in a badly-fractured elbow, for example, by simply putting the limb in a good supporting sling, is not better treatment than that of stiff splints and firm bandages. It seems rather absurd for a surgeon to have to excise an anchylosed joint in a bad chronic case in the hope of getting ligamentous union, and at the same time his dread of getting ligamentous union in a recent case will cause him to do his best to bring about anchylosis.

Anchylosis.

No.	Age	Sex	Nature.	Cause.	Treatment.	Result.	Remarks.
1	55	F.	False, of elbow-joint.	Old dislocation of radius, treated in extended position.	None.	....	Discharged for misconduct same day as admitted.
2	11	M.	False, of left hip.	Fall from wagon, 5 years ago, and a subsequent blow on the part.	Operation decided against.	Unrelieved.	
3	37	M.	False, of knee-joint.	Old strumous disease of knee.	None.	Unrelieved.	Was advised to have the limb amputated, but declined to submit.
4	36	M.	False, of ankle.	From a fall.	Manipulation and liniments.	Cured.	In hospital 9 days.
5	54	M.	False, of elbow.	From fracture.	Etherized and broken up, Stromeyer's splint, massage.	Cured.	In hospital 17 days.

Anchylosis.—(Continued.)

No.	Age.	Sex.	Nature.	Cause.	Treatment.	Result.	Remarks.
6	58	F.	False, of wrist.	Following fracture.	Manipulation, etc., liniments.	Stationary.	In hospital 14 days.
7	60	M.	False, of shoulder.	From dislocation which had been reduced.	Etherized and adhesions broken up, passive motion, etc.	Cured.	In hospital 37 days.
8	50	M.	False, of left shoulder.	From dislocation which had been reduced.	Etherized and adhesions broken up, galvanism and motion.	Cured.	In hospital 19 days.
9	45	M.	False, of shoulder.	Followed fracture of head of humerus a year since.	Etherized and adhesions broken up.	Cured.	In hospital 19 days.
10	16	M.	False, of right elbow.	Followed synovitis from an injury to joint.	Passive motion and constitutional treatment, cod-liver oil, etc.	Improved.	In hospital 3 months, 12 days.
11	32	M.	False, of right elbow.	From fracture, 10 years since.	Adhesions broken up under ether, passive motion, etc.	Improved.	There was an ulcer of same elbow from a recent injury. In hospital 97 days.
12	13	M.	Of right knee.	From strumous disease of 3 years' standing. No history of injury.	Etherized; patella was firmly adherent; Stromeyer's splint applied.	Improved.	In hospital 31 days.
13	40	M.	False, of knee.	Followed synovitis.	Chloral dressings, rest, etc., then passive motion.	Cured.	In hospital 12 days.
14	41	M.	False, of right ankle.	From a fracture caused by a fall.	Etherized and adhesions broken up; hypodermic injections of atropia, etc., passive motion.	Improved.	In hospital 42 days.
15	57	M.	False, of ankle and knee.	From old synovitis.	Hypodermic injections of atropia, tendo Achillis divided, passive motion, etc.	Improved.	In hospital 5 months.
16	9	M.	False, of knee.	No history of injury.	Excision. (See Excisions.)	Cured.	In hospital 2 months.
17	11	M.	False, of elbow.	....	Stromeyer's splint and passive motion.	Improved.	In hospital 16 days.
18	24	M.	False, of knee.	From fall, 10 years since.	Abscess and necrosis were found. Relieved by operation.	Cured.	In hospital 49 days.
19	19	F.	False, of right knee.	From fall when a child.	Tenotomy of hamstrings and forcible extension.	Improved.	In hospital 75 days. Walked well with apparatus.
20	20	F.	False, of elbow.	From rheumatism.	Adhesions broken up under anæsthetic.	Improved.	In hospital 37 days.
21	23	M.	False, of knee.	From synovitis, 8 years since.	Adhesions broken up under ether, and patella loosened by blows of hammer.	Stationary.	In hospital 45 days.
22	24	M.	False, of knee.	From synovitis, strumous (?).	Constitutional, cyst opened under knee.	Stationary.	In hospital 69 days.
23	62	M.	False, of elbow, with necrosis.	From old gun-shot wound.	Dead bone removed. Potass. iodid. and hyd. bichlorid.	Improved.	Reason to believe the patient was syphilitic.
24	18	M.	Of right knee.	From fall.	Declined operative measures.	Stationary.	In hospital 1 day.
25	30	M.	False, of elbow.	Injury on ship.	Adhesions broken up under ether.	Improved.	In hospital 5 days.
26	28	F.	False, of hip.	Rheumatism.	Movements under ether.	Stationary.	In hospital 8 days.

WILLIAM HUNT.

COXALGIA.

Comparatively few cases of coxalgia are treated in this hospital. Of the 27 recorded in the list, 16 had a positive history of previous injury. 1 was doubtful as to this matter, and from 10 no such history could be obtained. Many of the cases are far advanced. The average age of those admitted was twelve years. The greatest improvement that has been made in treatment for this disease consists in abandoning the old plan of keeping the patient continuously in bed for a very long time. So soon as the acute symptoms have passed away under the influence of extension and rest, the patient is allowed to get up, and with crutches, and the joint fixed or not in a splint, as the case may be, he is enabled to take air and exercise. The plan of elongating the sound limb by an iron stilt attachment to the sole of the shoe, so as to keep the foot of the diseased side from the ground, and allow the limb to hang while the patient uses crutches, promises good results. Two of the cases in the list were subjected to excision of the head of the femur. One of these recovered, the other died. Their history is more fully reported under the head of Excisions.

Coxalgia.

No.	Age.	Sex.	Nature.	Cause.	Treatment.	Result.	Remarks.
1	$3\frac{1}{2}$	M.	There was an external opening.	....	None.	....	Eloped before history could be obtained.
2	24	M.	Advanced. Open abscesses and caries of hip, also of ankle.	....	Rest; opening of abscesses from time to time. General, tonics, cod-liver oil, etc.	Unimproved.	Transferred to Philadelphia Hospital. In hospital 4 months, 11 days.
3	4	F.	In first stage.	Fell and injured hip a year ago, then fell again.	Rest, extension, apparatus, etc.	Improved.	In hospital 26 days. Removed to Orthopædic Hospital.
4	9	M.	In first stage.	Had a fall, 2 weeks before admission.	Rest, extension, iodide of iron, etc.	Cured.	In hospital 37 days. Was soon allowed to go on crutches, and then to go without them.
5	14	F.	Advanced disease; of 10 years' standing.	....	Excision of head of femur.	Improved.	In hospital 1 year, 1 month. (See Excisions.)
6	10	M.	Advanced disease.	Fell from a horse, 2 years before.	Etherized, adhesions broken up, limb straightened, and extension applied. In 3 weeks up on crutches.	Improved.	In hospital 44 days.
7	8	F.	Advanced disease.	Fell down-stairs, 2 years before.	Rest, lacto-phosph. lime and cod-liver oil, etc.	Improved.	In hospital 15 days.

Coxalgia.

No.	Age.	Sex.	Nature.	Cause.	Treatment.	Result.	Remarks.
8	5	F.	In first stage.	....	Rest, continuous extension for a short time, then up on crutches, and extension at night. Cod-liver oil, etc.	Greatly improved.	In hospital 6 months.
9	10	M.	In first stage, left hip.	From fall.	Rest and extension.	Improved.	In hospital 33 days.
10	4	M.	In first stage.	....	Rest, etc.	Removed by friends.	In hospital 1 week.
11	15	F.	In second stage.	From fall.	Rest, extension, cod-liver oil, aspiration, etc.	Cured.	In hospital 8 months.
12	21	M.	In third stage. Head of femur displaced.	No positive history of injury. Has had bone trouble from very early in life.	Smith's anterior splint, cod-liver oil, etc.	Cured, with anchylosis.	In hospital 5 months.
13	6	F.	Recurrent disease.	This attack was brought on by a severe fall. Patient had been in house in early stage of disease, from which she recovered.	Extension and rest for some time, then extension only at night.	Improved greatly. Removed by friends.	
14	4	F.	First stage, of right leg.	....	Extension, etc.	Cured.	In hospital 4 months.
15	5	M.	Disease; 9 weeks' standing; never sick before. Left hip.	....	Extension, cod-liver oil and lacto-phosph. of lime.	Cured.	In hospital 88 days.
16	23	M.	Of left hip.	Followed fall down steps; 3 weeks before admission.	Extension, rest, etc.	Greatly improved.	In hospital 60 days.
17	4	M.	Of left hip.	....	Extension, Sayre's short splint, cod-liver oil and lacto-phosphate, etc.	Greatly improved, to out-patient department.	In hospital 5 months.
18	21	M.	Admitted to await apparatus.	....	Apparatus adjusted.	Improved.	In hospital 40 days.
19	9	F.	Of right hip, old.	Fall from door-steps, 3 years ago.	Rest, extension for 1 month; up on crutches.	Improved.	In hospital 2 months.
20	24	F.	Of right hip, old.	Fell from bridge, 5 years since.	Extension.	Improved.	In hospital 2 months.
21	4	M.	Old; 10 months' standing; abscess.	From fall.	Rest, extension, etc.	Stationary.	In hospital 6 months.
22	11	F.	Recent.	From fall.	Rest, extension, then up on crutches.	Improved.	In hospital 3 months.
23	10	M.	Of right hip; abscess and sinus.	From fall.	Rest, tonics, cod-liver oil, etc., then up on crutches.	Much improved.	In hospital 4 months, 9 days.
24	17	M.	Of right hip.	From fall, 4 months since.	Excision of head of femur.	Died.	In hospital 15 days. (See Excisions.)
25	4	M.	Of right hip.	From fall, 5 months since.	Extension, rest, then up on wheel-chair. Night extension continued, etc.	Cured.	In hospital 4 months.
26	3	M.	Of left hip.	From fall, 6 months since.	Extension and rest.	Improved.	In hospital 28 days. Removed by request.
27	9	F.	Side not mentioned. Contractions, 2 years' duration. (Anchylosis.)	Supposed from fall.	Tendons of gracilis and adductor longus divided, extension and rest, then wheel-chair and night extension.	Cured.	In hospital 3 months.

WILLIAM HUNT.

HERNIA.

Hernia appears to be one of those surgical maladies in which the average practitioner feels tempted at least to try the effect of ether and taxis before sending for a surgeon or consigning the patient to a hospital. Taxis, too, often means simply vigorous attempts to push or force the knuckle of intestine into the abdomen without any regard to the anatomical relations of the parts; thus the chances of recovery, after subsequent operation, are often materially diminished, and cases are lost which, under more prudent measures, it is likely would have resulted favorably.

In using taxis, the patient being etherized, if the hernia cannot, after a fair effort, be replaced, and symptoms of strangulation exist, an operation should at once be performed, for at this time it can hardly be called a serious one, unless previous or repeated attempts at reduction have been made. It is the delay, and the bruising the intestine often receives, which so frequently seriously complicates the operation. Thus, after the recognition of serious incarceration or incipient strangulation, the operation should not be unduly postponed, for the risk in many cases of rapid inflammation, with the shock of peritoneal invasion, frequently destroys any possible chance the operation would otherwise afford. We therefore urge early herniotomy. Should simple incarceration exist, without symptoms of strangulation, the postural treatment and elevation of the patient's limbs, with the application of ice, etc., is first carried out, having the patient carefully watched for any unfavorable symptoms which would require operative interference.

The following will show the number of hernias treated since 1870, also those reduced by taxis, ether, cold, opium, etc., with the mortality following reduction by herniotomy, taxis, etc.:

Of the 72 cases of strangulated or incarcerated hernia admitted, 55 were males and 17 females. Of this number, 48 recovered, 2 were improved, 4 were unimproved, 18 died. Reduced by ether, taxis, cold, opium, castor oil, or rest, 39; did not yield to manipulation, but symptoms were relieved under treatment, 6. Of the 45, 40 recovered, 5 died. Of the latter, peritonitis was present in each case upon admission. Of these, 1 died within 3 hours; 3 within 24 hours; and 1 on the 3d day.

Number subjected to herniotomy, 27. Of this number, 14 recovered, 13 died.

Of the 13 deaths, six were within 24 hours. In Cases 7 and 22, taxis and ether had been used before admission; there was also peritonitis; Case 31 had delirium on admission; Case 48, stercoraceous vomiting on admission; Case 67, hernia was the size of a man's head; patient collapsed; Case 71, hernia was the size of a child's head; patient moribund. Two deaths were on the 2d day. Case 54, ether and taxis before admission, also gut gangrenous; Case 58, symptoms very severe; strangulation for 48 hours. Three were on the 3d day. Case 1 had peritonitis on admission; Case 3 had peritonitis on admission; Case 72 had peritonitis on admission, with stercoraceous vomiting. One was on the 4th day. Case 37 had peritonitis on admission, with stercoraceous vomiting. One which occurred on the 5th day. Case 44 had peritonitis on admission.

There were of the inguinal variety of hernia, 40; femoral, 17; scrotal, 14; umbilical, 1.

The youngest subject on whom operation was performed was 2½ years of age; the next youngest, 13 years of age.

The ages of those who died after herniotomy were: 1 at 36, 2 at 39, 2 at 40, 1 at 46, 1 at 47, 1 at 50, 1 at 52, 1 at 53, 1 at 55, 1 at 58, 1 at 66.

Ages of the 72 cases: 1 at 2½ years, 3 between 10 and 20 years, 15 between 20 and 30 years, 14 between 30 and 40 years, 15 between 40 and 50 years, 11 between 50 and 60 years, 8 between 60 and 70 years, 5 between 70 and 80 years.

Of the 6 cases not entirely reduced by taxis, etc., 4 were femoral (one a large tumor), 1 was inguinal, 1 was scrotal (an enormous tumor). All these were relieved. Hernias had existed a long time, but under rest, opiates, etc., symptoms subsided. It is probable that omentum remained, but the gut in each was reduced.

The youngest subject operated upon in the hospital for strangulated hernia was a male child, 30 months old; the oldest subject, a male, nearly 80 years of age. (In 1875, Dr. T. H. Andrews brought before the clinical class of the hospital a hearty male child, upon whom he had operated some months previously for strangulated hernia. At the time of the operation the infant was barely *forty-five hours* old, probably the youngest subject of herniotomy on record. The hernia was evidently congenital, and he reported that the case (although apparently a very unfavorable one) made a rapid recovery without a bad symptom.)

Method of Operating.—We use in operating a scalpel, forceps, a grooved director, and a blunt pointed bistoury. After the primary incisions the top of the forefinger of the left hand is carried to the point of stricture or constricted tissue, the edge of which is divided with either an ordinary knife or with the probe-pointed bistoury.

Dr. Levis has devised a dilator and director combined, which he used very successfully (see figure) for many years, in and out of the hospital.

The radical operation for the cure of hernia is not now practised in the hospital. It is true that several operations were performed some years ago, but, although there was no fatal result, serious symptoms followed the operation in more than one of the cases, and in each instance the hernia sooner or later reappeared.

The risk of the operation is so great, as also is the uncertainty of any permanent beneficial result following it, that we now prefer the well-fitting hard rubber truss. Even in adults who have had hernia for years occasionally we find, after wearing such a truss during a period varying from several months to two or more years, a complete cure follows, while in youthful subjects the probability of an early cure is reasonably expected. The histories of several of the hernia cases have been thought worthy of record, but the details of the majority would present no more than mere repetition.

CASE 4. *Strangulated inguinal hernia in a boy aged 13; operation; recovery.*—Charles A., aged 13, was admitted March 13, 1872, with an inguinal hernia, which had been strangulated for four days. Ether and taxis were used unsuccessfully. Herniotomy was then performed soon after his admission. He was discharged, cured, April 17, 1872.

CASE 9. *Strangulated scrotal hernia in a child aged 30 months; operation; recovery.*—Louis A. was admitted July 7, 1873, with strangulated inguinal hernia on the right side. The hernia was congenital, and nothing had been done for its relief until six weeks previous to the child's admission, when a truss was applied. It had remained reduced until July 5, when the child had a fall down-stairs. On the next day marked symptoms of strangulation appeared.

On the 7th of July restoration by taxis was attempted, but it was unsuccessful. When the child was brought to the hospital he was verging on collapse, very pale, with quick, feeble pulse, constant sick stomach, and general prostration. On examination a hernial tumor was found on the right side of the scrotum, very tender to the touch, with considerable abdominal swelling. After etherization, taxis having failed to reduce the intestine, the ordinary operation was performed by Dr. Morton, at 6 P.M., on the same day. A large piece of intestine was found of a dark maroon color, which was firmly constricted at the internal ring. It was impossible to reduce the mass, but, after steady dilatation with the finger, the ring yielded, and the intestine was replaced. Five deep silver sutures, clamped with shot, held the walls of the wound in good apposition. Quarter-grain opium suppositories were administered, and teaspoonful doses of brandy given every two hours. He was put upon a liquid nourishing diet. The symptoms immediately subsided, rapid convalescence followed, and the child was discharged, quite well, July 19, twelve days after the operation.

CASE 10. *Strangulated scrotal hernia of unusual size; great exhaustion; gut returned by taxis; death.*—John N., aged 67, was admitted July 27, 1873, with hernial strangulation. He had suffered with hernia for many years, but had always worn a truss until a year since, when this was left off. The scrotal tumor subsequently increased in size, and frequently was only with the greatest difficulty returned. When admitted the tumor was so large that it hung as low as the middle of the thigh, and measured in circumference twenty-seven by eighteen inches. Shock and exhaustion were prominent symptoms. After etherization the mass was successfully returned, but the following day the patient died, not having reacted since admission.

CASE 13. *Strangulated hernia; operation; recovery.*—John B., aged 39, was admitted August 16, 1873. The hernia he attributed to a strain produced while carrying a heavy load ten years ago. The tumor had always been readily reduced, but twenty-four hours before admission the gut became incarcerated. On entering the wards the tumor occupied the right groin, and measured eight inches in its long circumference. He complained of abdominal pain, but had no vomiting. Ether and taxis having failed, the usual operation was performed by Dr. Levis. A drainage-tube was left in the wound, along side of which a portion of gut some hours afterwards appeared. The wound now had to be reopened to reduce the gut, deep silver sutures were introduced, and the tube was left out.

August 17.—One grain of opium was given every three hours, and quinine in large doses.

August 18.—Wound slightly opened at lower part for drainage, and lead-water and laudanum dressing applied.

August 29.—Discharged, cured.

CASE 17. *Strangulated femoral hernia; operation; recovery.*—Bridget W., aged 30, was admitted January 29, 1874. The patient lifted a heavy wash-tub of water two days before. When examined it was found that a femoral hernia existed, which was the size of a pigeon's egg. Ether was administered, and then taxis by the house surgeon, who thought he completely reduced the gut. During the night the patient, against orders, got up and went to stool. At 10 A.M. next day the hernia was again discovered, and now refused to respond to taxis; and on the same evening, symptoms of strangulation appearing, the usual operation was performed. A small portion of the gut was found strangulated (which evidently had not been previously reduced), and was firmly held by the ring. After reduction the wound was closed by three deep and several superficial sutures. A firm, soft pad and a roller bandage applied.

February 17.—Discharged, well.

NOTE.—It is probable that the taxis upon the first attempt caused a diminution in the size of the tumor and gave some relief; a portion of the contents of the knuckle being forced out, but the bowel itself, unquestionably, was not returned, as shown at the time of the operation, by the firmness of its attachments to the ring.

CASE 18. *Incarcerated hernia; spontaneous reduction after the use of ice and opium.*—Male, aged 41, was admitted May 30, 1874. The hernia appeared suddenly while the patient was walking; the tumor was quite hard and painful and about the size of a hen's egg. A two-grain opium suppository was introduced. Ice was applied to the tumor, in two hours after a second opium suppository. The patient fell asleep, during which time the gut went up spontaneously.

CASE 27. *Strangulated hernia; operation; cæcum and mass of intestine replaced; recovery.*—Thomas S., aged 27, a sailor, was admitted June 3, 1875. Four years ago he fell from the rigging upon the deck, striking upon his feet, and at the time felt pain in right groin. Soon after he noticed a small tumor, which gave him no trouble for two years. He then, after lifting heavy boards, developed an inguinal hernia. This was readily reduced. Since then the hernia has frequently appeared, but he always could replace it. He had never worn a truss. The evening of his admission, while walking, his foot slipped into the gutter and the hernia appeared; this time it could not be replaced. At 11.30 P.M., when he was admitted, the hernia was as large as a cocoanut, somewhat tympanitic on percussion; pain was intense, especially at umbilicus; frequent vomitings; pulse weak; skin cold; somewhat delirious. Ether and taxis were employed without avail. Dr. Hunt then made an incision three and one-half inches long and opened the sac, from which a small amount of serous fluid escaped. Nearly three feet of intestine protruded. The internal ring was nicked, and the cæcum and appendix and ileum were then returned. The mesentery appeared thickened. Deep silver sutures were inserted after the return of the gut. The urine was drawn off and the knees were well flexed. Opium suppositories, two grains, were ordered every four hours.

June 4.—Passed a good night; vomited once; opium continued.

June 7.—Penis œdematous; suppuration at lower part of incision; tongue coated. Up to this time he had had forty grains of opium; one-grain suppositories now substituted for the two-grain, and a puncture was made at the lower part of incision, which gave vent to considerable fluid.

June 10.—Sutures removed; free discharge; opium now suspended; Rochelle salt, two drachms, gave free action of bowels. From this time the progress of the healing of the wound was quite satisfactory, and patient was discharged, well, July 23, 1875.

CASE 33. *Large inguinal hernia in a female; taxis; spontaneous reduction.*—Julia C., aged 42, admitted November 12, 1875. During childbirth, sixteen years ago, she had developed, by the violence of the labor, a hernia in the left groin. At first it was small, but during this long period it has continued to increase in bulk, notwithstanding the various forms of truss which she had worn, until upon admission the tumor was fourteen and one-quarter inches in circumference, and nine inches in length from the rounded extremity to its junction with the belly wall. The hernia was entirely reducible, and by invaginating the sac it was possible to pass two fingers through the ring into the abdominal cavity. Although the contents of the tumor could be returned to the abdomen while the patient was recumbent, yet as soon as she coughed or rose into the erect position, the intestines were protruded into the large sac. The finger, introduced into the opening when the bowel was returned and the sac invaginated, could distinctly feel the body of the pubis below, and the tendinous edges of the inguinal ring above and laterally.

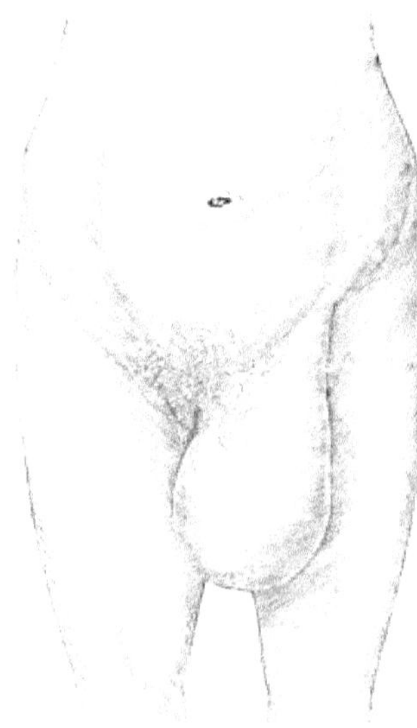

The patient had worn a number of trusses, but had not been able to obtain one that would keep the bowel within the abdomen. Before undertaking any operation for the radical cure of the case, however, it was thought proper to try a truss which would fill up the hernial canal. Accordingly, Dr. Levis had made a large triangular pad to fit the groin, having projecting from its centre a piece of wood like a finger, and about two and one-half inches long. The hernia was then reduced by taxis, the sac invaginated, and the projection of the truss passed into the hernial canal. Then the elastic band attached to the pad was buckled tightly around the waist, and two perineal bands adjusted to keep the apparatus from becoming displaced.

The following note was made upon her discharge: "This truss seems to fulfil the indications, and, if the patient can tolerate it, will probably make her much more comfortable than she has been without incurring the risk of an operation for the radical cure of the hernia. Even if a certain amount of inflammation should be produced it would not be regretted, because, if lymph were thrown out around the hernial opening, partial occlusion might result."

CASE 34. *Strangulated hernia; hydrocele; operation; recovery.*—John B., aged 55, was admitted March 25, 1876. He had had left sided hydrocele since childhood. The tumor was the size of a small cocoanut. On the right side there was a pyramidal swelling, running down from the groin into the scrotum. He stated that one year ago, while working, a lump appeared in the right scrotum, which could be pushed back into the belly, but on the day before admission he found that the lump could not be replaced. He had pain, and the tumor became enlarged in size. Upon examination distinct fluctuation was found; no impulse in coughing; it was not translucent; taxis unsuccessfully tried. He was then put to bed, and laxative enemata given. After two injections the bowels responded. Locally, lead-water and laudanum were applied to scrotum.

March 26.—Vomiting frequent; no passage; opium to relieve pain; poultices.

March 27.—A.M., no action of bowels; ice-water to scrotum; no sign of stercoraceous matter; pain continues in umbilical and *left* inguinal region. Later in the day taxis was again tried on the right or suspected hernial side, and the gut slipped into the abdomen, and thus established the diagnosis. The left side was then tapped, and twelve ounces of fluid were drawn off. Poultices continued on abdomen. Turpentine injections moved the bowels, after which a two-grain opium suppository was introduced, and two hours later a second. All pain now vanished. Discharged, cured.

CASE 37. *Strangulated hernia; peritonitis; operation; gut gangrenous; bowel unreduced; death.*—Mary H., aged 69, was admitted April 21, 1876, stating that she never had hernia before yesterday. When lifting a tub she felt a sharp pain in the bowels, which continued. In trying to defecate she found a small tumor in left groin; soon after this she began to vomit, and continued the rest of the day, the pain steadily increasing. Taxis was tried before she came into the hospital. Upon admission pulse was good; pain considerable, located near and on the tumor in the left groin. As the bowels had not been open for a week, several injections were administered. After the fourth a good passage was obtained, to great relief of the patient.

April 22.—8 A.M., vomiting of stercoraceous matter began. In the absence of Dr. Hunt, Dr. Hewson cut down on the tumor. The gut was found to be very dark in color, and so offensive that it was deemed inexpedient to return it. The wound was allowed to remain open; salicylic acid and emollient poultices were used. Beef-tea, whiskey, and morphia were given. Only one attack of vomiting after the operation.

April 23.—More comfortable; wound looks well; no change in the protruding gut; abdomen distended; vomiting returned, for which opium was given freely.

April 24.—Worse.

April 25.—Died.

Post-mortem showed that the hernial sac had not been opened by the knife; a pendulous mass of fat was found adherent to the mouth of the sac; an inch and a half of small intestine was found softened and strangulated, surrounded by a ring of lymph.

CASE 40. *Strangulated congenital hernia, complicated with hydrocele; operation; recovery.*—Benjamin B., aged 56, was admitted August 11, 1876. Has had congenital hernia on left side; also hernia in the right side for twenty years; rupture on right side came down while walking. For twenty-four hours had been irreducible; the symptoms were those of incarceration. Ether and taxis produced no relief; morphia was administered.

Dr. Hunt saw patient June 12; condition same as on admission. On the right side the swelling had increased and was very hard. A hydrocele being suspected with the hernia, an aspirator was introduced, and a large amount of straw-colored fluid was drawn off; this reduced the tension, but the hernia remained firm. Resisting taxis, the usual operation was performed. The gut was found firmly agglutinated to the sac; this was peeled off, and the hernial mass was returned; morphia ordered.

August 16.—Bowels acted on freely; been taking one grain of opium every three hours.

September 12.—Cured.

CASE 41. *Strangulated hernia; operation; death.*—Peter B., aged 36, admitted December 17, 1876. Had had rupture for eighteen years, and had worn a truss for ten years. The mass was always reducible. When straining at stool, six hours since, he had pain, and found he could not reduce it. He was "a hard drinker." Upon entering the wards he had vomiting and pain, and a tumor as large as a child's head. Taxis having failed, the usual operation was performed by Dr. Morton; at least eight inches of small intestine were out, filled with fæces and gas. Gut a deep maroon color; structure dark and unyielding; it was divided and the gut returned. A small vessel was ruptured in mesentery, which was enclosed in a catgut ligature. Opium, one grain every three hours, and stimulants were given, but serious symptoms continued to increase.

December 22.—Died.

Autopsy.—Portions of small intestine were found bound down by lymph; no general peritonitis.

CASE 63. *Direct inguinal hernia, strangulated; operation; recovery.*—John L., aged 27, admitted May 7, 1878. Patient believed that his hernia was congenital, although, as far as he knew, he first had trouble from the groin swelling about nine years ago. Two and one-half years before (October 14, 1875), after a great muscular effort, he found he could not reduce the mass, and came to Pennsylvania Hospital, where, under ether, it was reduced. Since then he generally had worn a truss. Yesterday, when without the truss, the hernia came down. Since then he had suffered great pain. When admitted ether was given, and taxis was not successful; ice was then applied, and a two-grain opium suppository given. These were ordered to be continued through the night.

May 8.—Not so well; some vomiting. Dr. Hunt made the usual operation.

May 10.—Temperature, 102½°; evening, 103°.

May 11.—Inflammatory action about edge of wound, but no abdominal tenderness. Bowels freely moved, naturally.

June 5.—Discharged, well.

Since the above cases were tabulated an interesting case of strangulated hernia, associated with an undescended testicle, was admitted and operated upon. The rarity of these cases is sufficient to warrant the introduction of the case in this paper.

Undescended testicle; symptoms of peritonitis: operation; portion of intestine found strangulated: return of the gut with a mass of mesentery: removal of the testicle.—Gabriel S., aged 22, single, a laborer, an Italian, was admitted November 28, 1879, with the following history: Had always enjoyed good health, but, as long as he could remember, he had a lump just above the right groin, which was moderately soft and painless. On admission the patient had an anxious expression, was pale, and in a condition of shock. In the right inguinal region there was found a lump somewhat larger than a goose's egg, which was very firm, painful, and fluctuated. The skin covering the mass was quite inflamed. There was no testicle in the right side of the scrotum. Two days before admission he had lifted a box of lemons, and at once experienced sharp pains, which gradually had increased in severity.

November 29.—Several attacks of vomiting, with pain in the abdomen, which, since twenty-four hours, had become tympanitic, led to the supposition that a portion of intestine was strangulated alongside the undescended testicle. After etherization an incision four inches in length was made over the tumor. Dissection revealed a tense, very dark-colored sac, which, when punctured, gave vent to a fluid similar to that found in ordinary hernias. When the sac was freely divided a portion of thin, spread-out mesentery was observed, the vessels in which were enormously distended, and the mass almost black from being strictured. On raising this up an atrophied testicle was brought into view, and directly under this a knuckle of strangulated intestine was found. It was very dark in color, and it had been tightly pressed down by the fluid and testicle above it, and firmly strictured by the ring. The intestine and mesentery were then with some difficulty replaced in the abdomen. The cord of the testicle was then seized with a tenaculum, but the testicle was so firmly held in its position that it was impossible to alter its position, except to a very limited extent. The cord was then pierced by a curved needle armed with a strong double ligature, and each side was then tied, and the testicle was removed. Silver sutures were then introduced, and made to include the lining membrane of the sac at several points, thus closing effectually the canal; an opening was left opposite the stump of the tied cord. Carbolized oil dressing was applied, and opium in half-grain doses given every two hours, with an occasional hypodermic injection of morphia. The abdomen was also covered with a soft towel, which had been thoroughly soaked in laudanum. The testicle when opened was very soft, quite black, and apparently disorganized.

The patient did well for several days. Gradually peritonitis developed, and on the eight day after the operation he died.

The autopsy showed diffuse peritonitis: intestines glued to each other and to peritoneum (abdominal surface); omentum rolled up along the transverse colon, except the small portion, which was returned with the hernia, this was glued at internal ring, and was gangrenous. There was about one fluidounce of pus near wound and over the fundus of the bladder.

Résumé of the Cases of Hernia treated in the Pennsylvania Hospital from 1870 to 1879.

No.	Age.	Sex.	Date of Admission.	Variety.	Duration.	Time strangulated.	Taxis.	Anæsthetics.	Operation.	Result.	Days in Hospital.	Remarks.
1	39	M.	March 3, 1871.	Inguinal.	5 days.	Many hours.	Failed.	Ether.	Usual.	Died.	3	Peritonitis at time of operation.
2	58	F.	July 8, 1871.	Femoral.	....	10 hours.	Failed.	Ether.	Usual.	Cured.	17	
3	58	M.	Sept. 3, 1871.	Inguinal.	....	30 hours.	Failed.	Ether.	Usual.	Died.	3	Peritonitis at time of operation.
4	44	M.	March 15, 1872.	Inguinal.	....	Some hours.	Failed.	Ether.	Usual.	Cured.	35	
5	35	M.	Dec. 9, 1872.	Inguinal.	....	24 hours.	Failed.	Ether.	Usual.	Cured.	50	
6	28	M.	Dec. 15, 1872.	Inguinal.	1 day.	1 day.	Failed.	Ether.	Usual.	Cured.	30	
7	50	M.	Feb. 28, 1873.	Femoral.	....	24 hours.	Failed.	Ether.	Usual.	Died.	1	Very great shock; the tumor was glued together by lymph, forming a solid ball.
8	20	M.	March 12, 1873.	Inguinal.	4 days.	23 hours.	Failed.	Ether.	Usual.	Cured.	26	
9	2½	M.	July 7, 1873.	Inguinal.	Congenital.	18 hours.	Failed.	Ether.	Usual.	Cured.	12	Gut dark maroon color.
10	67	M.	July 27, 1873.	Scrotal.	Long standing.	Several hours.	....	Ether.	....	Died.	1	Tumor was 27 by 18 inches in circumference. Sinking on admission.
11	54	M.	Aug. 5, 1873.	Inguinal.	Several years.	12 hours.	....	Ether.	....	Cured.	1	
12	48	F.	Aug. 7, 1873.	Femoral.	Several years.	24 hours.	Failed.	Ether.	....	Cured.	4	Pil. opii, gr. i; ol. ricini, ℥i.
13	39	M.	Aug. 16, 1873.	Inguinal.	11 years.	6 hours.	Failed.	Ether.	Usual.	Cured.	13	Tumor was 8 inches in long circumference. Drainage-tube left in; portion gut appeared; tube removed; deep sutures introduced.
14	75	F.	Sept. 5, 1873.	Inguinal.	Long standing.	24 hours.	....	Ether.	....	Cured.	6	
15	21	M.	Sept. 16, 1873.	Scrotal.	Several years.	12 hours.	....	Ether.	....	Cured.	1	
16	50	M.	Oct. 18, 1873.	Scrotal.	....	Several hours.	Succeeded.	Ether.	....	Cured.	13	Fell through a bridge, fracturing ribs and producing hernia; never noticed it out before; seems to be of long standing, for ring is large.
17	30	F.	Jan. 29, 1874.	Femoral.	3 days.	Several hours.	Failed.	Ether.	Usual.	Cured.	27	
18	41	M.	May 30, 1874.	Inguinal.	Several years.	24 hours.	Failed.	Ether.	....	Cured.	2	Cold and opium; spontaneous reduction.
19	52	M.	July 6, 1874.	Femoral.	24 hours.	12 hours.	Failed.	Ether.	....	Cured.	1	Tumor size of a hen's egg. Cold and opium; spontaneously, with slight effort of his own during night.
20	23	M.	Sept. 29, 1874.	Scrotal.	Several years.	12 hours.	....	Ether.	....	Cured.	1	Scrotum enormously distended; taxis and ether used before admission.
21	56	M.	Oct. 8, 1874.	Inguinal.	Several days.	....	Succeeded.	Ether.	....	Cured.	9	
22	47	F.	Oct. 13, 1874.	Femoral.	Several days.	18 hours.	Failed.	Ether.	Usual.	Died.	1	Gut seriously involved; taxis and ether before admission; continued exhaustion and peritonitis.

Résumé of the Cases of Hernia treated in the Pennsylvania Hospital from 1870 to 1879.—(Continued.)

No.	Age.	Sex.	Date of Admission.	Variety.	Duration.	Time Strangulated.	Taxis.	Anæsthetics.	Operation.	Result.	Days in Hospital.	Remarks.
23	30	M.	Oct. 25, 1874.	Inguinal.	3 years.	24 hours.	Succeeded.	Ether.	....	Cured.	1	Cold and opium, and taxis three times, which succeeded.
24	40	F.	Dec. 31, 1874.	Femoral.	8 years.	....	Failed.	Ether.	....	Unimproved.	8	Larger the last 4 days; ether and taxis failed.
25	23	M.	Jan. 2, 1875.	Scrotal.	6 years.	18 hours.	Failed.	Ether.	....	Cured.	4	Vomiting; cold and opium.
26	65	M.	Jan. 17, 1875.	Scrotal.	Long standing.	24 hours.	Failed.	Ether.	....	Died.	3	Taxis and ether before admission, sinking when admitted.
27	27	M.	June 3, 1875.	Inguinal.	4 years.	6 hours.	Failed.	Ether.	Usual.	Cured.	50	Tumor as large as a cocoanut.
28	49	M.	June 28, 1875.	Double inguinal.	....	Some hours.	Succeeded.	Ether.	....	Cured.	4	
29	44	M.	Aug. 12, 1875.	Scrotal.	41 years.	2 hours.	Failed.	Ether.	....	Cured.	19	Taxis failed with ice and opium.
30	27	M.	Aug. 13, 1875.	Inguinal.	2 years.	3 hours.	Failed.	Ether.	....	Cured.	1	Spontaneous reduction.
31	40	M.	Sept. 25, 1875.	Inguinal.	Several years.	3 days.	Failed.	Ether.	Usual.	Died.	1	Vomiting for 24 hours, and delirious on admission; a large portion of gut distending scrotum; gut lividly congested; mesentery ligated and cut off; blood in sac.
32	21	M.	Oct. 14, 1875.	Inguinal.	Several years.	Several hours.	Failed.	Ether.	....	Cured.	2	
33	42	F.	Nov. 12, 1875.	Femoral.	16 years.	4 hours.	Failed.	Ether.	....	Cured.	14	Tumor 11½ by 9 inches. On 12th, after passing water, she reduced it.
34	55	M.	March 25, 1876.	Scrotal.	Congenital.	3 days.	Failed.	Ether.	....	Cured.	11	Also hydrocele left side; constant vomiting of greenish mucus until reduced; lead-water and laudanum applied to scrotum; evening of third day went back under taxis.
35	71	M.	April 4, 1876.	Inguinal.	Several days.	Several days.	Failed.	Ether.	....	Died.	1	Peritonitis and stercoraceous vomiting on admission.
36	70	F.	April 18, 1876.	Inguinal.	10 years.	....	Succeeded.	Ether.	....	Cured.	7	
37	69	F.	April 21, 1876.	Femoral.	48 hours.	....	Failed.	Ether.	Usual.	Died.	4	Stercoraceous vomiting.
38	49	F.	June 8, 1876.	Femoral.	20 years.	36 hours.	Failed.	Ether.	....	Died.	2	Shock on admission. Taxis and ether before admission; continued shock.
39	52	M.	June 9, 1876.	Inguinal.	Several days.	....	Succeeded.	Ether.	....	Cured.	22	Complicated with abscess.
40	56	M.	Aug. 11, 1876.	Double inguinal.	Congenital and 20 years.	5 days.	Failed.	Ether.	Usual.	Cured.	18	Complicated with hydrocele. On 17th, during delirium, jumped over hospital fence.
41	25	M.	Sept. 7, 1876.	Inguinal.	....	....	Succeeded.	Ether.	....	Cured.	1	
42	51	M.	Oct. 17, 1876.	Double inguinal.	3 years.	....	Failed.	Ether.	....	Unimproved.	7	
43	37	M.	Nov. 16, 1876.	Inguinal.	Many years.	Several hours.	Failed.	Ether.	....	Cured.	5	
44	36	M.	Dec. 17, 1876.	Inguinal.	18 years.	6 hours.	Failed.	Ether.	Usual.	Died.	5	Gut dark maroon color; tumor size of child's head; hard drinker. Taxis and chloroform before admission; peritonitis.

45	70	M.	Dec. 26, 1876.	Scrotal.	15 years.	....	Succeeded.	Ether.	....	Cured.	18	Complicated with hydrocele.
46	24	M.	Dec. 29, 1876.	Inguinal.	10 years.	4 days.	Succeeded.	Ether.	....	Cured.	7	Some of the old irreducible portion remains out in scrotum.
47	16	M.	Feb. 15, 1877.	Inguinal.	4 years.	5 days.	Failed.	Ether.	Usual.	Cured.	40	Shock on admission and for 24 hours. Congestion of gut disappeared on stretching stricture. Pulse 44. Erysipelas on 18th.
48	66	M.	June 12, 1877.	Inguinal.	9 days.	9 days.	Failed.	Ether.	Usual.	Died.	1	Continued stercoraceous vomiting before admission. Gut gangrenous; some jaundice.
49	31	M.	June 21, 1877.	Inguinal.	Congenital.	30 hours.	Failed.	Ether.	Usual.	Cured.	71	Stercoraceous vomiting. Some of the gut fast to scrotum.
50	33	M.	July 7, 1877.	Inguinal.	....	24 hours.	Succeeded.	Ether.	....	Cured.	2	
51	37	M.	July 25, 1877.	Double scrotal.	Several years.	....	Succeeded.	Ether.	....	Cured.	8	
52	63	F.	Aug. 15, 1877.	Femoral.	Long standing.	Several days.	Failed.	Ether.	....	Unimproved.	6	
53	40	M.	Sept. 22, 1877.	Inguinal.	10 years.	Few hours.	Failed.	Ether.	....	Cured.	3	Opii suppository and cold; spontaneous reduction in 10 hours.
54	30	M.	Nov. 13, 1877.	Inguinal.	14 years.	24 hours.	Failed.	Ether.	Usual.	Died.	2	Ether and taxis before admission. Omentum dark-brown color; gut congested.
55	67	M.	Nov. 19, 1877.	Inguinal.	Many years.	Several days.	Failed.	Ether.	....	Cured.	2	
56	22	M.	Nov. 30, 1877.	Inguinal.	3 months.	1 hour.	Failed.	Ether.	....	Cured.	2	
57	40	F.	Dec. 12, 1877.	Femoral; incarcerated.	2 years.	10 days.	....	Ether.	....	Improved.	17	Nothing done. Ol. ricini and opii.
58	53	M.	Feb. 9, 1878.	Inguinal.	2 years.	2 days.	Failed.	Ether.	Usual.	Died.	2	Vomiting for some time. General peritonitis.
59	40	M.	Feb. 26, 1878.	Inguinal.	16 days.	....	Succeeded.	Ether.	....	Cured.	2	
60	28	M.	March 29, 1878.	Inguinal.	24 hours.	....	Failed.	Ether.	....	Cured.	10	Taxis before admission; taxis and ether failed second time. Ice and opii; leg flexed; hips elevated and ol. ricini; spontaneous reduction in few hours.
61	70	F.	April 5, 1878.	Umbilical.	2 days.	....	Failed.	Ether.	....	Died.	1	Shock on admission; death in 3½ hours.
62	70	M.	April 10, 1878.	Inguinal.	8 years.	24 hours.	Failed.	Ether.	....	Cured.	1	Spontaneous reduction in 2½ hours; ice and opii, and ol. ricini.
63	27	M.	May 7, 1878.	Inguinal.	7 years.	24 hours.	Failed.	Ether.	Usual.	Cured.	31	
64	68	M.	May 25, 1878.	Scrotal.	15 years.	....	Failed.	Ether.	....	Improved.	18	Tumor, 10 by 6 inches.
65	53	F.	Oct. 1, 1878.	Femoral.	Many years.	6 days.	Failed.	Ether.	....	Cured.	17	
66	28	M.	Oct. 12, 1878.	Scrotal.	Several years.	10 hours.	Failed.	Ether.	....	Cured.	3	
67	52	M.	Jan. 1, 1879.	Femoral.	25 years.	1 day.	Failed.	Ether.	Usual.	Died.	1	Ether and taxis before admission. Tumor size of man's head; gut much congested.
68	22	M.	Nov. 25, 1878.	Scrotal.	Several hours.	....	Failed.	Ether.	....	Cured.	1	Kicked in belly, and shortly after tumor in scrotum.
69	50	F.	March 7, 1879.	Femoral.	2 years.	1 day.	Failed.	Ether.	Usual.	Cured.	94	Ether and taxis before admission. Highest temperature, 103°.
70	46	F.	March 10, 1879.	Femoral.	10 years.	1 day.	Failed.	Ether.	....	Unimproved.	2	Would not submit to operation.
71	16	M.	April 14, 1879.	Scrotal.	25 years.	1 day.	Failed.	Ether.	Usual.	Died.	1	Scrotum black; size of child's head; gut gangrenous and sloughing.
72	65	F.	June 12, 1879.	Femoral.	6 years.	4 days.	Failed.	Ether.	Usual.	Died.	3	Stercoraceous vomiting on admission; gut gangrenous; mitral systolic murmur.

Thomas G. Morton.

GYNÆCOLOGICAL CASES AND OPERATIONS, INCLUDING OVARIOTOMY.

1.—*Vaginal Atresia, Congenital and Acquired.*
2.—*Ovarian Tumors, Ovariotomy, Paracentesis Abdominis.*
3.—*Vesico-Vaginal Fistula.*
4.—*Recto-Vaginal Fistula.*
5.—*Laceration of the Female Perineum.*
6.—*Amputation of the Cervix Uteri.*

1.—VAGINAL ATRESIA, CONGENITAL AND ACQUIRED.

PRIOR to the year 1873 the records of the surgical operations performed in the hospital were not systematically kept, and the clinical notes of many interesting cases are unfortunately missing. This fact will probably explain why it happens that the records are silent, up to the time mentioned, upon many topics that now receive proper attention. Operations upon the female genital organs for imperforate hymen, with retained menstrual secretion; atresia vaginæ, either from congenital malformation, or following injury received during parturition; vesico-vaginal and other fistulæ; laceration of the perineum, and similar conditions, have doubtless been more or less frequently performed in the past, and reports of a few of them have been included in the consideration of allied or associated conditions, under appropriate headings, elsewhere in these pages, but of others, particularly vaginal deformities, no previous record appears. It is altogether improbable that prior to 1873, during a period of more than a century, no such case should have been under treatment, whereas, since keeping a systematic record, five or six patients have applied for relief. Moreover, the great activity within the last few years of specialists in the field of gynæcology, and the increase in the number of hospitals, some of them devoted exclusively to the treatment of women's diseases, would naturally have tended to reduce the number of such operations in this hospital rather than to increase it.

The subject of vaginal atresia, or colpatresia, is naturally divided into the following forms: (1) the simplest variety of atresia, imperforate hymen; (2) congenital atresia, from imperfect development of that fœtal involution of the epiblast forming the vagina and neck of the uterus (the uterine fundus having been duly formed by the junction of the Müllerian ducts); (3) congenital atresia, with absence of uterus (failure of union of Müllerian ducts); and (4) acquired atresia, as a result of inflammation, sloughing, and cicatricial contraction, following a lingering labor.

Of the first series one case is given, that of Mary McC., 15 years of age. Of the second, we communicate the clinical history of two patients,—Sarah J. W., aged 14 years, and Mary C., aged 20 years. With the exception of Mary C., in all of these complete relief was obtained by operation, and a patulous vagina resulted. The third class of congenital malformation is represented by the history of Bertha D. Such deformities being outside the limits of reparative surgery, do not admit of operation, and the cases are discharged after a careful examination reveals the true condition of the parts. The final series includes those unfortunate cases in which cicatricial contraction and inflammation have completely or practically obliterated the canal of the vagina, two cases of which are appended,—Roxie B. and Sarah J. A.,—difficult and protracted parturition being the cause. These patients always suffer with dysmenorrhœa, or complete emansio mensium, with more or less serious disturbance of the general health, placing them in a truly pitiable condition. In the cases communicated, after a simple surgical procedure, dissecting the coalesced vaginal walls apart, carefully cutting, or rather tearing, through the cicatricial tissue until the uterus is reached, subsequently keeping the opening patulous by means of Molesworth's or other dilators and tents, a most satisfactory result in one case has been obtained, so that menstruation can be painlessly performed and the patient's health perfectly restored; but in the other the cure was less complete, on account of the refractory disposition of the patient, and the necessarily imperfect after-treatment.

(1) *Imperforate hymen (membrana vulvæ impervia).*

Mary McC., aged 15 years, was admitted into the women's surgical ward April 4, 1875. The patient stated that two months before entering the hospital she began to have severe tearing and bearing-down pains in the vagina and surrounding parts, but within the two weeks preceding her application for admission the suffering had returned, and was now almost constant. There was some irritability of the bladder, but no disturbance of the bowels. She had never seen any menstrual discharge, but had, for several months, exhibited all the attendant symptoms of menstruation. Her mother called attention to some enlargement of the lower portion of the abdomen, and also spoke of a swelling or a dark-colored tumor in the vulvar fissure. Upon examination, the patient was found to be in vigorous health, and well developed for her age. The hypogastric region was unusually prominent. The girl being fully etherized, the local cause of the amenorrhœa was sought for, and the projecting tumor found to be an imperforate hymen, bulging outward, as a result of retained menstrual discharge. Dr. Morton made a small incision, which gave outlet to eighteen ounces of dark-colored, tarry fluid. The opening was subsequently enlarged. The patient recovered, and was discharged, cured, on the fifth day.

(2) *Congenital vaginal atresia (vagina in sinum desinens): vagina a cul-de-sac.*

Case I.—Sarah J. W., aged 14 years, was admitted January 18, 1870, with complete congenital occlusion of the vagina. She had never menstruated. Two days later, under ether, Dr. Hunt made the vagina patulous, carefully dissecting the adjacent walls of bladder and rectum, up to the uterus; a good reaction followed, although twenty-four hours after operation there were some bearing-down pains, relieved by full doses of morphia. Four weeks afterwards a free discharge of menstrual fluid came away, and was followed by a healthy menstrual flow. After the operation the vagina was filled with a large tent of oiled lint. She was discharged, cured, February 19, 1870.

CASE 2.—Mary C., aged 20, was admitted December 15, 1873, with complete atresia of the vagina. This patient was well formed, had fair-sized breasts, a normal amount of hair upon the genitalia. The vulva, including the nymphae, was well developed, but the vagina was represented by only a very small cul-de-sac. On examination, through the rectum, and at the same time pressing the lower part of the abdomen well in, a small, rather firm body was detected, occupying a central position, and very movable, which was thought to be the imperfectly-developed fundus of the uterus. As the patient had monthly pains, with all the usual concomitant symptoms of menstruation, it was deemed best to make an exploratory operation, dissecting up to the hard body the supposed uterus. From the menstrual molimen it was believed that the ovaries were present, and were functionally active, especially since two very small oval bodies, it was thought, could be felt on either side. A preliminary operation had been performed six months before, in which Dr. Morton had divided the cul-de-sac, and after tearing the tissues, principally with the finger-nail and handle of the scalpel, he dissected the parts in the direction of the supposed uterus to the extent of four inches in depth. Since then the opening had closed again, the patient neglecting the after-treatment.

January 13, 1874.—Patient etherized, and a canal or artificial vagina was again opened by Dr. Morton to a depth of three and a half to four inches, and a large oiled tent was then introduced.

January 16.—Dilatation has been kept up, and a solution of permanganate of potassium used as a wash.

February 1.—The vagina has gradually closed, notwithstanding the best directed efforts to keep it open. Discharged, incurable.

(3) *Congenital vaginal atresia, with absent uterus and ovaries (defectio vaginae, uteri, et ovariorum).*

Bertha D., born in Hanover, aged 22, was admitted August 14, 1873. Although well developed and ruddy, yet has never menstruated; not anaemic; has had sick headache, pain in the back, occasional fever at menstrual periods, but no discharge. Pains were noticed first at the age of 16, and continued every four weeks, increasing in severity, keeping her in bed for several days each period until she was 20. At this time there was so much constitutional disturbance that her feet and limbs would swell, and the headache and dorso-lumbar pains were unendurable. Her family physician, on several occasions, applied leeches to the thighs with marked benefit. On making a vaginal examination it was found that the vagina formed a cul-de-sac, about one inch deep, and with the catheter and finger in rectum no uterus could be discovered. The external parts were apparently normal. The girl is well formed, full-blooded, and breasts in good state of development. She stated that an aunt of hers had never menstruated until after marriage, at the age of 28 years.

August 25.—Ether administered. No uterus or ovaries can be found, nor can any hard body be felt in the neighborhood. A fold of fibrous tissue resembling a broad ligament is felt, and at its termination, far to the right, an uncertain mass, smaller and less defined than a virgin ovary, and giving less resistance to the finger, and it slides away under touch.

September 5.—Patient states that for nearly six years she has had attacks of epistaxis every four weeks, with rush of blood to the head, vertigo, and loss of vision, and that a year ago she had an attack of hæmoptysis, losing a gill of blood daily for a week.

September 20.—Discharged, incurable.

In connection with the preceding case it is proper to note the interesting fact that the patient was markedly hysterical, and on several occasions, while under observation, had well-developed hysterical convulsions.

CASE 1. *Congenital vaginal atresia, with absent uterus and ovaries.*—Ellen E., aged 18, was admitted March 5, 1878. On examination it was found that the external parts were normal. There was no vagina, but simply a bare cul-de-sac. By the rectum examination neither uterus nor ovaries could be discovered. No treatment was adopted.

CASE 2. *Congenital vaginal atresia, with absent uterus and ovaries, and absence of hair upon the pudenda.*—In the spring of 1878 I examined Sarah B., aged 28, who had never menstruated. She had suffered immensely from monthly pains, so much so that at such times she was compelled to remain in bed. She was apparently in the most perfect health, was very well formed, quite stout, and had a fair color. No previous examination had ever been made. After etherization I found an entire absence of vagina, ovaries, and uterus. Just below the urethra, which opened regularly, there was a dense, unyielding tissue, about the size of a dime. In addition there was an entire absence of hair about the pudenda, but on the axillary region the usual amount was noticed. No treatment was adopted.

(1) *Vaginal occlusion by cicatrix (vagina occlusa).*

CASE 1.—Roxie E. B., aged 18, was admitted November 6, 1872. She had a very serious labor, was delivered with instruments, and the vaginal tissues had been lacerated, and from all accounts there had been sloughing of tissue. Following this the vagina closed almost entirely, and when first examined the finger could be passed into the vagina about half an inch.

November 28. Dr. Morton divided the tissues in a median line, and then separated the parts with the forefinger, and carried the division of the tissues as far as the uterus, which was readily felt. A large tent was then introduced, and the parts kept well cleansed with a permanganate of potassa solution. The daily use of the vaginal rubber dilator (see figure) kept the vagina patulous, and the patient was discharged greatly improved, and with instructions to continue the use of the instrument.

On the 26th of November, the following year, the patient was again admitted, the vagina having closed up nearly to the condition it presented at her first admission.

December 1.—The vagina was lengthened by incisions and tearings. The os uteri was seized with a tenaculum, and brought well down, and a very large tent of lint soaked in oil was introduced. This was changed daily, and the use of bougies was made part of the daily treatment. She was finally discharged as cured. Since leaving the hospital we have not heard of her condition.

CASE 2. *Contraction of the vagina, with vesico-vaginal fistula following difficult labor.*—Sarah J. A., aged 23, was admitted March 21, 1871. The cicatrices of the posterior wall were divided, and portions were excised. The vesical opening was freshened, and five sutures were introduced; a catheter was retained in the bladder. On tenth day the sutures were removed. Discharged, cured.

2.—OVARIAN TUMORS, OVARIOTOMY, PARACENTESIS ABDOMINIS.

THE records show comparatively few ovariotomy operations. This in all probability results from the fact that the opinion prevails that such operations have less chance of success when performed in a large hospital than they would have in a private house. However, since the year 1873, ovariotomy has been performed in four cases. Two of these operations recovered, and two died. Of the causes of death immediately inducing a fatal issue, one was from tetanus, on the ninth day, the other from suppression of urine, on the second day after the operation.

In all of the cases the ordinary clamp has been placed on the pedicle, and in one case a drainage-tube, as recommended by Dr. Atlee, was left in until the fourth day. In four instances ovarian cysts have been simply tapped, but quite a number of these tumors were emptied by paracentesis in the medical wards, which have not been included in the series of cases here presented.

CASE 1. *Simple ovarian cyst: ovariotomy: recovery.*—Anna M., aged 26, a Swede, was admitted February 28, 1873, at the recommendation of Dr. A. Frické, who wrote concerning the patient, who had been under his care, as follows. After referring to her abdominal tumor, he says, "The patient is in otherwise perfect health; has menstruated regularly; was married some months ago; has never been pregnant." Upon careful repeated examinations a unilocular, very movable, ovarian cyst, with fluid contents, was found. The patient said that she had never experienced any pain in the tumor, but it was rapidly growing. The other abdominal organs were apparently healthy.

On March 4, Dr. Morton, after etherizing the patient, made the usual incision, about an inch and a half in length, in the median line below the umbilicus. A cyst then at once presented, which was seized with a tenaculum, and firmly held while a trocar was introduced. As the contents of the tumor drained off the cyst walls were from time to time drawn out through the abdominal opening until the entire sac was extracted. It was found necessary to enlarge the incision slightly,—about half an inch. The clamp was applied, and the pedicle secured well outside of the abdomen, and the wound was brought together with deep silver sutures. An excellent and rapid recovery followed, without an unfavorable symptom.

March 27.—Discharged, well.

Some months afterwards Dr. Frické saw this patient, and reported that she had since menstruated regularly, and in all respects was in good health.

CASE 2. *Ovarian tumor: tapped: relieved: death from exhaustion: post-mortem.*—Mary D., aged 19, was admitted February 29, 1873, with a solid ovarian tumor, associated with considerable abdominal dropsy. There was very great discomfort, and some fever symptoms, the temperature being from 100° to 101°.

May 24.—Paracentesis abdominis was performed, and five pints were removed, which gave considerable relief. Gradually, however, symptoms of exhaustion appeared, and the patient died June 5, 1873.

Post-mortem.—The abdomen contained about a bucketful of serum, without evidence of general peritonitis. Springing from the site of the right ovary was a large solid tumor, which had firm attachments to the adjacent peritoneum. The surface of the tumor was irregularly lobulated. Some of the nodules were soft and doughy. Upon section, the tumor, in parts, was firm, principally whitish in color, with patches of red (ecchymotic). Other parts were yellowish-white, more cystic in character from central softening, and escape of fluid contents. The disease was evidently malignant (encephaloid) in character.

CASE 3. *Ovariotomy; death from tetanus, on the ninth day.*—Catharine G., aged 28, was admitted August 9, 1873, with an immense ovarian tumor. When married, at 19, she was healthy. Since that time she has had five children, the youngest being ten weeks old. After the birth of the second child, in May, 1867, Mrs. G. noticed that her abdomen did not diminish in size, but became more and more prominent. In December, 1872, it was tapped, and four and one-half gallons of liquid was drawn off; and on May 11, 1873, she was tapped again, and six gallons removed. A few months later, in July, 1873, four gallons were taken away in the same manner. The last two operations were performed while the patient was pregnant with her last child. No unfavorable symptoms followed.

Upon admission, a tumor, irregular in shape, nodular, freely movable, painless on pressure, extends from the lower edge of the sternum to the crest of ilium, dipping deeply into the abdomen. The other organs were healthy.

August 11.—The usual median incision was made by Dr. Morton; the main cyst was tapped. Afterwards numerous other cysts were evacuated, and the mass was then withdrawn. Everywhere firm adhesions were found, which had bound down the tumor to the peritoneum and intestines in all directions. The uterus was also firmly attached to the growth, and only with great difficulty was it freed from its connections with that organ and from the intestines, which were also glued to the mass in the same way. The tearing of the adhesions gave rise to profuse hemorrhage from a multitude of bleeding points, and death seemed imminent. One clamp was applied to the pedicle and one to some torn long adhesions, which bled freely. The usual dressings were applied; a full dose of morphia was given.

August 11, evening.—Pulse 140; skin soft; there was but little pain; takes nourishment freely.

August 12.—Some vomiting; pulse 140.

August 13.—Vomiting; pulse 120; irritable, but better.

August 14.—Stomach better; pulse 120; first clamp came away.

August 16.—Second clamp came off.

August 17.—Pulse 108; somewhat stronger.

August 18.—Pulse 140; slight trismus; half a grain of Magendie's solution of morphia ordered; fed by rectum; some delirium.

August 19.—Pulse 156; trismus marked; weaker; morphia every fourth hour by the skin.

August 20.—Died suddenly. No post-mortem was allowed.

CASE 4. *Ovariotomy; death from suppression of urine.*—Harriet B., aged 43, was admitted June 11, 1875. This patient was in the hospital some time ago with an ovarian tumor, which was then tapped, giving great relief. She now presents an abdomen of immense size, and the tumor increasing.

June 20.—Tapped, and eight quarts were removed, after which the more solid portions of the tumor were readily felt.

July 25.—Tapped; eight and a half quarts were removed.

September 10.—Tapped, and about the same amount as last operation were drawn off.

September 13.—Thirteen quarts removed.

September 30.—Advised to go to the country, and to return when in better health for an ovariotomy operation.

October 4.—Readmitted on account of abdominal distress, which is so great that she is unable to leave the city.

October 6.—Ovariotomy was performed by Dr. Morton; a multilocular ovarian cyst was tapped, after the usual abdominal incision, from which many quarts of fluid escaped. The liquid from the cyst was dark in color, like ale. This cyst, like the preceding case, was found exceedingly adherent to the abdominal walls;

it had no pedicle. The écraseur was applied; there was no bleeding; the wound was closed in the usual manner.

October 7.—Vomiting and exhaustion, but no peritonitis.

October 8.—In spite of every effort to produce reaction the patient now sank, and death resulted from suppression of urine.

No post-mortem was allowed.

CASE 5. *Ovarian tumor; tapped; improved.*—Lizzie J., aged 33, married, was admitted October 27, 1875. Four years before, after childbirth, she had first noticed a swelling in the right side of the abdomen: this swelling did not apparently increase for a year, when it began to grow rapidly. The pain was of a dragging character. Abdomen measured thirty-seven inches in circumference: the uterus was found to be normal.

November 15.—Tapped by Dr. Levis, and thirteen pints were removed. The fluid removed was viscid and coffee-colored in appearance.

November 24.—Discharged, improved.

CASE 6. *Ovarian tumor; tapped.*—Elizabeth W., aged 26, was admitted May 3, 1877. There was no history of hereditary disease. She commenced to menstruate at 14, and was married at 19, and had two children. Her present malady commenced one year before the birth of her last child. About five years ago she began to have pain and discomfort in the abdomen; then swelling was noticed, and a gradual decline in her general health. Her second gestation only lasted seven months. Upon admission there was an irregular abdominal enlargement; apparently there were two tumors, each about the size of a child's head.

May 12.—A cyst in the left side was tapped; three pints of a thick grayish fluid was removed.

May 18.—A cyst in the right side was tapped.

Dr. J. G. Richardson, who examined the fluid taken from the first tapping, reported that "it contains, besides red and white corpuscles, an abundance of compound granular corpuscles,—good examples of the peculiar ovarian cell."

CASE 7. *Multilocular ovarian cyst; ovariotomy; recovery.*—Mary G., aged 19, single, was admitted into Dr. J. H. Hutchinson's ward December 15, 1877. (For notes of case, see August 20, 1877, medical ward.) Patient's condition was about the same as after the previous tapping. Four months ago, however, her abdomen was not quite so globular as it was at that time. Over the umbilicus percussion was dull, and the tumor is readily forced from side to side.

December 26.—She having been transferred to the surgical wards, several days ago was put under the influence of ether, and Dr. Morton made the usual incision, two inches in length, in the median line. The contents of the sac were drawn off by the trocar. After a large amount had been evacuated the finger was introduced, and various cysts were broken up and their contents removed. There were no adhesions. A clamp was applied to the pedicle, and the tumor was then cut off. A portion of a small-sized drainage-tube was left in the very lowest part of the wound, and its edges brought together by interrupted silver sutures. Carbolized charpie was applied to the wound, and the whole retained in place by a roller and abdominal binder. A two-grain opium suppository was inserted at once on the completion of the operation. During the removal of the tumor the room had been kept at a temperature of 94°. The quantity of fluid drawn off from the cyst was twelve quarts and eight ounces, having a specific gravity of 1012. It was neutral in its reaction, or very slightly acid. A suppository of opium was ordered every three hours, with beef-tea, milk, and small doses of brandy. The urine to be drawn off, and the ordinary directions given for careful nursing.

December 27.—Abdomen soft and not distended; free discharge from the wound; urine has been voided naturally; opium suppositories have been given every three hours.

December 28.—Removed drainage-tube; doing well; appetite.

January 2, 1878.—Restless; had two movements of the bowels.

January 5.—Pins and sutures removed.

January 7.—Clamp removed; slight discharge; surface dressed with the oxide of zinc ointment.

January 22.—Up and about the ward.

February 3.—Menstruation appeared.

March 7.—Discharged, cured.

The temperature record shows that on the evening following the operation the thermometer indicated 100½°, dropping to 99° the following morning, the 27th. That evening it rose again to 100°, with the pulse 100, and respirations 25 in the minute. The pulse on the 28th was 106, and on the 29th 110, the temperature remaining at 100°. On the evening of the 30th temperature was recorded 101° and the pulse 120. The evening temperature on January 1 was 102° and the pulse 124; after this temperature fluctuated between 100° and 101°, and the pulse ranged from 100 down to 90 and 80 per minute, and so continued until January 15, when again the temperature reached 99°, with pulse and respirations normal.

CASE 8. *Ovarian tumor; tapped; relieved.*—Emily A., aged 24, was admitted May 22, 1878, with a large tumor of the right ovary; symptoms have existed for four years. Dr. Agnew tapped the tumor, and drew off eight quarts of fluid; great relief followed, and the patient, at her own request, was discharged, July 11, 1878.

The following cases were treated in the hospital previous to 1873, and, since they present some points of surgical interest, are thought worthy of a permanent record:

Cystic disease of the ovary; attempt made to remove the tumor found impracticable; cyst injected; death from suppurative inflammation.—Hannah P., aged 34, was admitted May 8, 1865. She had suffered for eight years from ovarian dropsy, with irregular attacks of pain, and febrile symptoms, and gradual increase in the size of the tumor. An operation for the removal of the cyst was begun by Dr. Hewson, July 20, 1865, by an incision three and one-half inches long in the median line. Dissection proved the cyst and the abdominal walls to be so firmly and continuously adherent that extirpation was abandoned, and the cyst was left open. During the operation a large amount of serum and dark grumous, semi-solid lymph was evacuated. Numerous injections of various fluids were introduced into the cyst, but the patient steadily sank from subacute peritonitis and recurring suppurative inflammation among the old adhesions of the tumor, and died August 14, 1865.

Unilocular ovarian cyst.—Alice M., aged 21, was admitted December 15, 1866. The patient was married, but had never borne children. The tumor appeared several years previously, and had already been tapped eight times. There had been numerous attacks of subacute peritonitis. On admission she was much exhausted, and colliquative diarrhœa set in, though the size of the tumor rapidly increased, and she died January 10, 1867.

The tumor sprang from the left ovary. The cyst was unilocular, and measured thirty-five inches in the circumference of its long diameter, and thirty inches in that of its short or transverse diameter, with moderately thick, fibrous walls, and contained several gallons of ropy, viscid, variously-colored fluid.

Piliferous ovarian cyst, communicating with the intestinal canal.—Bessie C., aged 52, was admitted March 11, 1867. No satisfactory history could be obtained of the development of the disease, which was a large tumor in the right side of the abdomen. For some days after admission she presented symptoms of intestinal obstruction, but for several weeks preceding her death an uncontrollable exhausting diarrhœa existed and much hastened the fatal event.

The tumor was rounded, quite firm, and seemed closely attached to the abdominal walls.

April 27.—Death.

At the autopsy evidences of old peritonitis were seen. The intestines were found closely glued together, and at the junction of the cæcum with the ascending colon there was a slough of the bowel. Fecal matter had been poured into the cavity of the peritoneum. Attached to the sloughing bowel, but lying free in the cavity, was found a true cyst, containing hair and sebaceous matter, with thin walls and in an advanced state of gangrenous decomposition. This cyst undoubtedly developed from the right ovary, became detached, gangrenous, and finally involved the adjacent portion of intestine in destructive inflammation.

Large pediculated uterine fibroid: abdominal section; removal of tumor: death.—Sarah R., aged 23, was admitted May 9, 1868, with an abdominal tumor, which had been growing for a number of years. On examination it was difficult to determine exactly its origin: the growth was very movable. The patient's health was failing so rapidly that an operation was decided upon. Dr. Agnew made a large abdominal median incision on the 28th of May, 1868, and found the tumor was pediculated, growing from the fundus of the uterus. The pedicle was ligated, and, after being divided, was cauterized by the actual cautery. Very little blood was lost, but she sank and died, apparently from nervous shock, in twenty-four hours.

The tumor removed was a dense fibroid, somewhat pear-shaped, and weighed rather more than 12 pounds. It was attached to the fundus of the uterus by a short pedicle two inches thick. There were hardly any adhesions between the tumor and the surrounding parts.

Cases of Ovarian Disease treated in the Surgical Wards of the Pennsylvania Hospital.

OPERATOR.	AGE	NAME.	ADMITTED.	DISCHARGED.	OPERATION.	RESULT.	CAUSE OF DEATH.
Morton.	26	Anna M.	Feb. 2, 1873.	March 27, 1873.	Ovariotomy.	Cured.	
Morton.	19	Mary D.	April 29, 1873.	June 5, 1873.	Cyst tapped.	Relieved.	
Morton.	28	Catharine G.	Aug. 9, 1873.	Aug. 11, 1873.	Ovariotomy.	Died.	Tetanus.
Morton.	40	Harriet B.	June 10, 1875.	Oct. 8, 1875.	Ovariotomy.	Died.	Suppression of urine.
Levis.	56	Lizzie I.	Oct. 27, 1875.	Nov. 24, 1875.	Cyst tapped.	Relieved.	
Hewson.	26	Eliza W.	May 3, 1877.	June 1, 1877.	Cyst tapped.	Relieved.	
Morton.	19	Mary G.	Dec. 15, 1877.	March 7, 1878.	Ovariotomy.	Cured.	
Agnew.	24	Emily A.	May 22, 1878.	July 11, 1878.	Cyst tapped.	Improved.	

3.—VESICO-VAGINAL FISTULA.

19 cases of vesico-vaginal fistula are recorded; of this number, 6 were cured, 6 were improved, 3 were unimproved, 2 died, and 2 declined treatment. Of the cause of the malady, we note in most of the cases tedious or hard labors; in one case a large wooden spool was introduced into the vagina, which, after several years, ulcerated into the bladder; in several of the cases the cause is not noted.

The operation which has been performed in each case was that which may be described as the "Agnew operation," which has been now performed with marked success, not only by the author of the method, but by others all over the country. Early in the history of this method of closing vesico-vaginal fistulæ Dr. Agnew introduced the operation into the Pennsylvania Hospital, and we have seen no reason for adopting any other. The account (condensed), which has already been published by Dr. Agnew, of the various steps of his operation we think can with profit be again introduced in this paper.*

"**Arrangements for the Operation.**—The bed on which the patient is to lie should be a firm mattress; over that part where the hips are to rest there should be spread a strip of oiled cloth, and over this a folded sheet, the object being to protect the bed. A low stool, turned upon its side and covered with blankets, forms an excellent support, over which the patient is to be turned.

"Instruments required are, a duck-bill speculum (Fig. 1); two long-handle scalpels (Fig. 2); a pair of long, rat-toothed forceps (Fig. 3), slightly curved, with an attachment at the end of the handle embodying the adjuster for running down the wires, and the crochet, to favor by counter pressure the passage of the needle through the distal side of the fistula; a needle-holder (Fig. 4), which can, with one hand, be detached from the needle, or again made to grasp it, and by which the needle can be introduced at any angle; one pair of long scissors (Fig. 5), curved a little on the flat; a shot-compressor (Fig. 6); a shot-perforator (Fig. 10); a sigmoid self-retaining catheter (Fig. 7). The needles (Fig. 8) should be constructed with great care, seven-eighths of an inch in length, slightly curved for one-fourth of an inch at the extremity, the cutting edge confined only to the extent of the curve, and sufficiently wide to allow the proximal part to pass without tugging and pulling, the eye should be well sunken; fine silver wire; No. 3 shot; and light gutta-percha tubing to attach to the catheter to convey the urine.

* *Lacerations of the Female Perineum and Vesico-Vaginal Fistula.* By D. H. Agnew, M.D. Philadelphia: Lindsay & Blakiston, 1873.

Fig. 9 shows the needle in operation, the forceps having been passed through the fistula, and the method of using the adjuster on the forceps; Fig. 11 the appearance of the shot and cut wire after removal.

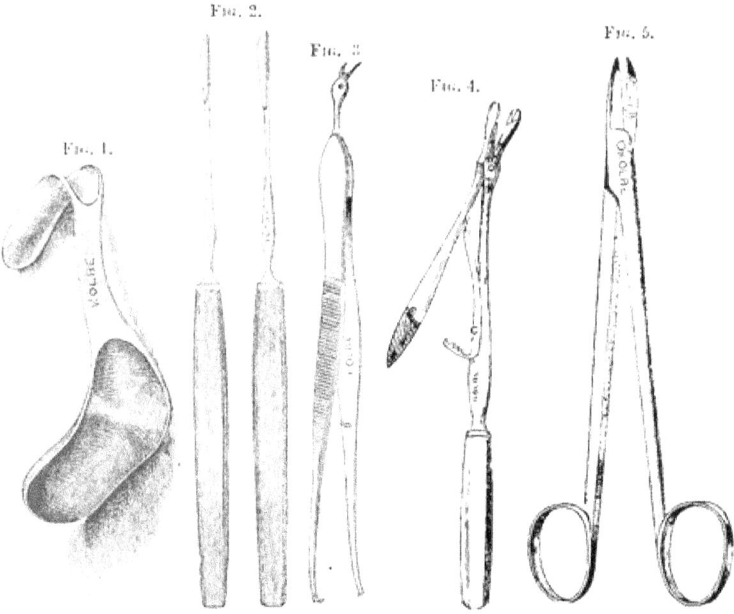

"**Operation.**—The patient having removed all her clothing, save a chemise and night-gown, lies down upon the bed, and is brought under the influence of the anæsthetic. . . . When sufficiently unconscious, the stool . . . (covered with blankets) is placed across the foot or side of the bed, and the patient carefully lifted and placed over it, resting on her abdomen, two or three pillows being laid under her breast and head in such a way as to form an inclined plane. . . . The legs being next flexed upon the thighs, are given over to assistants.

"The operator now takes the speculum, . . . and, introducing it into the vagina, commits it to one of the assistants having charge of the limbs, who draws it firmly toward the rectum, when the air, entering the vagina, expands the tube. . . .

"The surgeon now takes his seat in a position to command a full view of the fistula, and, seizing its lower margin with the forceps, enters the knife from three-eighths to half an inch from the opening, bringing it out just short of the vesical mucous membrane, and by successive sawing movements paring away until the entire circumference of the fistula has been freshened.

"Should the mucous membrane of the bladder protrude, a piece of sponge may be pressed through the opening to keep it out of the way. The greatest difficulty in executing this part of the operation will be experienced at the angles or commissures of the opening. . . . If it is properly done there should be at least three-eighths of an inch or more of oblique, raw surface, visible everywhere around the fistulous opening. . . . There will be cases . . . where the scissors come in more advantageously than the knife. . . .

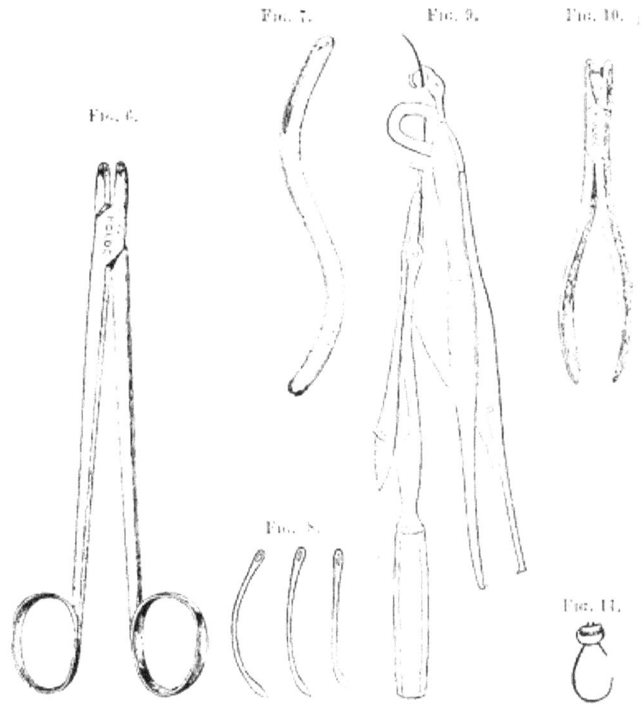

Fig. 6. Fig. 7. Fig. 8. Fig. 9. Fig. 10. Fig. 11.

"When the fistula is very small, receiving, for instance, only the end of an ordinary probe, some advise transfixing with a long awl-shaped instrument, and, raising the sides, by a single stroke of the knife cut out a sufficient amount of tissue. . . .

"Arrest of Hemorrhage.—The bleeding which follows the foregoing process is not generally very profuse, stopping under the application of cold (or hot) water. . . . Should this not succeed, the stitches should be inserted and the edges drawn firmly together. . . .

"The Direction of Approximation.—Most operators favor an approximation of the sides of the fistula transversely, yet there are no reasons why they may not be closed longitudinally. . . .

"Introduction of the Sutures.—The needle bearing the wire is placed in the grasp of the needle-holder, and whilst the proximal border of the fistula is steadied by the forceps, is entered at the middle of the wound three-eighths of an inch from the freshened surface, brought out at the mucous membrane of the bladder (not including it), carried across the opening, made to enter the opposite side, and emerge the same distance above its raw surface.

"The needle-holder is now disengaged from the needle by simply pressing the upper blade of the instrument while the spring is being pressed forward by the thumb, made to seize the extremity through the upper border of the fistula, and while the parts are supported, by applying to them the hook at the end of the forceps the needle is drawn through, turned, and brought out of the vagina. . . . In this manner the requisite number of threads are inserted, the distance between them being a trifle less than one-fourth of an inch.

"As each is deposited in its proper place the needle is to be removed, the ends of the wire twisted together, and given in charge of one of the assistants supporting the thighs.

"ADJUSTMENT.—In the important stage of the adjustment the wire first inserted is separated from the others, and the ends passed through the hole of the adjuster at the end of the forceps. As the latter is slid down the wire is drawn upon until the edges of the wound are brought into accurate contact. The set which the wire thus obtains is sufficient of itself temporarily to maintain the apposition. All of the threads are subjected successively to this process, and while being done care must be observed that the edges be properly everted so as to secure the contact of raw surfaces, and also that no clot be permitted to lie between them.

"The next step is to secure the sutures. . . . Perforated pellets of shot are run down the wires, then seized with the strong compressing forceps, and while the metallic threads are being drawn upon, pressed firmly against the line of adjustment, and then compressed so as securely to maintain their position. The sutures are next cut off close to the shot.

"AFTER-TREATMENT.—A catheter, frequently changed, should be kept in the bladder, and the bowels locked up until after the stitches are removed. Great care should be taken that the first movement of the bowels does not open the wound, and on this account it is better to give small doses of oil or salts, frequently repeated, and finally use an enema of warm water."

REMOVAL OF THE SUTURES.—On the eighth or ninth day the stitches should be removed.

Vesico-vaginal Fistula.

No.	Name.	Age.	Admitted.	Discharged.	Cause and Extent.	Operation.	Operator.	Result.	Remarks, etc.
1	Martha G.	25	Oct. 10, 1865.	Nov. 11, 1865.	Tedious labor; 48 hours elapsed before delivery; fistula extended from neck of the bladder to os uteri, and involved the cervix.	Agnew's. The opening was closed longitudinally with thirteen silver sutures.	Agnew.	Cured.	Two operations were performed prior to admission, and two in hospital. Sutures removed on ninth day.
2	Martha L.	28	Feb. 13, 1866.	Mar. 7, 1866.	Tedious labor with fourth child; instrumental on second day.	Agnew's. Fistula 2½ inches long; fourteen silver sutures; the approximation secured longitudinally.	Agnew.	Died.	The fistula had not united; septicæmia; no abscesses.
3	Annie A.	21	April 5, 1866.	July 25, 1866.	Tedious labor.	Agnew's. Four silver sutures.	Hewson.	Unimproved.	Flatulence; hemorrhage from breaking up of adhesions.
4	Cornelia A. H.	24	April 14, 1866.	May 1, 1866.	Prolonged second stage in first labor; fistula of enormous size.	Agnew's. Edges brought together antero-posteriorly by thirteen silver sutures. The two upper ones involved the anterior lip of the uterus.	Agnew.	Died.	Peritonitis.
5	Elizabeth L.	36	June 5, 1866.	July 27, 1866.	Tedious labor 2 days, also vaginal stricture.	Agnew's. Four sutures.	Agnew.	Improved.	
6	Rosanna K.	36	June 15, 1866.	July 25, 1866.	Tedious labor.	Agnew's.	....	Improved.	
7	Frances H.	30	June 28, 1866.	July 5, 1866.	Tedious labor.	....	....	....	Declined operative treatment.
8	Sarah M.	24	Mar. 11, 1868. Readmitted May 13, 1868.	April 14, 1868. Redischarged May 27, 1868.	....	Agnew's.	....	1st operation, improved, 2d do., cured.	
9	Catharine H.	45	June 24, 1868.	June 29, 1868.	....	....	....	....	Declined treatment.
10	Rebecca J.	18	Mar. 21, 1869.	May 3, 1869.	....	Agnew's.	....	Cured.	
11	Hannah L.	20	Nov. 4, 1869.	Nov. 27, 1869.	Tedious labor; fistula existed 11 mos.	Agnew's.	....	Cured.	A minute opening left.
12	Elizabeth H.	36	Aug. 10, 1871.	Dec. 3, 1871.	....	Agnew's.	....	Unimproved.	
13	Sarah I. A.	23	Mar. 21, 1871.	July 28, 1871.	Tedious labor.	Agnew's.	Hewson.	Improved.	Three operations performed; a very small opening existed when discharged.
14	Sallie S.	20	Feb. 13, 1874.	Mar. 5, 1874.	Tedious labor; the os completely occluded, and great loss of tissue; rent large.	Attempts made to open the vagina to the os by cutting and tearing.	Levis.	Unimproved. By request.	
15	Rebecca M.	35	April 10, 1875.	June 26, 1875.	Tedious labor; the fistula is two inches long, and thick edges.	Six silver shotted sutures; vagina closed, and the labia were united, leaving a small opening near urethra.	Hewson.	Cured.	Much blood was lost in the operation. The benefit consists chiefly in the support given to the bladder, preventing it from protruding as formerly.
16	Rachel G., colored.	20	May 20, 1875.	July 31, 1875.	Several years ago she introduced a spool into her vagina, and then could not extract it. 3 years ago the spool was removed, but it caused great ulceration. Vagina strictured at the upper portion and vesical opening an inch.	Stricture divided, and fistula freshened; four shotted silver sutures.	Hewson.	Improved.	
17	Eliza L.	46	June 19, 1876.	July 27, 1876.	Extensive opening; tedious labor.	Oblique incisions on each side of urethra; silver shotted sutures.	....	Improved.	Seven previous operations.
18	Rachel G., colored.	23	Jan. 17, 1877.	April 5, 1877.	Opening large enough to admit the point of a pencil.	....	Morton.	Cured.	After leaving the hospital a month or so, found some dribbling.
19	Ida H., colored.	22	April 25, 1879.	June 27, 1879.	Confined three times, all still-born; very hard labors; was with last child three days in labor.	Usual method; two operations.	Agnew.	Improved.	

4.—RECTO-VAGINAL FISTULA.

We have had under care 3 cases of recto-vaginal fistula; 1 case was improved; in 1 case no benefit was derived from treatment, and 1 case declined any operation.

Case 1.—Margaret D., aged 33, was admitted June 15, 1872, with a severe recto-vaginal fistule. She declined treatment, and was discharged July 6, 1872.

Case 2.—Elizabeth B., aged 45, was admitted May 5, 1875. The fistula is of several years' duration. The opening was three inches in length, and extended well up into the rectum.

May 14.—Edges well freshened, and then drawn together with shotted silver sutures.

May 24.—Five stitches removed.

June 2.—Good union, except a small portion.

June 10.—Discharged, much improved.

Case 3.—Nellie S., aged 31, was admitted Septembe r25, 1876. Her present trouble commenced about twenty-one months ago, and was the result of a tedious confinement. The opening into the bowel was not large.

September 27.—Operation consisted in dividing the tissues and introducing four deep shotted silver sutures.

October 17.—Some of the sutures gave way, and the operation was not thoroughly successful. Discharged by request.

7 cases of lacerated perineum have been admitted and operated upon. An excellent recovery followed in each instance without an unfavorable symptom.

The method of operating in all of the cases was that first suggested by Dr. Agnew, and introduced into the hospital by Dr. Hunt, under whose care the first four cases were received and operated upon.

The details (condensed) of these operations, as given by Dr. Agnew,* are thought worthy of a republication in this report.

Operation for Lacerated Perineum.

"POSITION.—The position on the back, or the lithotomy operation, is the one preferred.

"OPERATION.—The operator seizes one side of the laceration, commences the denudation from behind forwards, including a portion of the labium. In breadth it should extend inwards, so as to include a little of the vaginal mucous membrane, and outwards towards the buttocks. The paring should not extend deep, but merely to skim the surface, and when completed should be over an inch broad. The opposite side is to be treated in the same manner, the raw surfaces in form and extent being as near alike as possible. The recto-vaginal septum is next made tense, and its surface is then freshened to the extent of three-quarters of an inch. The denuded surface when finished much resembles a horseshoe. The approximation is effected by the interrupted silver suture, a deep and a superficial set being used.

"The first stitch is entered three-quarters of an inch from the margin of the wound, below its lowest point at the anterior part of the ischio-rectal fossa, and carried forwards and upwards until it appears on the middle of the septum, just above the line of denudation. The thread is then pulled out of the eye of the needle, the latter withdrawn and made to pass unarmed through the corresponding parts on the opposite side, emerging on the septum, close to the first. The wire is now passed through its eye, and, as the needle is withdrawn, makes the complete circuit of the wound, so that when it is tightened the parts are pursed together. Three or four deep sutures are then introduced. The ends of the suture first introduced are then approximated, following this the others. The sutures are clamped with shot, the ends cut close to the shot; between each deep suture superficial ones are introduced.

"The deep sutures may be removed from the fourth to the eighth day. The superficial ones can remain in a day or so longer."

* *Laceration of the Female Perineum and Vesico-Vaginal Fistula.* By D. H. Agnew, M.D. Philadelphia: Lindsay & Blakiston, 1873.

CASE 3. *Laceration of the female perineum.*—Mary A., aged 47, was admitted into the hospital September 16, 1872. Her first labor had been difficult and prolonged; an intense pain had expelled the head suddenly and unexpectedly at last,—the perineum being without support,—and produced a laceration, which extended through the sphincters into the bowel. There was prolapsus of the bowel. Three months after the accident Dr. Hunt thoroughly pared the edges, and united the parts by four silver threads, deeply inserted: the wires were clamped by shot: three intermediate silver threads, with their ends simply twisted, completed the operation. On the eighth day the sutures were removed, and she was discharged, quite well, October 14.

CASE 4.—Eliza A. F., aged 52, was admitted November 25, 1872, suffering from complete procidentia. Twenty years previously she had been delivered of twins, and at this time the perineum was ruptured. Five years afterwards she gave birth to another child, when the rent was increased. Dr. Hunt inserted four deep silver sutures, which were secured by clamps of shot, and intermediate to these two superficial sutures, fastened by twisting the ends about each other. On the fourth day the deep sutures were removed, and on the seventh day the superficial ones. A week later the patient was discharged, well, and free from all displacement.

CASE 5.—Mary I. H., aged 46, was admitted June 18, 1873, with complete procidentia of twenty-two years' duration. Since the rent of the perineum has had seven children. The uterus was entirely out of the vagina, and the walls of the bladder are drawn entirely over it. The perineum was freshened, three-fourths of an inch on either side being denuded. Three deep and three superficial sutures were introduced. The patient made an excellent recovery, and the sutures were removed on the seventh day. Discharged July 12, 1873.

CASE 6.—Violetta B., aged 70, was admitted November 10, 1874. The laceration of the perineum occurred fifteen years ago. The uterus was entirely out of the vagina. The usual operation was made by Dr. Morton, who introduced five deep sutures, and used the shot and four superficial sutures, merely twisting the ends. On the seventh day several of the sutures were removed: five days later all were cut away.

December 5.—Discharged, cured.

CASE 7.—Isabella E., aged 33, was admitted January 30, 1879. During her first labor had instruments used, which produced a laceration of the perineum.

February 13.—Usual operation by Dr. Agnew.

March 27.—Discharged, cured.

Laceration of the Female Perineum.

No.	Name.	Age.	Admitted.	Discharged.	Cause and Extent.	Operator.	Result.	Remarks.
1	Rose D.	28	Oct. 28, 1865.	Dec. 20, 1865.	Labor.	Hunt.	Cured.	Usual operation.
2	Sarah B.	19	Feb. 25, 1870.	May 13, 1870.	Labor.	Hunt.	Cured.	Usual operation.
3	Mary A.	47	Sept. 16, 1872.	Oct. 14, 1872.	Sudden expulsion of the head; no support to perineum.	Hunt.	Cured.	Sutures, removed on the 8th day.
4	Eliza A. F.	52	Nov. 25, 1872	Feb. 26, 1873.	During confinement with twins, 20 years before; had procidentia uteri.	Hunt.	Cured.	Deep sutures, removed on the 4th day.
5	Mary I. H.	46	June 18, 1873.	July 12, 1873.	Hard labor; duration, 22 years; procidentia uteri.	....	Cured.	Three deep and three superficial sutures, removed on the 7th day.
6	Violetta B.	70	Nov. 10, 1874.	Dec. 5, 1874.	Labor, 15 years ago.	Morton.	Cured.	Five deep and four superficial sutures, removed on the fifth day.
7	Isabella E.	33	Jan. 30, 1879.	March 27, 1879.	Labor, instrumental.	Agnew.	Cured.	Usual operation.

6.—AMPUTATION OF THE CERVIX UTERI

4 cases are recorded of amputation of the neck of the uterus. 2 of these operations were for cancer, involving the cervix, and 2 for hypertrophic elongation of the neck. All of the cases recovered.

The écraseur was used in the first two cases, and was applied after the cervix had been well drawn down by hooks. In the two cases of elongated cervix, in the first instance the galvano-cautery was applied, and no blood was lost. Marked pelvic pain followed, and on the eighth day a chill of some severity occurred, followed by vomiting and severe nervous symptoms. After this a large amount of pus was discharged by the rectum. On the tenth day there was a severe hemorrhage, which was controlled by a tampon and iron. Fever came on, but soon lessened, and gradually the patient regained her usual health.

In the second case, the neck of the uterus was drawn well down and transfixed by Dr. Ellwood Wilson's curved needle (see figure), and a stout double ligature was carried through. By this means the parts were securely held, then with the knife the neck was at once removed. Six waxed silk sutures were then introduced and the two sides carefully approximated. There was some hemorrhage, but it was at once controlled by hot-water applications. The patient recovered without a single symptom.

CASE 1. *Cancer of the neck of the uterus; amputation of the cervix; cured.*—Sallie B., aged 26, was admitted January 7, 1867, with cancer of the neck of the uterus. The disease seemed confined to the os, and the patient was otherwise in a good condition of health. The disease was of two years' standing. Dr. Morton, after drawing the os well down with hooks, applied the écraseur, and removed the growth. There were no complications, and the patient was discharged, cured.

CASE 2. *Carcinoma of os uteri; amputation.*—Mary E., aged 30, was admitted November 19, 1874. She first noticed her malady five months ago; came on with pain in the back and lumbar region; had excessive leucorrhœa, the discharge gradually getting more and more offensive; has had two children, and family history good. The os uteri was transfixed with a long pin, and then was removed by the écraseur. There was very little hemorrhage. The wound healed up kindly; November 30 she was discharged, cured.

CASE 3. *Hypertrophic elongation of the cervix uteri: amputation: recovery.*—Sarah G., aged 27, colored, married, was admitted August 21, 1874, with an elongated neck of the uterus. Healthy until thirteen years old, when she first menstruated. She then took notice that the uterus protruded from the vulva when urinating. Subject to sick headache for many years. Five years ago was sick for five weeks with fever and night-sweats and vomiting. At this time there was a good deal of induration around the neck of the uterus. Soon after this had a discharge of a large quantity of matter from rectum, with lessening of the uterine induration. Four years ago married: previous to this was regular every three weeks, after marriage every four weeks, and until lately always regular. Never has been pregnant. After first six months of married life coitus gave her intense pain. Was treated at Women's Hospital last winter. Was well enough to work for five or six months afterwards, but had to give up. Constipated for the last three or four years, and has had leucorrhœa. On admission rather feeble. Appetite fair, but constipated. Examination shows cervix one and one-half to two inches elongated, lying just above vulvar orifice, not more than one-half inch within. Right and left cul-de-sac filled with semi-elastic mass, anteriorly and posteriorly. Uterus can be moved a little. Fundus felt through abdomen, a little to right of median line. A hard mass can also be felt above Poupart's ligament, seeming to project from right side of fundus. Os uteri pointing forwards to centre of vulva from a broad base, tapering down to a sharp point. Mucous membrane of cervix appears rough, as though it had been exposed to air. White and dry. No discharge from canal of cervix.

September 11.—Menstrual flow commenced.

September 21.—Removed half an inch of the cervix by the galvano-cautery in five and one-half minutes. No blood lost.

September 22.—No pain. No difficulty in passing water, except over a small spot near the urethra, that was accidentally burned.

September 25.—Solution of chloral injected into the vagina after washing it out with tepid water.

September 27.—Discharge much less, and not so offensive. Bowels constipated; ordered injection. There is some fixed pelvic pain, but none upon coughing or taking a full breath.

September 28.—Bowels opened by injection. Some tympanitis; gave morphia.

September 29.—Tongue coated and dry; groans constantly: abdomen soft. No chill, but is chilly at times. Some vomiting; chloral injection, diluted one-half. Ordered beef-tea, ℥i, with acid muriat. dil., gr. x, q. s.; also brandy, lime-water, and milk every two hours, and valerian, f℥i, for nervousness, and morphia, gr. ¼, q. s. to sleep.

October 1.—A large amount of pus discharged from the rectum; felt more comfortable afterwards. In the afternoon had a severe hemorrhage from vagina (it is time for period). Vagina tamponed with cotton soaked in persulphate of iron. Vagina to be washed with a weak solution of permanganate of potash.

October 3.—Much better: still a good deal of discharge from rectum, but little or none from vagina.

October 8.—Uterus red and slightly bloody; touched with Monsel's solution and water, equal parts; is convalescent; quinine reduced to gr. viii daily.

October 15.—Slight discharge of pus from rectum yesterday: vomited; bowels constipated; has had fever for some time.

October 22.—Has a quick, feeble pulse, and a peculiar sensation in cardiac region; tinct. digitalis, gtt. x, t. d.

October 29.—Slight fever again.

October 30.—Abdomen soft and relaxed; surface of uterus smooth, but not yet perfectly cicatrized; probe enters two and three-quarters inches. The menstrual flow is again upon her, and the discharge appeared normal.

November 5.—Flow stopped; usual symptoms.

November 9.—Discharged, cured.

CASE 4. *Hypertrophic elongation of the neck of the uterus; amputation; recovery.*—Mary I. R., colored, aged 19, single, was admitted into the medical ward, and subsequently, at Dr. John F. Meigs's request, was transferred to Dr. Morton's care, for surgical treatment and amputation of the cervix, November 5, 1879. Menstruation was established between 13 and 14 years of age; was regular for two years. Became irregular, and continued so until fifteen months ago, since which time has been regular. Has worked very hard at farm and general housework. About five years ago began to have dragging pain in back after exertion. About the same time, while lifting a large tub filled with water, she "felt something give way inside," and was seized with great pain in back, and more especially in abdomen. Remained in bed three weeks, but did no work for three months. At time of strain she examined her vagina with her finger, and could feel something she had never felt before. Was like a lump. When recumbent it would recede; upon arising would return. Upon resuming work the uterus became more and more prolapsed, and very painful. Finally the uterus protruded from vulva when stooping, and would recede just within labia majora when erect. At times was compelled to cease work for a few days at a time from severe pain. Has had more or less leucorrhœa during entire five years.

November 6.—On examination, the patient seems to be well nourished; is very nervous. Cervix uteri, in stooping, is considerably protruded, and when lying down emerges at the vulva. On digital examination, the neck is one and three-quarters inches longer than normal.

November 12.—After complete anæsthesia, Dr. Morton, with the assistance of Drs. Hunt and E. Wilson, introduced a well-threaded needle through the neck of the uterus, just above the place of intended division; the uterus was well drawn down, and one inch and a half of the neck was removed. After this Dr. Morton used the curved needle devised by Dr. Wilson, and then introduced six sutures, drawing the edges of the wound in close approximation.

November 18.—Sutures removed; wound well.

December 5.—Menstruation came on to-day; not quite so free as usual.

December 8.—Discharge ceased; had epistaxis to the amount of two ounces. All pain and weight in pelvic region has disappeared.

December 11.—Discharged, quite well.

THOMAS G. MORTON.

DISLOCATIONS.

In the 80 luxations reported, the following was the order of frequency as to the parts involved: humerus, 23; clavicle, 14; femur, 13; elbow, 8; ankle, 5; wrist, 3; tibia, 2; knee, 2; ulna, 2; fingers, 2; toes, 1; metatarsal, great toe, 1; cervical vertebræ, 1; fibula, 1; thumb, 1; radius, 1. 10 of the cases were compound, and 14 were complicated with fractures, either of the same bone that was dislocated, or of some other part. The diagnosis is put down exactly as it occurs upon the case-books. In a few this is, no doubt, open to criticism, as, for example, the luxations of the head of the humerus with fracture; for, while this accident now and then occurs, it is, we think, true that in most that are reported as such the upper fragment, or some of it, in cases of comminution, maintains its position in the glenoid cavity. It is the displaced upper end of the lower fragment projecting under the skin that gives rise to doubt sometimes. The true way to clear up this doubt, not only in these cases, but in all others of a questionable character, where the joints are concerned, is to etherize. A double purpose may thus be accomplished,—first, diagnosis, for it is very rare that a correct conclusion cannot be arrived at; and, second, if the case is one of luxation, it may be reduced at once. It is doubtful whether there is any hospital surgeon of long experience who has not had occasion to regret that he did not at once put this or that case under anæsthesia when it was one of obscure injury to a joint.

There is another important matter in this connection. If examination under anæsthesia should still fail to reveal the true condition, which sometimes happens, or if, after revealing it, the result of treatment, however skilful, should not be satisfactory to the patient, although that result was all that the surgeon could reasonably look for or promise, the latter is in the impregnable position of having done all that science, art, or skill could do in the case. The directions taken by the various luxated bones are not fully recorded, except in those of the femur, clavicle, and knee. Of the 13 femoral luxations, 8 were dorsal, 3 thyroid, and 2 sciatic. There was no case of luxation upwards and forwards on the pubis. The writer saw a case of this kind in private practice a few years since, and he was surprised afterwards to find how rare an accident it was; for, excepting the surgeons who saw the same patient, no one, old or young, whom he asked (many of them of large hospital and private experience) had ever met with a case of the injury. The clavicle furnished 8 acromial and 6 sternal luxations. These injuries are notoriously difficult to treat,—that is, with the expectation of bringing about a perfect result. Nothing, perhaps, contributes to a good cure (true also as to fractures of the same bone) so much as absolute dorsal decubitus for two or three weeks. No matter what form of dressing or apparatus is used, either before or after the patient is allowed to rise, the result of these cases, as to use of the part, is nearly always satisfactory. 2 cases of backward luxations of the elbow are reported. As to the humerus, the direction is stated in 11 cases; of these, 7 were axillary, 3 anterior, and 1 backward. It is fair to assume that 7 at least of the remaining 12, about which the direction is not given, were axillary. In the 2 knee cases, the tibial surface was thrown outwards in one and forwards in the other. In the whole list there is no record of a dislocation of the patella. This accident probably occurs oftener than is supposed, as in simple cases the efforts of the patient or bystanders to move the limb may easily cause the bone to slip into its place. The writer has met with the accident in private practice, and has readily restored the bone to position. There is also no case of dislocated lower jaw reported,—an accident which many think to be quite common. The same remark made as to the patella it is probable applies to the maxilla. It may be spontaneously reduced, or, if not, immediate assistance is sought from the nearest physician. We are disposed to think, however, that the accident is by no means of frequent occurrence. We have had cases of it in the hospital which were quite difficult to reduce, but none within the range of time (five or six years) of the accompanying table.

The results in the 80 cases are interesting and instructive,—57 were cured, 13 relieved, 7 died, in 2 the luxation was unreduced, and 1 eloped. All of the femoral luxations were reduced except one, who, on account of other injuries, did not react, and died shortly after admission. It is comparatively rare to find old luxations of the femur, the helplessness of the patient forcing him to seek early relief. The contrary is the case with the humerus, about which mistakes are often made, not only by the patient and his friends, but by medical men. In two cases of ancient humeral luxations it will be noticed that the bone was broken in the attempts to reduce it. These both recovered promptly from the fractures, and further trials at reduction were declined. The operation of subcutaneous section of the bone somewhere in the surgical neck promises to be of great service in this class of cases, as has been illustrated in a very successful one, in which the operation was performed by Dr. J. Ewing Mears, of this city.* Of the 13 relieved cases, 5 were of the clavicle. These are probably more correctly reported than the 9 other luxations of the same bone set down as cured. Under this head come 1 of dislocation of the hip, followed by arthritis; 3 of the wrist (?); 3, old, of the humerus; and 1 of a metatarsal bone. The 7 deaths comprised cases of a very severe character. 4 of these were compound luxations, and all were complicated with other serious injuries.

The great risk of attempting to save limbs in severe compound injuries of

* *Transactions of College of Physicians, Philadelphia*, 3d Series, vol. iii., 1877.

the ankle-joint is strikingly illustrated. We think there is not one of the surgical staff who does not agree in saying that the more he tries the experiment the more he has occasion to regret it.

The following cases of simple luxation of the knee- and ankle-joints are of such interest that the histories are thought worthy of a reproduction in this paper:

CASE I. *Lateral dislocation of the knee, followed by recovery, with good use of the limb.*—A strong, healthy German, aged 40, was caught in the belting of some machinery, and sustained, in addition to a scalp-wound and severe laceration of the elbow, a dislocation of the left knee-joint. When admitted the luxation was found to be partial, as all lateral dislocations in this situation, unless compound, are; but the deformity was very marked. The tibia was displaced outwards, and at the same time rotated *inwards*, so that the crest of the tibia and the top of the foot looked to the right. The leg was partially flexed on the thigh, and bent outwards, so that it formed an angle with the thigh: the patella was drawn outwards from its normal position, and the internal condyle was seen as a great protuberance on the inner aspect of the joint. The deformity is well shown in the cut, made from a cast taken at the time. Over the internal condyle the skin was ruptured, and the wound, which was about half an inch long, led directly to the bone; but future developments seem to show that the joint was not opened.

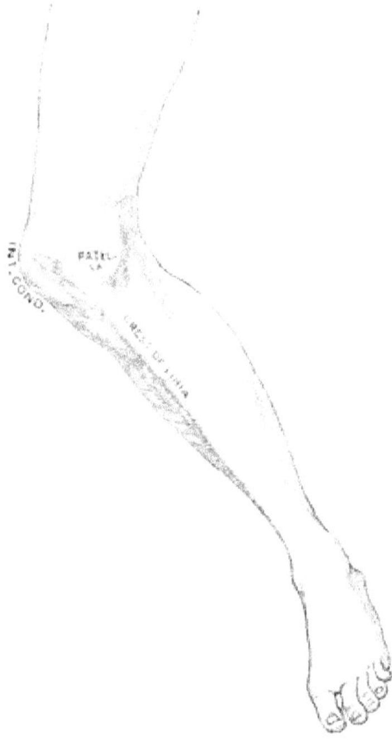

From a cast of the limb before reduction.

The reduction was readily effected under ether by extension and counter-extension, with direct pressure applied to the head of the dislocated tibia. The limb was then placed in a double-inclined-plane fracture-box, because there was less tendency to a repetition of the dislocation when the leg was partially flexed over the ridge made by the two inclined planes. The wound was dressed with carbolized oil.

The knee did not show the least tendency to synovitis, and the wound healed rapidly by granulation. About two weeks after the injury the leg was dressed with a long straight posterior splint, with a pad under the popliteal space to prevent any strain coming on the joint. After the lapse of seven weeks this splint was removed, and it was found that the patient, although he had been walking about for a considerable time without crutches when the splint was in position, could not bear any weight on the leg, when the splint was removed, without pain. He could flex and extend the leg to a limited extent, but there was preternatural lateral mobility of the tibia on the femur.

A silicate dressing was applied, small strips of wood incorporated in the dressing around the knee gave additional firmness. He wore this about ten days, and was then discharged, cured. He was able to flex the joint so that the leg moved through an arc of nearly 135°; he could walk without much limping, bending the knee a little with every step, but he soon became tired, and had pain in front of the joint. The patella appeared very prominent, as though the tibia were displaced a little posteriorly.

For some time after his discharge he used an apparatus to give lateral support to the knee, but when last seen he was able to dispense with it, and walk about and attend to his work for a great portion of the day without any marked inconvenience.

CASE 2. *Forward dislocation of the knee, with recovery at the end of six weeks.* —The patient was a strong, vigorous colored man, aged 46, who was struck upon the top of his head by a falling bucket of sand, which had broken loose when about eight feet from the hold of the vessel, from which he was unloading ballast. Upon admission, patient was quite conscious, and seemed to have sustained no serious head-injury, but there was found a partial displacement of the tibia and fibula forward upon the lower extremity of the femur of the right leg; the patella had entirely lost its prominent position, and was quite lost to sight behind the protruding head of the tibia. The popliteal space had also lost its characteristic appearance, and was almost bulging, owing to the presence of the condyles of the femur, the outer especially being so prominent that it could be distinctly seen just beneath the skin. The leg was extended and the foot slightly everted, but on lifting the foot the leg would bend considerably forward, owing to the increased mobility at the knee-joint. The patient had, in addition to the above injury, a clean incised wound of the left leg, about four inches long.

Under ether, having made a plaster cast of the knee-joint, the luxation was readily reduced by extreme flexion of the leg upon the thigh, the limb being placed across the knee of the operator; a Smith's anterior splint was then applied. At the end of eighteen days, no unfavorable symptom having occurred, the splint was removed for the first time, and there was found good antero-posterior motion, with little or no lateral movement. A silicate dressing, reinforced behind the knee by a piece of cardboard, was substituted for the wire splint, and the patient was allowed to get up and walk about on crutches. In six weeks from the time of the accident the silicate dressing was removed, and the patient found himself able to walk without aid, and indeed without a limp. Two days later he was discharged from the hospital, cured, when the final note was made that when his thigh was flexed at a right angle with the body he could flex his leg at the same angle with the thigh, and that there was no deformity or preternatural movement at the knee-joint.

CASE 3. *Internal dislocation of the os calcis and scaphoid from the astragalus.* —J. B., aged 36, was admitted July 29, 1876. This injury, more commonly, though incorrectly, spoken of as dislocation of the astragalus externally, occurred to a large, powerfully-built man of 36 years of age, who was a driver by occupation. While rolling a barrel of shellac down a wagon-plank it slipped from his hold and struck both his legs in front. He was brought immediately to the Pennsylvania Hospital suffering from considerable shock. Examination showed a compound fracture of the left tibia and fibula, at a junction of the middle and upper third of the leg, almost transverse in direction, and showing very little tendency to displacement. The right foot was much deformed, but there was no break of the skin. There was a large prominence on the external aspect of the foot, about an inch and a half in front of the external malleolus, and on a line with it, which was readily felt to be the smooth anterior articular surface of the astragalus, with apparently nothing covering it but tightly-drawn integument. The foot was inverted, and turned upwards, so that its plantar surface looked almost to the opposite foot.

The patient was at once etherized, and, after taking a plaster cast of the parts concerned in the dislocation, from which the accompanying drawings were made, the reduction was easily effected by extension applied to the foot, while the thumbs of an assistant pressed with moderate force upon the head of the astragalus. After the bones were restored a more careful examination was made, which showed no evidence of fracture, no disturbance of the articulation of the astragalus with the tibia and fibula, and no alteration in the contour of the plantar portion of the foot. The deformity was quite overcome, and the foot and leg were placed in a silicate dressing. No serious symptoms occurred, the compound fracture of the leg, as well as the foot, which was the seat of dislocation, being well some time before he left the hospital, his stay being pro-

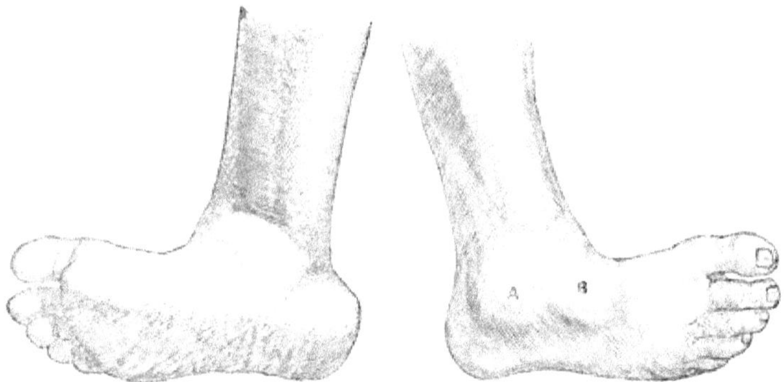

longed on account of an operation for convergent strabismus. Was discharged September 18, 1876. He was seen several months later, when he was found to be at work, and suffering no inconvenience.

Dislocations.

No.	Age.	Sex.	Nature.	Cause.	Treatment.	Result.	Remarks.
1	40	M.	Compound luxation of toes of left foot.	Large stone fell on foot.	Amputation through metatarsus.	Cured.	(See Amputations and Injuries of Nerves.)
2	30	F.	Luxation of elbow and fracture of head of humerus.	Tripped and fell backwards down-stairs.	Luxation reduced.	Cured.	(See Fractures.)
3	20	M.	Of hip, on dorsum ilii.	Fell from a high-seated wagon while driving.	Reduced by manipulation, under ether; knees bandaged together.	Cured.	In hospital 16 days.
4	29	M.	Of hip, into thyroid foramen.	Run over by loaded wagon.	Etherized and reduced by manipulation.	Cured.	In hospital 13 days.
5	39	M.	Of hip, into sciatic notch.	Fell into hold of vessel.	Etherized and reduced by manipulation.	Cured.	In hospital 14 days.
6	15	M.	Of humerus, with fracture.	Fell from cherry-tree.	Reduced.	Cured.	(See Fractures.) In hospital 25 days.
7	48	M.	Of humerus; 10 days before admission.	Was thrown against side of engine-room while at sea.	Etherized and reduced; third bandage of Desault.	Cured.	In hospital 37 days.
8	25	M.	Compound, of left ankle-joint.	Lump of coal falling on foot, from a height of 30 feet. Patient was at bottom of a shaft.	Amputation.	Cured.	(See Amputations.)
9	54	M.	Of left hip, on dorsum.	Fell through bridge.	Etherized and reduced by manipulation; arthritis followed, for which extension was applied.	Relieved.	Some arthritis when left house. In hospital 81 days.
10	56	M.	Of left humerus, also fracture of seventh rib.	Struck by pole of a wagon.	Etherized and reduced by manipulation.	Cured.	In hospital 28 days. (See Fractures.)
11	45	F.	Of elbow, backwards.	Knocked down by horse.	Etherized and reduced by manipulation; right-angled splint.	Cured.	In hospital 11 days.
12	27	M.	Compound, of ankle, tarsus.	Heavy stone fell on foot.	Posterior tibial artery tied; fragments of scaphoid, semilunar, and metatarsal bones removed; rest on inclined plane.	Died.	(See Pyæmia, No. 2.)
13	35	M.	Of right clavicle; fracture of left clavicle.	Fell, and was jammed between canal-boat and wall of lock.	Position, and retained by third bandage of Desault.	Cured.	In hospital 25 days.
14	27	M.	Of sternal end of right clavicle.	Struck by a locomotive.	Rest in horizontal position.	Cured.	Was complicated with serious laceration of face and fracture of coracoid process. (See Fractures.)
15	15	F.	Of left tibia, with fracture of fibula.	Fell down-stairs.	Fracture-box and compresses; pasteboard splints; silicate bandage.	Cured.	In hospital 10 days.
16	55	M.	Of sternal end of clavicle.	Knocked off of coal-cart, and struck shoulder against a post.	Third roller of Desault, and arm in sling.	Cured.	In hospital 18 days.
17	67	M.	Of acromial end of clavicle.	From a fall on deck of a vessel.	Third roller of Desault, and sling.	Cured (?).	In hospital 11 days. Accident happened 3 weeks before admission.
18	23	M.	Of acromial end of clavicle.	Fell from wagon.	Compress and adhesive strip.	Relieved.	In hospital 5 days. Bone in good position when left.
19	19	M.	Of acromial end of clavicle.	Struck by bridge while passing under it, on a railroad train.	Compress and broad adhesive strip.	Relieved.	In hospital 5 days.
20	23	M.	Of elbow, backwards.	Fell from tree.	Reduced, and right-angled splint applied.	Cured.	There was also fracture of condyles. In hospital 25 days.
21	17	M.	Of femur, on dorsum.	Caught between ship and wharf.	Etherized and reduced by manipulation.	Cured.	In house 31 days.

Dislocations.—(Continued.)

No.	Age.	Sex.	Nature.	Cause.	Treatment.	Result.	Remarks.
22	15	M.	Of femur, on dorsum.	A gravel-bank fell on him.	Etherized and reduced; head of bone again came out, when knee was allowed to fall inwards; reduced again, and knees bound together and extension put on.	Cured.	This injury was combined with a fracture through acetabulum of same side. In hospital 39 days.
23	32	M.	Of wrist (?).	Fall on pavement.	Etherized and luxation easily reduced.	Relieved, to out department.	In hospital 7 days.
24	32	M.	Of left elbow, backwards.	Fell down-stairs.	Etherized and reduced; laudanum and lead-water applied; sling.	Cured.	In hospital 19 days.
25	27	M.	Sciatic luxation of right femur.	Knocked down by falling wall.	Reduced by manipulation, under ether; knees tied together.	Cured.	In hospital 23 days. Had also a laceration of scalp.
26	12	M.	Compound luxation of left ankle.	Caught in belting.	Amputation.	Died.	Was complicated with other severe injuries. (See Amputations.)
27	23	M.	Of acromial end of right clavicle.	From fall on right shoulder.	Third bandage of Desault; Moore's sheet-dressing.	Cured (?).	
28	56	M.	Of humerus, with fracture of surgical neck.	Fell a distance of 25 feet on to pavement.	Palliative.	Died.	Was complicated with fracture of thigh and other injuries. Was 3 hours in hospital.
29	38	M.	Of acromial end of clavicle.	Fell down hatchway.	Third bandage of Desault, and compress; porous plaster.	Improved.	In hospital 18 days.
30	9	F.	Of humerus, forwards, with fracture of neck.	Fell while sliding down banisters.	Dislocation reduced under ether, and fracture dressed.	Cured.	In hospital 25 days.
31	11	M.	Of radius and ulna, backwards.	Hurt in a fight.	Reduced, and wet clay dressing; passive motion.	Cured.	In hospital 24 days.
32	17	M.	Of finger.	From a fall.	Reduced, and wet clay dressing.	Cured.	In hospital 1 day.
33	45	M.	Compound, of ankle, with fracture of radius.	Thrown out of a wagon.	Dupuytren's splint; Bond's splint to wrist; suspension of leg, etc.	Died.	Died of delirium tremens, 7th day after admission.
34	30	M.	Of humerus, subglenoid.	Thrown down by a wagon.	Reduced under ether.	Cured.	There was also a laceration of perineum, which was stitched. In hospital 14 days.
35	42	M.	Of humerus, with fracture of neck.	Fell down hatchway.	Luxation reduced, and fracture dressed with right-angled splint and cap to shoulder.	Improved.	Discharged for misconduct. In hospital 35 days.
36	10	M.	Of radius at wrist, with fracture of humerus of same arm.	Not stated.	Rectangular splint and pasteboard and wet clay.	Improved.	Removed by friends. In hospital 12 days.
37	31	M.	Of acromial end of left clavicle.	Fell down-stairs.	Compress and Velpeau's bandage.	Improved.	There was also fracture of rib. Continue treatment in out department. In hospital 7 days.
38	27	M.	Axillary of humerus.	Fell from main-yard of ship.	Reduced with difficulty by manipulation, under ether, on the 3d day.	Cured.	There were also symptoms of concussion of brain. In hospital 35 days.
39	55	M.	Of humerus.	....	None.	Eloped.	Said it had been reduced before, but slipped out; could not be found when wanted. In hospital 1 day.
40	34	M.	Of humerus, into axilla.	Fell down while drunk.	Reduced by heel in axilla and extension.	Cured.	In hospital 17 days.
41	40	M.	Of sternal end of clavicle.	From caving in of bank of earth.	Compresses, recumbent position, adhesive plaster, etc.	Cured.	Some prominence, but good use of arm obtained. In hospital 25 days.
42	19	M.	Of ulna, outwards.	Fell from wagon.	Reduced; rectangular splint at first, followed by Stromeyer's.	Cured.	There was also fracture of condyles of humerus. In hospital 47 days.

43		M.	Of femur, on dorsum.	From fall of a building.	Reduced by ether and manipulation.	Cured.	There were other severe injuries,—fracture of radius, lacerations, and contusions. In hospital 42 days.
44	30	M.	Of femur, on dorsum.	Run over by wagon.	Reduced by ether and manipulation.	Cured.	In hospital 15 days.
45	80	M.	Of humerus, backwards.	Fell from a stool.	Reduced by manipulation.	Cured.	In hospital 18 days.
46	50	F.	Of humerus, downwards.	Fell from cellar-door.	Ether; reduced by foot in axilla and extension.	Cured.	In hospital 7 days.
47	75	M.	Of humerus.	....	Ether and manipulation.	Cured.	Luxation was of 20 days' standing when admitted. In hospital 7 days.
48	71	F.	Old luxation of humerus; 10 weeks.	A fall.	Ether, manipulation, and extension.	Improved.	There was fracture of neck of femur. Patient removed by friends. In hospital 17 days.
49	35	M.	Dislocation of ankle, with fracture (Potts').	A fall.	Reduction and fracture-box.	Cured.	In hospital 41 days.
50	17	F.	Of sternal end of clavicle.	Fell on ice.	Recumbent position and compress.	Cured.	Discharged by request.
51	19	M.	Thyroid dislocation of hip.	Run over by loaded cart.	Palliative.	Died.	There were other severe injuries, and patient did not react.
52	31	M.	Compound, of left thumb, anteriorly.	....	Reduced, but did badly, and thumb was amputated.	Cured by amputation.	In hospital 50 days.
53	35	F.	Of left elbow.	Fell downstairs.	Reduced by manipulation.	Cured.	In hospital 2 days.
54	19	F.	Of right shoulder into axilla.	From a fall.	Etherized and reduced by manipulation.	Cured.	In hospital 1 day.
55	48	M.	Of metatarsal bone of great toe, upwards on tarsus.	Caught foot under floor while going up in an elevator.	Reduced, under ether, by manipulation.	Improved.	Caries followed abscess. In hospital 4 months.
56	42	F.	Of shoulder into axilla.	From a fall.	Reduced by manipulation, under ether.	Cured of dislocation.	In hospital 27 days. There was paralysis of arm remaining when discharged.
57	37	M.	Dislocation forwards of fifth cervical vertebra on sixth.	Fell from a fence and struck back. Fell only 3 or 4 feet.	Palliative.	Died.	There was also a fracture of spinous process of second dorsal vertebra. There was paralysis, priapism, etc. In hospital 1 day.
58	29	M.	Compound, of left elbow.	Run over by heavy wagon.	Palliative.	Died.	There were other serious injuries, comminuted fracture at knee, etc. Died same day.
59	18	M.	Compound, of right fibula.	Caught between elevator and floor.	Reduced; rest; in fracture-box; suspended; dressed with moss and salicylic acid; abscess evacuated under rapid respiration, etc.	Cured.	In hospital 5 months. There was partial anchylosis of ankle-joint.
60	37	M.	Compound, of second finger of right hand.	From fall on shipboard.	First attempt to save after reduction; this failing, the finger was amputated.	Cured.	In hospital 111 days.
61	40	M.	Of left elbow.	From a fall.	Reduction and anterior angular splint.	Cured.	In hospital 2 days. To out department.
62	...	M.	Of humerus, anteriorly.	Fell from scaffold.	Reduction under ether.	Cured.	In hospital 25 days. There was also laceration and contusion of lip, and a fracture of the humerus at lower end.
63	31	M.	Of radius.	Fell through joists of a building.	Angular, lateral splint.	Cured.	In hospital 29 days. There was fracture of ulna of same arm, and the parts were difficult to hold in position. To outpatient department.
64	50	M.	Of right hip, thyroid.	Fell while running, and was run over by cart.	Hip reduced under ether.	Cured.	In hospital 36 days. There was also a fracture of fifth rib and contusions.
65	40	M.	Compound, of left knee, tibia outwards.	Caught in belting of machinery.	Reduced, dressed with carbolized oil, and placed on an inclined plane; then to straight position and silicate dressing.	Cured.	In hospital 66 days. There was also severe laceration and contusion of elbow and a scalp wound.

Dislocations.—(Continued.)

No.	Age.	Sex.	Nature.	Cause.	Treatment.	Result.	Remarks.
66	46	M.	Of right knee-joint; tibia forwards, femur backwards.	Knocked down while under hatchway of vessel. A bucket of ballast weighing 400 pounds fell about 8 feet, and struck him directly on the head. Was chiefly standing on right foot.	Reduced by flexion, under ether.	Cured.	In hospital 44 days. Had other contusions and lacerations.
67	36	M.	Of sternal end of right clavicle, forwards.	Run over by wagon.	Compress and retaining bandage; recumbent position.	Cured (?).	In hospital 6 days.
68	38	M	Compound, of right foot, outwards; internal malleolus projecting through broken skin; fracture of fibula.	Was trodden upon in a brawl.	Reduced by manipulation; wound stitched; dry carbolized oil dressing and fracture-box.	Cured.	In hospital 58 days.
69	40	F.	Of acromial end of left clavicle, upwards.	Fell from street car.	Compress and adhesive plaster dressing.	Cured (?).	In hospital 1 month. There was also a bad fracture of lower jaw.
70	66	F.	Of right shoulder, axillary, old.	From fall, 5 weeks ago.	Attempted reduction while under ether.	Unreduced.	In hospital 47 days. The humerus was fractured high up during the attempt to reduce. The limb was placed in sling, and good use of arm resulted.
71	23	M.	Of femur, upwards and backwards.	From bank of earth caving in and falling on him.	Reduced by manipulation, under ether.	Cured.	In hospital 18 days.
72	57	M.	Of humerus, axillary.	From fall.	Reduced by manipulation.	Cured.	In hospital 4 days.
73	7	M.	Of ulna, backwards, with fracture.	From fall off steps.	Luxation reduced by forced flexion; splint.	Cured.	
74	69	M.	Old, of right humerus; 9 weeks.	From fall.	Reduced under rapid respiration.	Cured.	In hospital 32 days.
75	72	M.	Of right humerus.	From fall.	Reduced in out department.	Improved.	In hospital 11 days. Patient admitted for other injuries.
76	51	M.	Old, of humerus.	From falling on ice, 10 weeks since.	Patient was very weak and anæmic. Attempt was made to reduce under ether. After adhesions were broken up the humerus broke about insertion of deltoid, no extra force being applied. Limb was dressed for fracture, and did well. There was more freedom of movement at joint than before.	Unreduced.	In hospital 1 month.
77	42	M.	Of sternal end of clavicle.	From fall through hatchway.	Rest in bed.	Improved. Eloped.	In hospital 18 days. Also had concussion of brain.
78	54	M.	Of right humerus, axillary.	Was thrown from wagon.	Reduced by manipulation.	Cured.	In hospital 6 months. At same time there was a compound dislocation of tibia at ankle and fracture of fibula.
79	45	M.	Dorsal, of right hip.	Large rock rolled against him.	Reduced under ether.	Cured.	In hospital 46 days. At first was thought to be complicated with fracture, but the symptoms were owing to accompanying contusions.
80	36	M.	Old subluxation of wrists.	Large lump of coal fell upon the part, 2 years since.	Under ether the actual cautery was applied, and dressed with poultice.	Improved.	In hospital 4 months.

WILLIAM HUNT.

HEMORRHOIDS.

During the past five years 40 cases of hemorrhoids have been under treatment; of this number, 33 were males and 7 were females. 3 cases were under 20 years of age; 9 were between 20 and 30; 15 were between 30 and 40; 9 were between 40 and 50; 4 were between 50 and 70. 35 were cured; 5 were improved. In most of the cases the hemorrhoids were internal, but in several instances internal and external tumors were combined. In twenty-two cases the hemorrhoids were ligatured; the tumors were seized with a tenaculum, and, after being well drawn down, were encircled by a strong, well-waxed silk cord. In ligaturing external hemorrhoids, the skin is always very carefully divided before the ligature is tightened, so that the skin is not directly included in the strangulation; but the ligature is made to fall into the incision at the base of the tumor. When a number of hemorrhoids are ligated simultaneously, the anus being considerably involved, the division of the sphincter ani muscle may be made, either immediately before or after the operation, with excellent effect. This simple procedure not only diminishes the chances of irritation of the bladder, which so often is accompanied by retention of urine, a symptom not unfrequent after operations for hemorrhoids, but also lessens the painful spasmodic contractions of the anus, which so frequently occur. The divided muscle readily heals, and no ill effects are ever observed from the section.

In two cases, the hemorrhoids after removal by the knife, have been treated, in one instance with the actual cautery, in the other with nitric acid; an excellent method for applying the latter is with the glass brush (see figure), which is kept immersed in the acid, and is always ready for use. In one case the actual cautery alone was used; in three cases external hemorrhoids were successfully removed by the knife. After the operation, as soon as any odor is observed, the parts are washed with carbolized water, or with a solution of permanganate of potassa. Patients who submit to an operation for hemorrhoids are frequently much depressed, dyspeptic, full of notions, sallow in appearance, and often weak from loss of blood. These symptoms, which are the result of drain, often extending through years, require attention, and for this purpose we use anodynes, milk in very large amounts, full doses of quinine before and after the operation, and occasionally, stimulants. There is no minor operation which affords probably such immediate relief, and one which is followed by such improvement in the patient's general condition, as that for the cure of bleeding piles. The change is at once not only physical, but mental. We recommend the ligature because the operation is readily and quickly accomplished, and there is no danger from hemorrhage.

It may be well to remark that, after ligaturing the hemorrhoids, we *never* excise any portion of the constricted masses, for occasionally the tumors shrink so much that the ligatures become loosened, and hemorrhage may then occur. After the operation a large, soft, flaxseed-meal poultice is applied to the perineum; this is found to be very soothing. An opium suppository should also be introduced.

In twelve cases, attention to the functions of digestion, with tonics, laxatives, and cold applications, with the daily use of a suppository, was all that seemed to be required. One of the very best formulas for allaying the irritation incident to hemorrhoidal affections consists of the following combination:* Acetate of lead and tannin, each, fifteen grains; carbonate of lead and extract of stramonium, each, thirty grains; creasote, five drops. This mass, made up with a sufficient quantity of cocoa butter, should then be moulded into fifty suppositories; these not only allay congestion, but frequently induce an absorption of the more mild forms of hemorrhoidal disease.

THOMAS G. MORTON.

* Original formula of Dr. Paul B. Goddard, and furnished by Mr. Geo. McKelway.

RUPTURE OF THE LIVER.

INJURIES which are of such a severe character as to cause a rupture of the liver are not unfrequently associated with laceration of one or more of other viscera, so that we find, not uncommonly, the lungs, kidney, intestines, and sometimes the bladder, occasionally involved in the general crush. Such injuries of the trunk are generally from railway or heavy wheel crushes, or from falls from a great height, and from the very nature of the injury primarily fatal. Occasionally rupture of the liver occurs from direct violence without any other organ being implicated. Hemorrhage in any case is most alarming, and is often immediately fatal, and if not, peritonitis usually follows within a brief period. The most profound shock is always present, and the most severe pain ensues; of course decided doses of morphia by the skin are indicated, with absolute rest, and stimulants in guarded doses.

Of the 9 recorded cases of injuries involving the liver, 4 were incident to the passage of heavily-loaded wagons or cars over the body. In all of these cases a fatal issue resulted in a few hours after admission. 2 cases are recorded of punctured wounds of the liver; in one instance the injury was inflicted upon the upper, in the other upon the under surface of that organ; both cases made excellent recovery. In one case, an emery-wheel broke when making two thousand revolutions per minute, and a fragment of the wheel, which weighed upwards of four pounds, struck a workman directly over the region of the liver. The case progressed favorably for several days, when peritonitis developed, and on the sixteenth day the patient died. At the autopsy it was found that the wound of the liver, which was more than four inches in length, had fairly united. In two cases of lacerated liver,—in one instance from a blow from a locomotive, the other from a fall from a scaffold,—each terminated fatally in a few hours from hemorrhage and shock.

CASE 1. *Rupture of the liver; death from hemorrhage.*—Nathaniel S., aged 30, was admitted April 25, 1865. He was run over by a cart, the wheels passing over his body diagonally and upwards. Crepitus was distinct over the third and fourth right ribs, and there was bloody expectoration. He sank, and died within twenty-four hours, with symptoms of internal hemorrhage. At the autopsy, the liver showed a rupture on the convexity of the right lobe, and the rent involved at least one-half of the thickness of the organ. The second, third, and fourth ribs on the right side were fractured just inside their angles; the lungs were collapsed, and the pleural cavity was filled with blood. The peritoneal cavity was likewise filled with blood.

CASE 2. *Rupture of the liver; death from internal hemorrhage.*—A German, aged 30, was admitted with intense shock. He had been crushed by a heavily-laden wagon an hour previously. He died forty-eight hours after being brought to the hospital. At the autopsy, a laceration of the convexity of the right lobe of the liver was found, four inches in length, which extended deeply into the substance of the organ. Three quarts of blood were found in the peritoneal cavity, and the peritoneum was roughened by patches of recent lymph.

CASE 3. *Penetrating wound of the liver: recovery.*—Colon H. F., aged 23, was admitted into Dr. Morton's ward August 29, 1867. The patient, an upholsterer, while in a wrestling match, fell upon his long pair of shears: the blade cut between the seventh and eighth ribs, seven inches below the right nipple, passing inwards and slightly upwards for a distance of several inches. The external wound was two and a half inches long; the hemorrhage, which was profuse, came from the deep portion of the wound. Was much prostrated, with considerable shock. The finger, when passed in the wound, came directly in contact with the liver, which was found cut an inch in length, and to a considerable depth. After passing the abdominal covering, the liver was wounded upon the convex surface, and when the finger was introduced the liver was forcibly pressed upon the finger. Pain of a violent character was experienced in the right shoulder. The wound was closed by three sutures. One-quarter of a grain of opium was given every two hours. On the third day, suppuration, swelling, and redness necessitated the removing of the sutures; two ounces of pus were then discharged. From this time the patient continued to improve. An evident depression of the cicatrix showed that the liver was attached to the abdominal tissues. The pain continued in the shoulder at intervals during his hospital residence. The pus examined was found filled with broken liver-cells, pigment, and portal blood. Was discharged, cured, September 19, 1867.

In the summer of 1879 the patient presented himself for examination; he has been quite well since leaving the hospital.

CASE 4. *Rupture of the liver from a blow from a broken emery-wheel; peritonitis; death.*—Charles S., aged 25, was admitted September 28, 1875. An emery-wheel, weighing about 12 pounds, and revolving at a velocity of 2000 times a minute, burst, and a fragment weighing 4 pounds struck him over the cartilages of the false ribs, upon the right side, throwing him back about four feet. When brought to hospital he was suffering greatly from shock, and was in great pain. The cartilage of the seventh rib was exquisitely tender upon pressure, but no dislocation or fracture of the rib could be made out. Respiration was also attended with great pain. The skin was cold and clammy, pulse weak and fluttering. He was given brandy, and hot-water cans were applied to feet, and the chest was strapped, after which he felt much relieved. Hypodermics of morphia, ♏x, were given, to be repeated, if necessary; urine drawn by catheter. Great pain in abdomen; has recovered from shock.

Evening.—Belly becoming tympanitic and painful upon pressure; some vomiting; respiration costal. Taking pulv. opii. gr. i, every two hours, and applied flaxseed poultice to abdomen.

September 29.—Risus sardonicus. Vomiting much more than yesterday: it seems to be merely a regurgitation, and without much effort. Some fever: he is barely under the influence of opium.

September 30.—Suppository of opium, gr. ii, instead of pill. Tympanitis increasing; conjunctiva slightly icterode; urine drawn by catheter brownish in color, sp. gr. 1031; acid, contains a good deal of albumen (no casts), and displays the change of colors when tested with fuming nitric acid. A specimen of urine drawn five hours later contains more bile pigment but less albumen.

P.M.—Skin comparatively cool and moist; pain much less. The face has lost most of its anxious look.

October 1.—Much easier this morning; has been passing flatus quite freely; conjunctiva still jaundiced.

P.M.—Hydrarg. chlor. mit., gr. ¼, suppository of opium reduced to gr. i.

October 2.—Not nearly so well this morning; pain all over the abdomen; skin jaundiced; cheeks flushed; countenance very anxious; has not been passing nearly so much flatus.

P.M.—Gr. ij, opium suppository; pupils are but slightly contracted, but patient complains greatly of thirst.

October 3.—Looks much better. Last evening he passed a large amount of fæces, with great relief. Conjunctivæ are more jaundiced than ever; no pain.

October 4.—Friction heard over right chest; is slightly delirious; wants to get out of bed.

October 5.—Delirium more marked. Ordered whiskey, ℥i, every two hours, in milk.

October 6.—Somewhat better.

October 7.—Passed large quantities of fæces, with great relief.

October 8.—Treatment continued, except to increase whiskey to every hour.

October 9.—Removed from receiving ward to upper surgical, and a bath of temperature of 115° given, after which he felt much better.

October 10.—Feels much better. Able to pass his water.

October 13.—Sinking.

October 14.—Died.

Autopsy.—The specimen showed a laceration, commencing on the anterior border of the right lobe, at the fundus of the gall-bladder, and extending an inch and a half upon the convex surface, where the laceration divides in two directions, running two inches farther to the right and one inch towards the left. On the anterior border the laceration affects the whole thickness of the organ, as well as the laceration running towards the left. The tearing on the right side extends obliquely downwards and backwards half way through the thickness of the organ, reaching the main branch of the hepatic vein to the right lobe. On the under surface the rupture extends along the line of the gall-bladder as far as the transverse fissure. The anterior portion of the injury is not united, but is covered over by thick layers of lymph, deeply stained with blood. The tearing in the central portion of the organ is found to be pretty firmly united, and along the line of union there is a line of whitish, cicatricial-looking tissue. The surface of the liver at many parts is covered with recent lymph, deeply stained with blood. The abdominal cavity was greatly distended; the intestines were firmly adherent to the abdominal walls. The abdominal cavity contained a large amount of very dark fluid, large clots of blood lying between the coils of intestines, which were generally bound together by inflammatory adhesions; no other lesions were found.

CASE 5. *Rupture of the liver; death.*—Joseph H. W., aged 23, was admitted September 29, 1876. Was struck by a locomotive; was in immense shock from the blow, and internal hemorrhage. There was no reaction, and death occurred same day.

CASE 6. *Punctured wound of the abdomen, involving the liver and gall-bladder; great effusion into the peritoneal cavity; paracentesis; recovery.*—Amelia M., aged 23, was admitted May 7, 1877. Intoxicated on admission. Small wound, three and one-half inches to right of median line and three and one-half inches above umbilicus. Very little external hemorrhage; severe pain all over abdomen; given morphia hypodermically.

May 8.—Passes urine; tongue coated.

May 10.—Distention of abdomen, pain, etc.; kept under morphia; whiskey, ℥i, daily.

May 14.—Small injection of hot water and soap, producing slight operation; kept under morphia.

May 21.—Belly more distended; bowels have become regular once daily, and her appetite is good. Opii suppository, gr. ii, twice, and whiskey, ℥iii; wound healed.

May 28.—Very little change; abdomen more distended; fluctuation felt on palpation above the umbilicus; stools are clay color.

May 31.—Dr. Hunt made an incision two inches above and to the right of the umbilicus; then introduced trocar, and drew off seven pints of a dark, yellowish-green fluid, which contained large quantities of bile pigment, sp. gr. 1020; neutral reaction.

June 5.—Filling up again slowly; bowels constipated. Ordered injection, which operated freely; appetite improving rapidly; opium suppository every four hours.

June 21.—Sits up in bed; swelling not any larger; occasional vomiting spells since 10th; ordered bismuth.

July 3.—Up for first time; is weak, but has no pain.

July 10.—Gains strength; is out in yard.

July 21.—Discharged, cured.

CASE 7. *Rupture of the liver and kidney from a fall; death.*—Aaron J. S., aged 18, was admitted July 11, 1879. He fell from the fourth story of a scaffold and sustained a compound fracture of the thigh and leg, also a fracture of one rib, and a fracture of the liver and right kidney. Was, on admission, in a state of profound shock, and never rallied. Died in a few hours.

CASE 8. *Rupture of the liver; death.*—James F. S., aged 6, was admitted September 16, 1879. Patient was run over by a heavy dray at 4 this P.M. 7 P.M., brought to hospital in the most profound shock, very restless, and calling for water. No pulse at wrist whatever, and none could be felt except in femoral, which was a mere thread,—140; respiration 56; heart very weak. There was fracture of sixth and seventh ribs. With these signs of internal hemorrhage, face, lips, and conjunctivæ blanched and cold. Hot-water cans to extremities, and enema of hot water—118°—was given, but not retained. In one hour died. Autopsy revealed fracture of liver and ribs.

CASE 9. *Rupture of the liver; death.*—Jacob S., aged 37, was admitted November 22, 1879. While intoxicated fell from a street car, and a four-wheeled coal-cart, loaded with two tons of coal, passed diagonally across his abdomen, from left to right. On admission into hospital he was unable to be aroused; pupils dilated; respiration heavy; breath cold; skin cold to the touch; face blanched; pulse almost imperceptible. There were several bruises over right side, and the tenth and eleventh ribs were fractured. He was given, hypodermically, ʒij brandy, also ʒij ether hypodermically, and brandy, ℥ss, by mouth administered. Mustard and hot-water cans to extremities, and morphia, gr. $\frac{1}{6}$, hypodermically. This was about 11.30 A.M. At 2 P.M. skin grew warmer, pulse grew stronger, and he rallied, and became conscious. Hot coffee, milk, and beef-tea given at regular intervals of one hour.

6 P.M.—Drew off urine. Abdomen painful upon the slightest pressure. Applied flaxseed poultice. Pulse jerking; belly tympanitic; morphia, gr. $\frac{1}{4}$, hypodermically. Complains of great thirst. Ribs strapped.

November 23.—Suppression of urine. Abdomen more distended. Diarrhœa, with bloody stools.

November 24.—Sank rapidly, and died at 2.30 A.M.

Autopsy showed peritonitis and rupture of liver, left lobe being lacerated obliquely along its entire length posteriorly. Kidneys congested, the right one black, and considerably bruised.

Rupture of the Liver.

Injury.	Cause, etc.	Complications.	Treatment.	Result	Remarks, Cause of Death, etc.
1. Rupture of liver.	Wheels of a loaded cart passed diagonally across the body.	Fracture of ribs; internal hemorrhage.	Morphia hypodermically.	Died.	Convexity of the liver involved, also fracture of second, third, and fourth ribs.
2. Rupture of liver.	Crushed by heavily-laden wagon.	Hemorrhage.	Morphia.	Died.	Three quarts of blood found in abdominal cavity.
3. Penetrating wound of liver.	Fell on a long pair of upholsterers' shears.	Hemorrhage.	Wound closed by silver sutures; afterwards abscess opened.	Cured.	Pus examined showed broken liver-cells, pigment, etc.
4. Rupture of liver.	Struck by a broken emery-wheel.	Shock.	Morphia.	Died.	Peritonitis.
5. Rupture of liver.	Struck by a locomotive.	Hemorrhage and shock.	Morphia.	Died.	No reaction.
6. Incised wound of liver.	From a knife.	Shock.	Morphia, and paracentesis abdominis.	Cured.	Collection of fluid in abdominal cavity.
7. Rupture of liver.	Fell from the top of a high scaffold.	Compound fracture of thigh and rib; laceration of kidney; shock.	Morphia.	Died.	In few hours from shock.
8. Rupture of liver.	Run over by a dray.	Fracture of rib and shock.	Morphia and brandy by skin.	Died.	Shock.
9. Rupture of liver.	Run over by street car.	Shock.	Morphia and brandy by skin.	Died.	Peritonitis and hemorrhage.

THOMAS G. MORTON.

PHOSPHORUS NECROSIS.

During the past five years 4 cases of phosphorus necrosis have been admitted. The disease in 2 of the cases was confined to the superior and in the 2 others to the inferior maxilla. An entire recovery followed in each instance. All of the patients were adult males, and had been employed in a lucifer-match manufactory. Two of the patients had worked in the same establishment, three had worked as "dippers," and all had been exposed directly to the phosphorus fumes more or less for years. Case 1 had been sixteen years in the business, and was a "dipper" for eight years. Case 2 had been working at match-making for eighteen years, and for three years had been a "dipper." Case 3 had been many years in the business. A peculiar pustular eruption on the gum, which is one of the first and earliest signs of the disease, according to the statement of several patients, was very marked in this case. Case 4 had been a "dipper" for three years. In all of these cases the poisonous influence seems undoubtedly to have attacked the jaw through a carious tooth.

In Cases 1, 2, 3 the disease seems to have followed immediately after the extraction of decayed molar teeth, yet it is probable that the disease had been insidiously at work for a long time before, causing possibly the toothache which induced the extraction of the teeth, this removal allowing the fumes to act more directly upon the already diseased alveoli.

In the history of these cases the disease seems to have been slow in its progress, a long period elapsing before the necrosed portion of the bone becomes sufficiently detached from the periosteum to allow of its removal, but in one instance detailed to the writer by Case 3 the disease ran its course to a fatal termination with terrific rapidity. The subject, a girl aged 20, who was working in a lucifer-match manufactory, was seized with the usual symptoms of periostitis of the inferior maxilla, after the extraction of a tooth. She had a chill, followed by fever. Rapid suppuration ensued, and the great drain, with the intense suffering incident to the entire jaw being at once involved, was sufficient to cause a fatal issue within a week. The treatment in the hospital cases has been to get rid of the source of irritation as soon as practicable, for the excessive drain, with the great suffering, is productive of immense prostration.

In some cases the separation of the bone is wonderfully rapid. In others this is much more slowly accomplished. In one case a large portion of the jaw was quite loose at the time the patient came in, so that an external incision was not required. In Case 3 the suffering of the patient required an immediate operation. Since in some cases the disease had been arrested by a partial excision, this expedient was thought advisable; and at the first operation two inches of the diseased bone, on the left side, were cut out. Then, at a second operation, all the front of the maxilla to the ramus on the right side, and finally the ramus of the left, were removed.

Morphia hypodermically is usually required in very large doses; and, as the only food the patient can take is necessarily of a fluid character, it should be of the most easily digested and nutritious description. The most important local treatment consists in the free use of a disinfecting mouth wash.

The subject of phosphor-necrosis is of such interest that a short account of the duties of the "dippers," who seem generally to be the sufferers, may not be deemed irrelevant. The facts I have obtained from Chas. J. D., Case 3, whose father was for forty years, and himself many years, in the business.

The "dippers," who are always men, are more likely to be affected than women, who only pack the matches; yet several instances have occurred where they have likewise been affected. The "dippers" are constantly obliged to be over the fumes of the phosphorus, which hangs about like a cloud. In making matches, the glue which is used is slowly dissolved, afterwards the phosphorus is added, and the mass is rapidly agitated for an hour or so; the heat is gradually increased to 150° (during all this time the fumes are very great, and the "dipper" is obliged to be bending over it, and necessarily inhales it constantly); the mass is then allowed to cool. It is again made liquid by heat, when whiting is added. The mass is then spread out on a marble slab, and the bundles of little sticks, having been previously immersed in sulphur, are "dipped." The same person attends to the entire process. The health of the "dippers" is generally very poor, they have miserable appetites, usually a haggard appearance and cough, although many have been able to work for years. It seems that after any one working in the match-factories has had teeth extracted, the rule now is not to allow such a one to resume work for a week or ten days. Amorphous phosphorus, it is said, will not produce the phosphorus disease, and attempts have been made to introduce it in making the article known as the "Safety Matches." The phosphorus is applied to the box sides, and the sticks have some chemical compound applied to them, and only are ignited by friction upon the box. With the amorphous phosphorus there are no fumes. But I understand that the sale of this match is very limited, and is not increasing. The call for the old-style match has not diminished. Unfortunately, those who have to work in this business of common sulphur-matches will, until some other invention is made, be liable to suffer from this most terrible disease.

CASE 1. *Phosphorus necrosis; extirpation of whole right half and part of left half of lower jaw without external incision.*—Frederick C., aged 22, single, native of England, was admitted January 11, 1865. Was a lucifer-match maker, and had been in the business for sixteen years, being a "dipper" half that time. Family and personal history good. Patient had been well until the

Phosphorus necrosis of the inferior maxilla; removed by Dr. Hunt. Case I.

beginning of present trouble. Teeth were sound, except that there had been one slightly carious molar of right side of lower jaw. From this he habitually picked small particles of food after eating. Soon it pained about the root, and was extracted. From this time (seven months since) the peculiar phosphorus necrosis may be said to have fairly begun. It progressed rapidly, and about a fortnight before admission he was compelled to cease work. Upon admission he was pale, thin, and weak; pulse 90 to 100; face very much deformed by indurated swelling, particularly conspicuous on right side and about chin. Three fistulous openings, surrounded by flabby granulations, were upon right side of neck. From these fetid pus exuded freely. A horribly-offensive odor came from mouth, also saliva and pus dribbled constantly. Gums spongy on margins, but general induration at sides. Four sound but loose teeth occupied the right alveolar cavities. Upper jaw seemed healthy. Patient could not separate jaws sufficiently to introduce a finger with facility. Probe detected necrosed bone. Portion of left lower jaw only seemed clear of disease. Symphysis was completely involved. The *morale* of patient was extraordinary. Ordered milk-punch, beef-tea, chloride of iron, quinia, and anodynes at night. Potassii permang., gr. ij to f℥viii water, as mouth-wash.

January 28, 1865.*—The patient was thoroughly etherized. The teeth that were in the way were extracted. By means of a cheek-holder, held by an assistant, the mouth was widely opened, and with a stout scalpel Dr. Hunt made an incision directly along the base of the alveolar processes, at the margin of the gum, from the root of the coronoid process to the symphysis. The soft parts, including the periosteum, were easily separated from the horizontal ramus, and the jaw was divided at the symphysis with a pair of strong cutting pliers. A blunt-edged cranial elevator was now used and inserted behind the angle of the jaw. By careful and patient working he was able to separate the attachments of the pterygoid muscles and the internal ligaments; he then broke off and detached the coronoid process, leaving but small and crumbling portions of it attached to the temporal muscle. He then inserted the elevator behind and above the angle, and tilted the condyle forwards. The ascending ramus was now grasped with a pair of strong bone forceps from the angle to the neck, and by means of twisting movements the rest of the attachments gave way, and the whole of the right lateral half of the bone, with the exception of a small portion of the crumbling coronoid, was removed.

* Case reported in the *American Journal of Medical Sciences*, vol. xlix. p. 353, April, 1865.

The left half now claimed attention. About an inch and a half of this, beyond the symphysis, seemed to be diseased. The periosteum was readily recognized on this portion. The tongue was secured by passing a double ligature through its body, and, by means of a loop, it was held by an assistant. This proved to be a wise precaution; for, although it is considered unnecessary by some authorities, we found that in this case, as soon as the genio-hyoid muscles were divided, all control of the tongue was lost, and it would certainly have fallen back on the glottis had it not been for the loop. In Dr. Wood's well-known case the patient swallowed her tongue, and was very near being suffocated.

This precaution it appears is much more necessary when anæsthetics are used, for after the patient has his senses, and his head moderately elevated, there seems to be less danger from the slippery organ. This was proved in this case, for when the loop was removed on the day after the operation the patient had a moderate degree of control of, and no tendency to swallow the member. To return, the tongue muscles having been divided and the tongue and periosteum separated as far back as an apparently healthy portion, a chain-saw was inserted, and the piece was readily cut out on a line corresponding with the anterior border of the first molar tooth. The hemorrhage was slight: but a few ounces of blood were lost, and no vessels required ligature. Discharged, cured, March 28, 1865.

This patient, as well as Case 3, recovered completely; a hard rim, apparently of ossific, or very dense fibrous tissue, was generated by the periosteum in each instance, which took the shape of the original bone, although much contracted. Case 1 was seen several years later, and Case 3 a year later; both were able to use the inferior maxilla very well, and, with the beard full grown in each case, there was no trace externally of the loss sustained.

CASE 2. *Phosphorus disease, involving the superior maxilla.*—William M., aged 29, was admitted October 28, 1874. Working in a match manufactory since 11 years of age, but was very little exposed to the direct fumes of phosphorus until he came to this country three years ago, since which time he had been employed as "dipper" in a factory, almost continuously. For the first two years his health was as good as ordinary match-dippers, some days feeling well and others quite badly. The past year his health had not been at all good, suffering from a very bad cough, and feeling quite weak. Three months ago, for the first time, he suffered from toothache, and he had a tooth extracted, which was much decayed; on being removed, a small piece of the alveolar process came away with it. Within a month from this he had four teeth drawn, but continued at his work all the time. On admission he was pale and weak, although his appetite was fair and he slept well. Bowels were regular; urine acid, sp. gr. 1022, and free from albumen; temperature, 97° F. The upper jaw on diseased side bulged outwards. Dead bone could easily be detected; of the teeth the two middle incisors were loose. Ordered lacto-phos. lime and cod-liver oil.

October 30.—Dr. Morton removed the whole of the diseased portion of the upper jaw by external incision, using principally Lister's bulldog bone forceps. Hemorrhage not specially marked, and controlled by pressure.

November 9.—Continues to improve; some fetor; parts touched with sol. argent. nit., gr. xx, ad ℥j. Wound granulating, but looks tardy.

November 23.—Was discharged, cured.

CASE 3. *Phosphorus necrosis; removal of the entire inferior maxilla except the ramus of the right side; recovery.*—Charles J. D., aged 30, was admitted July 19, 1878. He was a match-maker by trade. Four months before had a left molar tooth extracted. Two or three days after this he experienced so much pain and tenderness in the part that he called upon the dentist, who he feared had injured the jaw in removing the tooth. A swelling could be felt at the place where the tooth was extracted, and there was pain, which gradually increased. On admission there was external redness, which extended from the chin to the ear, with great tenderness on pressure. The swelling was so great that the mouth could only be opened one-half an inch. The odor was very offensive, pus exuding freely from the socket of the extracted tooth, which was carious, the adjacent bone being also exposed. The gum was red

Phosphorus necrosis of the inferior maxilla; removed by Dr. Morton. Case 3.

and swollen. Goulard's extract of lead, with laudanum, was applied externally, and quinine, gr. xv, ordered daily, with morphia, as required, to relieve pain. Diet of milk and eggs.

July 25.—*First Operation.*—After etherization the parts were more readily examined, and the necrosed condition was found to be more extensive than at first appeared. With the handle of the knife and a finger the periosteum was easily peeled up, and the bone was found bathed in pus. A curved incision, corresponding with the edge of the inferior maxilla, was made. After the bone was reached the tissues were easily dissected off, the facial artery requiring ligation. A chain-saw was now carried around, and two inches of the bone removed, in which section, apparently, all of the disease was included.

July 26.—Treatment continued as before; the morphia had to be slightly increased.

Second Operation.—Dr. Morton removed that portion of the inferior maxilla which extended as far as the ramus on the right side. Convalescence was slow but steady, and on December 12 he was discharged, at his own request, with a fistulous tract, and some pain on the left side, about the ramus.

December 16.—Was readmitted. Had had severe pain on the left side of his face, which is more swollen; discharge quite free from point of last excision.

December 20.—*Third Operation.*—An incision was made to-day five inches in length, and all the remaining portion of the jaw on the left side was removed. No vessel of any magnitude required ligation. From this time the improvement was steady, and on February 3, 1879, he was discharged, well. In these three operations the entire inferior maxilla was removed except the ascending ramus on the right side.

December, 1879.—D. is in excellent health. There is a rim of dense fibro-ossific material, which is in place of the original bone. Movements of the jaw are perfect, and the mouth can be opened to the usual extent.

CASE 4. *Phosphorus disease of the malar bone.*—Robert D., aged 25, match-maker, was admitted September 5, 1879. On admission, patient states that he has never had any specific disease, and that he has always been a strong, healthy man. He had worked many years in a match-factory as a "dipper," and had once suffered from necrosis of the upper jaw on one side, with some loss of structure, but now the malar or cheek-bone is affected. Last March he first noticed a slight swelling of the chin, attended with some pain. In April he found one of his teeth loose, which he pulled out, bringing a small piece of bone with it. In a few weeks after this the posterior part of the alveolar border of the left side of the upper jaw came away. About the same time local abscesses formed, which were lanced over the malar bone. On admission the cheek is red, swollen, and painful; two sinuses are observed on the cheek, and a probe readily comes in contact with necrosed bone. The glands about the neck are a good deal swollen, and painful. Ordered citrate of quinine, gr. x daily, and iron, gtt. x t. d.

October 15.—A curved incision, two inches long, was made by Dr. Levis, exposing malar bone, which was very much necrosed. The bone was removed with forceps, and wound plugged with carbolized lint.

Later there was very little discharge; granulating nicely.

November 2.—Cavity nearly closed up.

November 11.—Discharged, nearly cured.

THOMAS G. MORTON.

THE TREATMENT OF PLEURAL EFFUSIONS BY PARACENTESIS THORACIS.

In quite a number of cases aspiration for pleural effusions, whether the result of inflammation or existing as mere transudations, has been performed in the medical wards of the hospital, but these will not be considered in the present place. The histories of the cases of paracentesis thoracis here presented only include those which have been operated upon in the surgical wards during the past five years, and which were mainly cases of empyema. The disease in each of the instances given had followed some injury of the chest, and it is likely that we may thus account for the fact that no matter what had been the original design of treatment, a free incision was, as a rule, finally resorted to, because pus, and not serum, was ascertained to be the fluid to be evacuated. Whenever practicable the aspirating needle should undoubtedly be used, not only for examination of the fluid, but for its evacuation; but we have not unfrequently seen the needle discarded after ineffectual efforts to remove the effused fluid, and a large trocar compelled to be substituted. The minute puncture, the exclusion of air, the slow evacuation of the contents of the chest-cavity, are advantages which would, whenever possible, give the aspirator the preference; but if this instrument for any reason cannot be used, the efforts to relieve the chest should not be abandoned. Not only does the patient run a great risk from slow suffocation and general failure of the powers through hectic fever, but sudden death has, on several occasions, followed pleural effusions, and that when the patient seemed in fair condition, due, it is probable, to either pressure of a large effusion rapidly thrown out directly on the lungs, interfering with respiration, or paralyzing the heart, or else to that twisting and bending of the large vessels which must always more or less take place when the heart is from any cause displaced. Occasionally sudden death occurs, apparently as a result of the operation of paracentesis, but it is probable that in these unusual instances the patients have been either extremely feeble, or else that the operation was undertaken when but a faint hope of life remained.

Some months since I had under my care a young man, aged about 21 years, who was suddenly seized with pleuro-pneumonia. He had contracted syphilis a year previously and had suffered severely with ulcerated mucous membrane, and was greatly broken down. He was also at times a hard drinker, and had always a feeble constitution, so that when stricken down with the chest inflammation he required stimulation from the very commencement. At the close of the sixth day he was gasping for breath, with a dusky face, and his respirations were 50 in the minute. In consultation with Dr. J. Forsyth Meigs, we concluded that the only chance was in tapping the chest, which was immensely distended. A small trocar was introduced, and when a pint of serum was drawn off the patient suddenly turned his head to one side, gasped, and expired. Another very similar case will be found in our hospital notes. The cause of death in these cases does not seem to be definitely settled, but it is likely that the slight shock incident to the operation, and the apparently trifling re-distribution of blood, due to the removal of the pressure, may be sufficient to induce the fatal issue. Œdema of the lungs and paralysis of the heart from over-distention have also been suggested in explanation. In the whole domain of surgery there is no more wonderful exhibition of the peculiar tolerance and recuperative power than that seen when, in empyema, the chest is safely opened and detergent antiseptic washes used, although the system occasionally resents the removal of the fluid. In Case 4, the patient, before the operation, was cachectic and deadly pale, except for a marked hectic flush. He had also what seemed to be characteristic consumptive cough, the opening in his chest pouring out pints of offensive pus. After a free outlet was secured, he soon rallied from this condition under good food, tonics, and stimulants, and from the time that his pleural cavity was washed out daily he gradually gained in flesh and strength, and from his greatly-improved condition, when last observed, we believe is now entirely well.

In regard to the use of the varied forms of instruments recommended for tapping the chest, we can simply say that in all cases where serum is believed to exist the aspirator is certainly the instrument which ought to be used. If pus or broken blood-clots are present, as is usually the case after gunshot or other wounds, a free opening is unquestionably demanded. It then becomes a question of the daily use of the double catheter, or the introduction of a drainage-tube. The latter has many advantages, the principal being in the fact that the chest is kept constantly emptied, or nearly so. This, with the daily use of the syringe and double catheter, makes the most efficient form of local treatment.

In Case 5 a drainage-tube was introduced. As I experienced some difficulty in securing a counter opening for the exit of the drainage-tube posteriorly, I devised the trocar and canula (see figure, page 226), which, after being introduced at the most dependent point anteriorly, is then carried around the next lower rib, and brought out far posteriorly, so that, whether the patient is lying upon his back or sitting upright, the drain is always from the most dependent point, which is a matter of considerable importance. The trocar is now removed, and through the canula the perforated tube is introduced, and when a sufficient length has been introduced the canula is removed. The ends of the tube are tied together.

Case 1. *Pneumo-hydrothorax following pleuro-pneumonia; aspiration: death during operation.*—Fannie McB., aged 34, was received into the ward December 7, 1873, with the following history: Three months ago patient had a chill, followed by fever, prostration, pain in right side, and the usual symptoms of pleurisy. On admission, the patient was anæmic and weak, with a bad cough and very rapid respirations. Friction heard over right side.

On December 9 the patient was much worse. She had great dyspnœa, severe pain in right side, respirations 34 per minute. There was high-pitched, tympanitic resonance over entire right side, save at base, where the note was dull. This condition increased until the next day, when there were signs of collapse, when the patient was aspirated, and a considerable quantity of air drawn off. Ten minutes after, the needle was introdnced again over point of greatest dulness, and a few ounces of fluid drawn out, when patient suddenly died.

Post-mortem examination was refused.

Case 2. *Gunshot wound of back: empyema: paracentesis; recovery.*—George E., aged 19, was admitted October 7, 1874, with two gunshot wounds, entering his back, one going to the left of the sixth dorsal vertebra, and passing upwards and inwards, the other passing directly inwards near the lower angle of the left scapula. The accident was almost immediately followed by pneumothorax of left side, and there was great shock and dyspnœa.

On the 15th the percussion-note had become dull on the left side, and there was much constitutional disturbance. On the 17th the aspirator was used, the needle being passed through the sixth intercostal space, two and a half inches to the outside of the nipple, and forty-eight ounces of a dark, bloody fluid were drawn off. This was followed by decided relief. On the 24th a free incision was made in the seventh interspace, and thirty-eight ounces drawn off. A rubber tube was now inserted, from which there was free drainage, and subsequently a great discharge of pus. The cavity was, during his stay in the hospital, often washed out, carbolized water being used. Immediately after second operation patient was very ill, and suffered for a long time from fever, chills, and great prostration, and was also attacked by erysipelas and diarrhœa; but on April 23, 1875, he was discharged as practically cured. The sinus in his side had not entirely closed, but there was no discharge, and the lung in the upper part of the cavity was freely inflated.

In autumn of 1876 patient appeared at hospital "perfectly well."

Note.—In the operation of washing out the chest of this patient, which was performed daily for a long time, a silver female catheter was used, to which was attached a rubber hose. When the catheter was fully inserted, it was frequently noticed that it could be made to touch the pericardium, so that it was moved by each contraction of the heart. This was not accompanied by any pain or sensation to the patient, who watched the experiment with considerable interest, and felt no inconvenience from it whatever, showing that ordinary sensation does not exist in this situation.

Case 3. *Stab wound of the chest: empyema; paracentesis; recovery.*—Jacob R. P., aged 40, single, a farmer, born in the United States, was admitted into the ward in September, 1875, having been stabbed with a carving-knife, which entered the second intercostal space, on the left side, close to the sternum. Much blood was lost; the hemorrhage being finally controlled by pressure. Pain and tenderness afterwards set in, with bulging of the left side and difficult respiration. When admitted, on September 22, the patient was pale and debilitated. Pulse 130, and feeble. Respirations 48, temperature 99½°, orthopnœa marked. There was bulging of the left side of the thorax, with complete flatness on percussion below, and decided dulness above, third rib. On the right side, as was to be expected, there was increased resonance and puerile breathing. Paracentesis thoracis was performed between the fifth and sixth ribs, in the anterior axillary line, and sixty-four ounces of altered blood were drawn off with the aspirator. There was an immediate amelioration of all the symptoms, and the breath-sounds became more normal. On the 29th the temperature rose to 104½°, with pain and signs of a reaccumulation of fluid. The patient was put on stimulants. On the 5th of October the case became complicated with phlebitis of right leg. (There had been more or less œdema of legs for some time.)

October 6, Dr. Levis reopened aspirating wound by incision, and allowed the escape of a purulent fluid. The pleural cavity was washed out with carbolized water, and this was repeated from time to time. On the 17th there was a chill. On the 31st diarrhœa had appeared. But the patient now began to grow stronger, though pus still continued to drain from the tube, and the injections were kept up. On January 3, 1876, the patient had gained seven pounds in two weeks. On the 13th there was a drawback, and temperature rose to 101°; but on February 1 it was noted that the wound had closed, the tube having been withdrawn about nine days previously. On February 4 the upper part of left side was somewhat depressed, but clear on percussion to the fourth rib, and below that somewhat impaired. There was some dulness in the line of the axilla. The heart, which during the progress of the case had been greatly pushed out of place, was now found to have its apex-beat about one inch below and two and one-half inches internal to the left nipple. The patient was discharged, cured.

The patient was readmitted in August, 1876, for ulcer of the leg, when it was found upon examination that the respiratory sounds and percussion-note of the left side very closely resembled those of the right. Occasionally a slight friction-sound was audible, and there was some retraction at the left apex, but practically the left lung was as sound as the right. The apex-beat of heart was in about normal position, and there was a faint systolic murmur, which might have existed before original injury or might have been due to some pericardial roughening.

CASE 4. *Pistol-ball wound of thorax; paracentesis thoracis.*—Richard P., aged 29, married, a stonecutter, of American birth, was admitted on July 3, 1876, in a state of shock, having just been shot with a small pistol, the ball from which entered three inches below left nipple. There was great pain and tenderness on pressure in the neighborhood of the wound.

His treatment was principally for the relief of symptoms, supporting his strength, and relieving pain by whiskey, morphia, and beef-tea given from time to time.

On the 6th the face was jaundiced; there was pain on pressure over the liver. Morning temperature, 101½°; pulse, 100; respirations, 28.

July 8.—Great dyspnœa; pain increased by pressure over left lung; rapid rise of temperature to 103½° in the evening.

July 13.—The severe diarrhœa, which had lasted for several days, seemed to be yielding to opium. Temperature about 100°, with slight evening rise; pulse 100.

July 23.—Still some diarrhœa; complained much of chest, but examination deferred on account of its causing him so much pain. Morning temperature, 100°; pulse, 112; respirations, 28. The patient seemed now to have fallen into an irritable hectic state, attended by diarrhœa, emaciation, and want of appetite.

On the 30th, Dr. Morton took charge of the ward, and the ball was removed from the man's back, where it was found imbedded in the muscles, having passed directly through his body.

On August 4 the diarrhœa had ceased. There was also a decrease of tenderness, but complete dulness on percussion over left lung from apex to base; intercostal spaces obliterated, and ribs bulged out. The apex of the heart was found directly under the right nipple. Dr. Morton made a slight incision with a bistoury in the fifth intercostal space, three inches back of the left nipple. After repeated attempts to use an aspirating needle, which brought away only a few drops of pus, an ordinary trocar was thrust in, and one quart of thick fluid, consisting of broken-down blood, pus, and serum, was evacuated. In the evening the temperature was 101°, which was lower than it had been at that time for the four previous days.

On the 6th the canula, which had been left in, was removed, and a drainage-tube substituted. The apex of the heart was still found under the right nipple. The patient's pulse had risen to 120, but his temperature remained at from 100° to 101°.

August 10.—The patient seems weaker. The discharge of pus, now very fetid, continues. The tympanitic sound over the left lung, which followed the operation, has given place to a more healthy resonance, and a slight respiratory murmur is heard. The heart has moved three inches farther towards the right. The temperature has decreased.

From this time the patient gradually improved until October 2, 1876, when he was discharged by request. At this time he had a cough, with want of appetite, and a hectic flush, and was quite thin, his side still discharging. Being more or less neglected after leaving the hospital, his condition grew worse, and he had the appearance of a man in the last stages of pulmonary consumption when he again applied for treatment. Being readmitted into the hospital, January 3, 1877, he rapidly improved under tonics, iron, rest, and good food, the chest cavity being at the same time washed out with a solution of sulphate of zinc, gr. ii to ℥i of water.

On May 5, 1877, the patient was again discharged, now comparatively well, although there was still a small opening in his side, and he had some cough, from which we have since been informed he entirely recovered.

This case is a very remarkable one, as showing the tolerance of the system to such a great and long-continued drain, and marks very clearly the strong tendency to recovery observed in almost all these cases, the pleural cavity having been once safely opened, and a free opening secured for drainage and antiseptic irrigation.

CASE 5. *Stab wound of chest: empyema; paracentesis: recovery.*—James F., aged 34, married, a seaman, was stabbed with a knife a short distance above the right nipple. When admitted, on August 28, 1877, there was no external hemorrhage or bloody sputum. Suffered from extreme pain and weakness, followed by dulness on the lower part of the chest.

On September 3 the patient's temperature rose to 104°, from which it fell in a day or two to 99°, but on the 11th it rose to 103½°, and patient had now a large effusion, great dyspnœa, and an exhausting cough. These symptoms increased until October 22, when the chest was tapped with an ordinary trocar, and 192

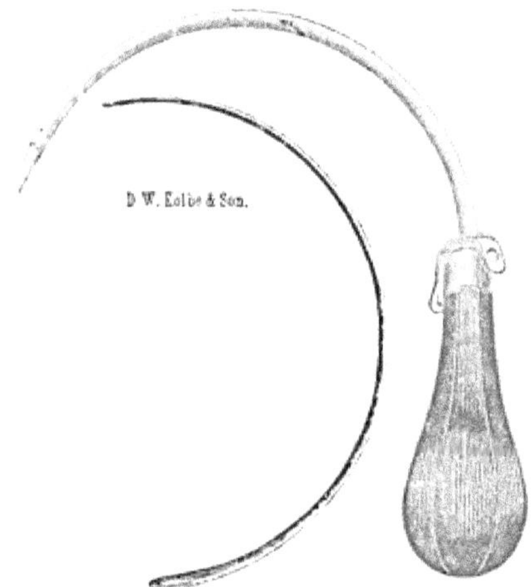

ounces of very offensive pus were drawn off. For a time the patient experienced great relief, but the pus soon collected again, and Dr. Morton inserted a drainage-tube through his canula (see figure). There was now constant drainage, whether the patient was in the recumbent or erect posture. From this time the patient improved steadily, although very slowly, his temperature ranging between 102° and 99°; and on January 10, 1878, there being very little discharge, the tube was removed.

On February 28 the sinus in his side had healed, and the patient was rapidly gaining flesh, and on April 2 he was discharged as cured.

In cases of this description, the graphic temperature record, enabling a ready means of noting and comparing the daily morning and evening observations, is of considerable service, since it may be taken as a guide to the time for the performance of the operation, and the tolerance of the system. It is remarkable how uniformly the bodily heat falls from a fever temperature to near the normal after the evacuation of the pus and the establishment of free drainage, and the use of the carbolized water injections.

Paracentesis Thoracis.

No.	Age.	Sex.	Admitted.	Discharged.	Disease or Injury.	Treatment.	Result.	Remarks.
1	31	F.	Dec. 7, 1873.	Dec. 19, 1873.	Idiopathic pleuro-pneumonia; pneumo-hydrothorax (phthisis?).	Aspiration.	Death	Needle at first drew off a little air. Introduced ten minutes after, drew off little fluid. Patient suddenly died.
2	19	M.	Oct. 7, 1874.	April 23, 1875.	Gunshot wound; empyema.	Aspiration. After free incision and drainage, cavity washed out.	Cured.	48 ounces, and afterwards 38 ounces, drawn off, then free drainage of pus.
3	40	M.	Sept. 22, 1875.	Feb. 4, 1876.	Stab; empyema.	Aspiration, free incision, drainage-tube.	Cured.	64 ounces altered blood aspirated. Lung fully expanded after cure.
4	29	M.	July 3, 1876.	May 5, 1877.	Gunshot wound; empyema.	Aspiration failed, then trocar, free incision, drainage-tube.	Cured.	At first opening 1 quart of blood, pus, and serum discharged. Patient had diarrhœa, great cough, continued fever, and loss of appetite. Slow, steady convalescence, under drainage, good food, etc. In this period was discharged once and returned.
5	34	M.	Aug. 28, 1877.	April 2, 1878.	Stab; pleurisy; empyema.	Ordinary trocar, afterwards drainage-tube.	Cured.	On first tapping drew off 19½ ounces of very offensive pus.

THOMAS G. MORTON.

CYSTITIS.

Inflammation of the urinary bladder, whether in the male or female, is a disease capable of causing so much annoyance, pain, and physical distress as to claim the sympathy, as it very often severely tries the patience, of the medical attendant. In many cases, moreover, the disorder is so little amenable to ordinary remedies as to demand surgical interference, the symptoms being so urgent and imperative that the patient is not only unfitted for his ordinary pursuits, but life itself becomes almost intolerable. Such cases, unfortunately, are by no means rare, and therefore, not unfrequently, find their way into the surgical wards of a general hospital, or appeal for relief to the skill of the surgical attendant outside the walls of the institution.

Cases of acute, non-specific cystitis, from chilling of the surface, contusions from horseback-riding, and other simple causes, generally yield in a week or ten days, either terminating by resolution or by passing into the subacute or chronic form. When such a case is received into the wards, in the early stage, there is generally marked constitutional disturbance, with nervous chills, followed by more or less fever and quickened pulse. When the disease is fully developed, there are frequent attempts at micturition, with strangury and severe pain in the perineum and neck of the bladder. The urine may at this time contain some blood and small fibrinous clots, but it always exhibits a decided amount of mucus and pus, and deposits the earthy phosphates upon standing. The scalding at the neck of the bladder, and the pain radiating through the pelvis and shooting down the thighs, may partially paralyze the bladder by reflex action, so that the viscus is not entirely evacuated, and may even contain a large accumulation of fluid, though stillicidium be present. Vomiting and decided systemic disturbance are rarely absent from among the symptoms of the disease when fully established. The occurrence of marked chills in the course of the affection, however, with an aggravation of the symptoms, or the onset of delirium with general prostration, may indicate the occurrence of suppuration or gangrene of the vesical walls, while symptoms of sudden collapse generally accompany rupture of the inflamed bladder.

The treatment of cases of simple acute inflammation of the bladder in the hospital has been by a thorough leeching of the perineum, the injection of warm medicated solutions into the bladder three or four times a day, also laudanum and starch enemata, morning and evening, with hot sitz-baths to relieve urgent symptoms. Counter-irritation over the pubes is often resorted to after the symptoms of local distress have been measurably relieved. The maintenance of the supine position is insisted upon, and the diet carefully regulated. Undoubtedly one of the best, if not the very best of all methods for the treatment of both acute and chronic cystitis, is the milk diet absolutely adhered to. Skimmed milk has been generally used in Dr. Morton's wards, and should be regularly given in small quantities every hour or two, the total during the twenty-four hours amounting to about two or three quarts. Should it disagree with the stomach at first, a half-ounce of lime-water is added to each glassful of milk, with a small piece of ice in warm weather, which makes it very acceptable and grateful to the patient, and prevents coagulation in the stomach. This milk treatment has proved extremely beneficial in the hospital, and, in fact, the beginning of the recovery in several cases was observed to date from the commencement of this course of treatment. Gonorrhœal cystitis does not require special attention in this connection, as it will be discussed among the specific accidents of gonorrhœa. We may say, however, that in general terms it demands the same care and the same general management as ordinary acute, non-specific inflammation of the bladder.

We would repeat, that by far the most important part of the local treatment consists in the abstraction of blood from the perineum, the leechings being repeated every second or third day. Washing out the bladder by means of a double catheter, with an emollient infusion, such as bran-water, or with a weak solution either of nitrate of silver (gr. ¼–½ to f℥i) or of sulphate of zinc (same strength), using at least a quart of warm fluid (temperature 100°) for each injection, is a very useful, and often an indispensable adjunct to the other measures employed. Care should be exercised to keep the patient well covered in bed and to protect him from draughts of air.

When the disease has become chronic the symptoms are less urgent, but at the same time less amenable to ordinary treatment, on account of the marked structural changes that take place in the bladder, which even extend along the ureters to the kidneys, adding chronic nephritis, or pyelitis, to the original malady. These changes in the bladder consist in ulceration and thickening of the mucous membrane, hypertrophy of the muscular bundles of the middle coat, and contraction of its fibrous envelope, so as to greatly change the appearance and reduce the capacity of the viscus, and thus seriously interfere with its functions. The general health is often impaired from the constant suffering and loss of rest due to frequent calls for micturition; the appetite fails, and emaciation and inanition may mark the cases of long-standing cystitis.

The treatment in this latter form would first be directed to the diseased organ. Since the disorder generally is produced by some obstruction to the outflow of the urine, an effort should be made to discover the nature of this mechanical interference, and, if possible, remove it, which may readily be accomplished if it be due to vesical or prostatic calculus, stricture of the urethra in its membranous or spongy portions, or to a narrow meatus, or a urethral polypus; but it is less amenable to treatment when caused by malignant disease, ulceration, paralysis, or other structural affection of the bladder, or to senile enlargement of the prostate gland. In the latter case, which is a common cause of bladder-disease in old men, the fluid extract of ergot, in teaspoonful doses, several times repeated during the day, has often a wonderful salutary effect, as recommended by the late Dr. Washington F. Atlee, who, being a sufferer from this disease, had used the remedy himself with great benefit.

Cases of chronic cystitis generally require tonics, more particularly quinine and iron, and a supporting treatment. The mineral acids are sometimes administered for their tonic effects, but, as a rule, where there is much irritability present, alkalies are required to relieve the extremely sensitive bladder. Bicarbonate of sodium, in infusion of buchu or uva ursi, or comparatively large doses of liquor potassæ, given in broom-tea or flaxseed-water, have often relieved the most distressing symptoms as if by magic. Opium or morphia injections are also required to aid in building up the patient's strength, by giving him good sleep at night. The oil of sandal-wood has been used in several cases in the wards of the hospital, with apparently excellent effect in diminishing the amount of pus and relieving the local symptoms.

Since 1873 there have been received into the surgical wards of the Pennsylvania Hospital eighteen cases of cystitis, in which the disease had existed for varying lengths of time. In one case all the rational symptoms of stone in the bladder were thought to be present, and an operation for median lithotomy was accordingly performed, but no calculus was discovered by the attending surgeon. Although no stone was found the patient was entirely relieved of his cystitis by the operation, and was discharged, cured. Division of the neck of the bladder for cystitis in the male is a rare operation, but it is justifiable in selected cases, since great relief is immediately experienced by the patient, and an opportunity is thus offered to inspect the mucous membrane and to make applications directly to its surface if desired. Inasmuch as no urine can be retained in the bladder as long as the opening is kept patulous, the parts are put in the best position for recovery. This operation in the female is now an acknowledged procedure, and two cases have come under Dr. Morton's care, in which he performed the operation of opening the base of the bladder through the vaginal wall, or colpo-cystotomy, with most gratifying results. One of these was performed upon a patient who had previously been in the hospital, the other was made out of the city, upon a patient of Dr. B. H. Detwiler, at Williamsport, Pa. The lady had suffered for years, and had become a confirmed invalid, as a result of chronic cystitis. She was in a truly lamentable condition before the operation, but since then has very much improved.

In a letter recently received by Dr. Morton, he was informed that "she is in good health now, is able to attend to her ordinary duties about the house, or to visit her friends." In this latter case, the wound in the base of the bladder has been kept permanently open by a hard rubber button, with a wide flange, so as to retain it in the bladder (see figure).

The history of the former case is as follows: In October, 1878, a married woman, 32 years of age, was admitted into the woman's surgical ward with chronic inflammation of the neck of the bladder, which had rendered her life miserable for several years. The symptoms were most distressing, and had not been relieved by any treatment. Upon admission, ether was administered, and the urethra well dilated with the fingers. Some improvement was experienced for a few days, but the symptoms reappearing, a second dilatation was performed, and the patient left the hospital much benefited. A few months later, however, she presented herself a greater sufferer than before, and she was clamorous for any operation that promised relief.

In May, 1879, with the assistance of Dr. Wm. Hunt, Dr. Morton performed the operation of colpo-cystotomy. The patient being anaesthetized, he inserted his left forefinger through the urethra, and making the anterior wall of the vagina prominent, cut directly down upon his finger, making an incision three-fourths of an inch in length. The opening was then distended, and the ordinary self-retaining rubber instrument inserted. The success attending the operation was marked from the first day. The distressing symptoms were at once relieved, an opium habit was broken up, and the patient's general health rapidly improved. The spasm of the bladder entirely ceased, and the character of the urine underwent a complete change, so that it no longer irritated the parts. The tendency to closure of the opening was so great that it was found necessary to constantly wear the instrument. It is true that she has the comparatively slight inconvenience of incontinence of urine, but this is infinitely preferable to allowing an accumulation of the urine, and thus invite a recurrence of a most painful malady.

In the following table, which is unfortunately incomplete, the causes of the cystitis are set down as far as could be ascertained. Two of the cases had introduced foreign bodies into the bladder. One was a male, who had inserted a tube from a glass thermometer, the other a female, who "sat down" on a long bone piercer, which slipped into the urethra beyond her reach, producing considerable irritation of the bladder. In the other cases nothing of special moment was observed. Of the entire eighteen, it will be seen that sixteen were male and two female.

Cystitis, with Results of Treatment.

No.	Age	Admitted.	Name.	Sex.	Duration of Disease.	Character of Attack.	Days under Treatment.	Result.	Remarks.
1	40	Jan. 27, 1869.	Edward A.	M.	3 years.	Chronic.	163	Cured.	Perineal section. Median operation for lithotomy.*
2	60	April 28, 1873.	J. B. F.	M.	2 years.		81	Improved.	
3	27	July 7, 1873.	Mary T.	F.	6 weeks.		72	Cured.	Leucorrhœa for weeks; had retention of urine three days.
4	64	Jan. 13, 1874.	Robert W.	M.	10 years.	Gonorrhœal.	35	Cured.	
5	32	July 2, 1874.	Edward G.	M.	Few days.	Acute.	5	Improved.	
6	64	Aug. 28, 1874.	William B.	M.	2 months.		27	Improved.	
7	44	Nov. 4, 1874.	Adam N.	M.	4 years.		14	Cured.	Carbolic acid injections.
8	36	Dec. 21, 1874.	Charles B.	M.	4 years.		14	Cured.	
9	66	Jan. 9, 1875.	Francis McC.	M.	Several years.		21	Improved.	Enlarged prostate.
10	72	Dec. 5, 1875.	James S.	M.	2 years.	Chronic.	19	Improved.	
11	52	Dec. 21, 1875.	H. J. W.	M.	17 years.		74	Improved.	Urethral stone removed six years before.
12	47	Dec. 31, 1875.	William B.	M.	Many years.		73	Improved.	
13	46	March 28, 1876.	John C.	M.			63	Improved.	Complicated with pyæmic rheumatism.
14	29	July 24, 1877.	Samuel W.	M.			56	Cured.	
15	25	Nov. 8, 1877.	Jos. D. W.	M.	Several years.	Chronic.	212	Died.	Was under treatment for cystitis four years before; had gonorrhœa five years before.
16	52	Nov. 13, 1877.	Walter L.	M.	5 years.	Acute exacerbation.	7	Improved.	
17	25	Nov. 8, 1877.	James D.	M.			211	Died.	Renal calculus.
18	32	Oct. 18, 1878.	Margaret W.	F.	For years.		8	Improved.	Colpo-cystotomy subsequently performed, with entire relief.

* See table of lithotomy operations, page 139.

FRANK WOODBURY.

VENEREAL AFFECTIONS.

Of the 351 cases admitted for venereal disease, 91 had syphilis,—that is, were constitutionally affected at the time of admission. The local lesions for which the others entered were as follows: chancre and chancroid, 91; bubo, 15; gonorrhœa, acute, 63; gonorrhœa, chronic, or gleet, 30; epididymitis and orchitis, 25; balanitis, 2; venereal warts, 3; syphilophobia, 1. Many of the cases were mixed in character, and complicated with the ever-varying troubles belonging to the disease; for example, from those of an insignificant but annoying kind, such as phimosis and paraphimosis, to those of a serious nature, such as ophthalmia, iritis, and brain affections.

How many of those having chancre and bubo were subsequently affected with general symptoms cannot, of course, be told, for the "cures" with which they are credited necessarily refer to the local lesions only. Most of them were sailors, subject to the temptations, changes, and exposures of their class. They leave at will, or are frequently remanded to duty by their superiors at the first available opportunity, and are often in the best possible condition for preserving and spreading the disease. The advocates of Lock Hospitals and Contagious Diseases Acts find their strong and, it appears to us, unanswerable arguments in the history of such men and their female companions. Why the principle of "stamping out" should not be applied to venereal as well as to other contagious diseases it is difficult to comprehend. Partial success can only be looked for at present as to all of them, and there is no reason why this excepted one should be allowed to run riot, with no attempt whatever in law to check its ravages.

There are no female venereal wards in the hospital. Women are never, except by accident, admitted for the acute affection, but a few get in from time to time with constitutional disease. In treatment the standard remedies are relied upon for the specific trouble. In small doses, the mercurials may be given for a very long time, and most patients grow rapidly better and stronger under their use. Should this not be the case, with some the line of treatment must be changed, and great improvement will often follow the use of tonics and cod-liver oil. The other great specific, the iodide of potassium, is used in most cases. A directly opposite course is now generally followed in its use as compared with the administration of mercury. It is given in enormous doses. To explain satisfactorily how it acts as favorably as it unquestionably does would puzzle, it seems to us, the most gifted therapeutist, for a very large part of what is taken may be recovered from the urine of the patient, and might easily be again given to him to repeat its rounds. To save trouble we will call it katalysis, and continue treatment. We think that the views of the most eminent authorities on syphilis as to the differences between the soft and hard sore, or chancroid and chancre, are fully confirmed. It is no unusual thing to see cases of the former produce a very great amount of local destruction, from which the recovery and repair are most remarkable and complete, while the latter is generally characterized by what appears at first to the patient to be an insignificant matter, but which he soon learns may be the source of a life-long contamination to himself and a plague to his descendants. When syphilis kills, it does it by slow degrees. There is a large number of individual cases of organic disease for which it is held responsible, and hence the purport of questions in diagnosis as to syphilitic history. There is no organ or tissue of the body that syphilis may not invade, in some producing rapid, and in others slow, destructive changes. Under the name of the local affection the patient is often credited with dying, and aneurism, brain tumor, laryngitis, etc., serve as well to state facts as to save reputations.

A more enlightened view as to the treatment of syphilis of late has given rise to the belief of some authorities that it is not only a curable, but a self-limited disease. It would be rather hazardous at present to practise upon the latter idea by the entire avoidance of specifics. It is best to work a cure by the judicious use of these, combined with nutritive tonics and good hygienic surroundings.

The cases of venereal were all tabulated, but as there is a great sameness as to treatment and no difference as to cause, it was thought better not to publish the wearisome list, but to summarize it, as follows:

Gonorrhœa, 63 cases. Average age, about 24 years.

Treatment.—The patient takes a bath, and is ordered to bed. Saline cathartics are at first given. It is an excellent plan to combine these with small doses of tartar emetic, especially when the appetite is good and the patient is vigorous. Weak urethral injections are ordered and repeatedly given. If the disease does not show rapid improvement, copaiba mixtures, generally containing liquor potassæ, are prescribed. Under this treatment the great majority of cases get well in from four to six weeks.

From time to time almost all the different plans in vogue have been tried, but most of the surgeons return essentially to the above methods. There is no encouragement, we think, to adopt any sort of heroic treatment with the idea of cutting short the disease. The attempts are too often painful failures and the consequences disastrous.

Gonorrhœal ophthalmia is treated by strong or weak solutions of nitrate of silver, according to the severity of the case. When weak applications are made, they are frequently repeated. It is not unusual to begin with moderately strong solutions, to apply them once a day, and to use weak ones in the interim. Should the cornea be threatened with perforation, atropia is applied to keep the pupil as much under control as possible. Great attention is paid to cleanliness; the discharges are wiped away and the syringe with warm water is frequently used. The patients mostly require a generous supporting diet during the progress of the case, and sometimes quinine and stimulants are necessary.

Gleet, 30 cases. Average age, 35 years.

These cases exhibit the usual ordinary obstinacy in yielding to treatment. Cures, however, are often accomplished. The internal remedies for gonorrhœa are sometimes given with good effect, but it is the local treatment that is by far the most important. This consists in injections, and dilatation by bougies. It is often found that a stricture is the source of the trouble, and where this is the case rapid improvement takes place under the use of instruments. These may be used also for the purpose of applying medicated ointments to the canal. The evil effects of the officious pursuit of direct remedial or specific measures in these cases is often illustrated, for by stopping them altogether for a time and attending to the general health of the patient, the disease disappears.

Epididymitis and orchitis, 25 cases. Average age, 25 years.

Most of these are of the first-mentioned kind, and originate from attacks of gonorrhœa. There is no reason to think that they are caused by the local methods of treating the clap, which is so commonly believed, for many cases may be found without injections having been used at all.

There is no doubt that an injudicious use of very strong injections might excite the disease, but what we mean is that injections are not necessary to produce epididymitis.

The orchitis cases, when chronic, are often found to have originated in syphilis, and require the treatment adapted to that condition. Epididymitis is generally treated by placing the patient in bed and having leeches and cooling lotions applied to the parts, which are well supported. Saline cathartics, with or without tartar emetic, are given. A method sometimes prescribed is to rub a solid stick of nitrate of silver over the scrotum after having it shaved. This treatment is painful, but often very rapid resolution takes place under it. After the acute symptoms have subsided, strapping with adhesive strip is resorted to should the case seem to require it.

Chancroid and chancre, 91 cases. Average age, 25 years.

The chancroid is treated locally by the standard applications,—*e.g.*, nitrate of silver, acid nitrate of mercury, carbolic acid, sulphate of copper, dry calomel, etc. The dressings are black-wash, weak carbolic solutions, lime-water, zinc, etc. The patients recover rapidly, as a rule, and often need no constitutional treatment other than placebos, which are sometimes given.

The chancre cases, on the other hand, besides undergoing the above local treatment, if requisite, are at once put under the mercurials, combined with the iodide of potassium. The corrosive chloride and the biniodide are the favorites, although the blue pill and calomel are sometimes given. The method of excising the chancre so highly advocated by some authorities is not followed. As stated in the preliminary remarks, most of this class leave the hospital on recovery from the local lesions, and there is no doubt that nearly all of them become subjects for classification under the following head.

Syphilis, 91 cases. Average age, 31 years.

These are cases where, by some form or another, the general system of the patient has been invaded by this protean disease. The mildest and most recent, as well as the severest and most chronic, forms are included under this heading.

Although many are much broken in health, it is found that they will generally bear small doses of the mercurials most admirably. When these are combined with tonics, they often act very favorably, and in some patients there is rapid improvement. It is not found necessary to push the treatment to salivation. In a few cases, where a quick impression is wanted, inunction with blue ointment is resorted to, and in some of the skin affections medicated vapor-baths are used.

The secondary local lesions, such as those affecting the fauces, the palate, larynx, and eyes, are promptly treated by appropriate remedies, while at the same time the more important (with one exception) constitutional treatment is continued.

The nitrate of silver, the sulphates of copper and zinc, astringent gargles, gargles containing the cinchona barks, the preparations of iron, etc., are all used, the individual patient being subjected to constant changes as this or that remedy is found to be most appropriate for him. Besides the brush, the douche and atomizer are used for the nostrils, larynx, and fauces.

When syphilis attacks the iris the local remedy precedes the constitutional in importance. This remedy is atropia frequently instilled. The strength used is generally gr. iv to f℥i. If a very rapid effect is required it may be applied at first in greater strength, but after full dilatation is obtained the weak solutions will be sufficient to keep up the impression. What constitutional remedies shall be used at the same time will depend upon the general condition of the patient. Mercury, the iodide of potassium, turpentine, quinine, etc., are prescribed in accordance with the judgment of the surgeon under whose charge the case may be.

The extreme advanced cases (tertiary?), as bone affections, gummata, etc., are treated with large doses of iodide of potassium, with or without mercury, according to the views of the practitioner under whose charge the patient may be. Tonics and cod-liver oil are important adjuvants, and these, combined with good diet, fresh air, and cleanliness, often bring about marvellous changes for the better.

Bubo, 45 cases. Average age, 25 years.

These were mostly of the suppurating variety, and were locally treated as cases of ordinary abscess,—*i.e.*, by rest, emollient dressings, and the knife. An old way of treating suppurating bubo in this house is sometimes still resorted to. The swelling is punctured in several places by a knife, but not slit up. Moist sponges are then placed over the part and firmly pressed against it by a spica bandage. In this way the matter may be discharged and the walls brought together without loss of substance or the usual scar being left.

The hard varieties of bubo come under the head of syphilis, and are treated accordingly.

Balanitis, 2 cases. Venereal warts, 3 cases. Syphilophobia, 1 case.

The simple forms of balanitis are easily cured by soap and water. The more obstinate require astringents and the separating of the surfaces of the prepuce and glans by lint or linen.

Venereal warts, if bad, are snipped off with the scissors. If not, they are touched with chromic acid or with collodion, or with this combined with corrosive sublimate.

Of the whole number of venereal cases, 223 are set down as cured, 114 relieved, 5 unimproved or declined treatment, 5 eloped, and 4 died. The average time in hospital was 29½ days.

William Hunt.

CONTRACTIONS; CLUB-FOOT; TENOTOMY.

11 cases of divisions of tendons are reported. Of this number there were 3 cases of congenital talipes equino-varus, 5 cases of acquired talipes equinus, 2 cases of severe contraction of the flexor thigh muscles following coxalgia and inflammation of the knee, and 1 case of (rheumatic) contraction of the palmar fascia of the little finger (Dupuytren's disease). 9 of the cases are regarded as cured, 2 were only improved.

It not unfrequently happens, in straightening limbs contracted either from joint-inflammation (false anchylosis) or from the healing of wounds of muscles, where shortening more or less serious has taken place, that it is often impossible to make such limbs resume their normal functions without dividing permanently shortened tendons.

This condition is often seen in the deformities which occur from the contractions in old diseases of the hip and knee-joints, when due attention has not been directed to extension, and keeping the extremity in a straight position, but the patients have been permitted to keep the limb in what appeared to them to be a most comfortable position. In one instance of sloughing from a pistol-ball, and in another from severe contusion involving the calf muscles, the subsequent structural changes produced so much contraction that division of the tendo Achillis in each instance was found necessary in order to restore the function of the part.

In cases of congenital talipes equino-varus, the treatment should commence at birth. The nurse or attendant of the infant should be instructed how to daily manipulate the foot, pressing in the prominent tarsus with the thumb, while the rest of the hand securely grasps the inner side of the foot. Steady outward movement and partial rotation should be made, and the foot thus should be gradually brought into a normal shape. At the time that this manipulation is performed the foot should be strongly flexed on the leg so as to bring the heel well down. By thorough, patient, and well-directed efforts, constantly kept up, it is wonderful how soon a most marked congenital case of talipes varus can be rectified; but it very often happens that although a varus can be cured (*i.e.*, all disposition to inversion disappearing), not unfrequently the very best directed efforts of not only intelligent but very anxious parents fail to correct the equinus. The deformity, however, is a small matter as compared with the varus. It is useless to apply any form of so-called brace upon an infant. Indeed, until the child is six or eight months old, a brace often does more harm than good, because the medical attendant or parents simply take it for granted that the brace is all-sufficient, when in reality the foot is so tender that it is impossible to make in this way any decided impression upon it.

Indeed, if pressure could thus be successfully made, there would be more or less atrophy resulting from the constantly constrained position the foot would be subjected to. It requires, not an inflexible brace or shoe which is supposed to hold the foot in position, but the brace should be the hand of the mother or the nurse applied when the child is asleep in the lap, or at other times when nursing, frequently twisting the foot into a correct position. Undoubtedly it is an arduous task: the infant at first makes a great outcry, but it soon gets accustomed to the manipulations and finally makes no objection; still, it is a work involving many months of care and constant attention. A light, well-padded splint of tin or moulded pasteboard may be judiciously applied at night and left off during the day, when the child is a few months old. When the child is ready to walk, a suitable apparatus should then be applied, but not, as a rule, the ordinary rigid Scarpa's shoe, which cramps the foot and only allows the hinge motion of the ankle, but a modification of it. To overcome the principal objection, I have used for several years past a brace which is constituted essentially as follows:

Fig. 1.

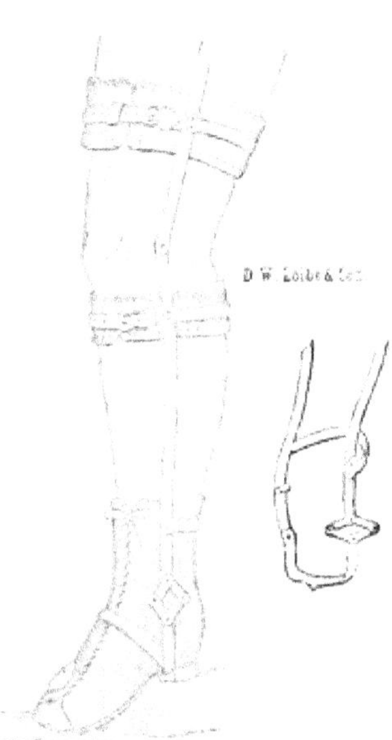

Taking the ordinary shoe, with lateral steel supports running up above the middle of the thigh, with a transverse brace and a belt or girdle above and below the knee to hold the apparatus in position, and a hinge at the knee- and ankle-joints, I have had an additional hinge placed opposite the internal malleolus, allowing outward flexion of the foot, and opposite this point I have taken out a portion of the outer steel rod and replaced it by a double antero-posterior hinge (see Fig. 1), which enables it to yield when pressure is made upon it, while the hinge in the inner support allows the turning out of the foot each time it is brought down to the ground. The weight of the body resting upon it with each step gives, by this means, an outward movement or partial rotation of the foot, which is consequently communicated to the tarsus, so that this portion of the articulation is made more pliable.

If it should be asked when we perform tenotomy, we answer *never* in very early life, and believe that in childhood it should be delayed as long as possible, until, at least, the tendency of varus is quite overcome. It may be asserted that, with ordinary care, in all cases, congenital varus should and can be cured without tenotomy; but not unfrequently the equinus remains more or less marked, and division of the tendo Achillis is sometimes required. This operation need not be performed, however, until the child is four or five years old at least, unless the heel is so much elevated as to cause marked lameness and difficulty in walking. After tenotomy we rectify the malposition at once. In all cases of congenital club-foot there is more or less atrophy of the limb; and consequently massage of the muscles of the leg from the birth is a most important factor in developing and increasing the usefulness of the member. There are no cases which do so well, if properly cared for, as club-foot, and scarcely any which, with neglect, turn out so badly. Especially is this the

Fig. 2. Fig. 3.

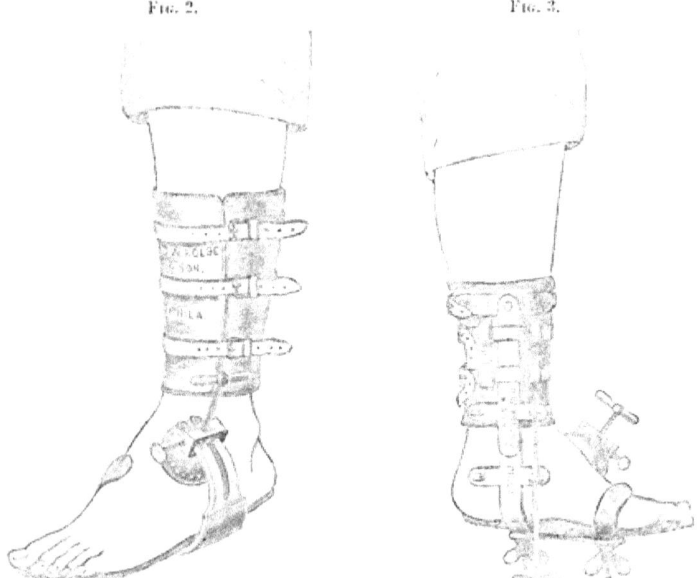

Apparatus for foot-stretching, devised by Mr. Louis Kolbe, Sr.

case after division of tendons, which, unfortunately, is so often practised at surgical clinics, where the operator, after dividing tendons, consigns the patient to the hands of the instrument-maker for a brace, and the patient is then intrusted to his parents or guardian, and not again seen, perhaps, for days or weeks. We have known, in not one case only, but many times, serious sloughs, erysipelas, and abscess as a result. The consequence is that, even with a slight excoriation, the shoe is left off for some days, or perhaps weeks, contraction takes place, and we have as a result a rigid, stiff foot, far worse than it was before the operation, and more difficult, if not impossible, in some cases, to rectify. This is more especially so if the tissues at the heel are cut too freely with the tendo Achillis. How frequent it is to have cases of club-foot return for advice in after-years with the deformity much worse than it was before the operation! Such cases are often the result of negligence on the part of the parents, but often from the want of proper instructions from the surgeon, while the cost of apparatus, and the necessary daily labor on the part of the friends, with a very natural aversion to giving the patient any suffering, all militate against the child's well-being.

When we come to consider the stiff, unyielding, rigid equino-varus of children who have in after-years a recurrence of the deformity after operation, it becomes a very serious question as to the propriety of operative treatment. When we examine such a case we find the feet are like stumps, with but very little motion in the tarsus, and but slight hinge-motion in the ankle. It is useless to think of tenotomy, for the rigidity is not due to tendon contraction or involvement, but to a general condensing of all the tissues about the joint. In such cases we apply flaxseed poultices day after day for weeks sometimes, until the skin and subcutaneous tissues are softened; then the poultice is only used at nights, and during the day the foot is subjected to a stretching process, the principal pressure being applied upon the tarsus. (See Figs. 2, 3, and 4.) This may be done several times a day, and must be kept up for months. No brace applied for the purpose of correcting deformity can or will be able to do *any* good in these cases of recurring deformity, for the apparatus only holds the foot in one position. The bones of the foot must be gradually forced into a normal position, and this can be better done by a "foot-stretcher," in addition, the occasional stretching and moulding with the hand, which also accomplishes very much. After a time it may be necessary to divide tense plantar fasciæ, the tendon of the anterior or posterior tibial, or the tendo Achillis; but at best, a club-foot which has been neglected after operation and become rigid with a recurrence of varus and equinus presents one of the most difficult cases to rectify in orthopædic surgery. Of course, the mal-nutrition of the parts, which is the result of diminished nerve-supply, and which always exists in this form of deformity, adds greatly to the embarrassments of the case. In adults with varus or equino varus, when no operation has ever been performed, a more favorable result may be expected; but in these cases, after the division of the tendons and while the patient is still under the influence of an anæsthetic, the foot should be thoroughly stretched; and this latter manipulation will generally be required to be kept up as a part of the regular

FIG. 4.

Apparatus for stretching, devised by Dr. Morton.

treatment for a long time afterwards. Occasionally the immense atrophy and distortion in neglected adult club-foot is so great, that amputation of the leg is preferable to any attempt at correction of the deformity, and two such cases have come under our observation. In one instance, amputation was performed; in the other, the removal of the foot was advised, but temporarily declined by the patient.

In most of these cases there is not only atrophy of the foot, but atrophy as regards the development of the thigh- and leg-bones, so that the shortening of the deformed limb may amount to an inch or more.

Contractions; Club-foot; Tenotomy.

No.	Admission.	Age	Sex	Cause, Disease, or Injury.	Tendons divided and other Treatment.	Result.	Discharged.
1	Jan. 29, 1870.	7	M.	Talipes equinus as a result of ulcer of the heel, following compound comminuted fracture of calcaneum.	Tendo Achillis; posterior splints.	Cured.	April 17, 1870.
2	June 22, 1875.	37	M.	Inflammation of the knee and ankle; contraction of flexor thigh tendons; contraction of tendo Achillis.	Tendo Achillis and internal hamstring tendons; foot stretched.	Improved	Oct. 25, 1875.
3	Feb. 17, 1876.	18	M.	Injury of right foot. Run over by a hand-car; slough followed and some bone discharged; talipes equinus acquired.	Tendo Achillis; plantar fascia; foot stretched.	Cured.	June 15, 1876.
4	Nov. 23, 1877.	36	F.	Congenital talipes equino-varus; foot rigid.	Tibialis anticus and posticus; tendo Achillis; foot stretched and at first placed in fracture-box, then a club-foot shoe.	Cured.	Feb. 20, 1878.
5	Oct. 25, 1877.	9	F.	Coxalgia; great contraction of tendons about the hip; anchylosis of joint; flexion of thigh and great adduction.	Adductor longus and gracilis; the adhesions broken up; sixth extension; brace applied limb, three-quarter inch shorter than the other.	Cured.	Jan. 20, 1878.
6	July 27, 1877.	10	F.	Fracture of femur, with contusions of the leg posteriorly, which sloughed, involving the tendo Achillis.	Tendo Achillis divided; posterior splint.	Cured.	Nov. 8, 1877.
7	March 8, 1878.	24	M.	Contraction of tendo Achillis from gunshot injury, 2 years ago. Ball passed through gastrocnemius.	Tendo Achillis divided; fracture-box and then posterior splint.	Cured.	May 2, 1878.
8	Oct. 1, 1878.	19	F.	Anchylosis of knee-joint, with extreme flexion at a right angle; atrophy of leg.	Biceps, semitendinosus and membranosus; Stromeyer splint applied.	Improved	Dec. 13, 1878.
9	Dec. 6, 1878.	16	M.	Congenital talipes equino-varus; 10 years ago an attempt was made to correct the deformity.	Tibialis anticus and plantar fascia, and tendo Achillis; club-foot shoe.	Cured.	May 15, 1879.
10	Oct. 28, 1878.	9	M.	Congenital talipes equino-varus, was operated on 4 years ago, but neglected treatment; deformity returned.	Plantar fasciæ and tendo Achillis; foot stretched.	Cured.	Dec. 27, 1878.
11	Jan. 19, 1879.	60	M.	Digital contractions; Dupuytren's disease; little finger only involved; for 20 years gradually contracting.	Palmar fascia; finger extended at once; passive motion.	Cured.	Feb. 22, 1879.

THOMAS G. MORTON.

STRICTURE AND URETHRAL ACCIDENTS.

UNDER the general heading of stricture are included, it will be noticed, those of the rectum, as well as the urethra. Of the former there were only two cases, one of these having a syphilitic origin and the other coming from an accident. The first one died, the other is now in good health and actively engaged as a nurse. There were 34 urethral strictures arising from gonorrhœa, 2 from excess in drinking, 1 accompanying enlarged prostate, 1 is put down as irritable, and of 4 there is no history.

13 were traumatic cases, most of them from falls received in such a way as to injure the perineum. These accidents are of a very serious character. Two of them were fatal. But one death arose among the cases having a venereal origin, unless the two from whom no history was obtained belonged to that class.

Most of the patients having ordinary strictures who are admitted into our hospital are seafaring men, and, having all the peculiarities of the class, are reckless as to themselves. They seek temporary relief because they are forced to do it, and then disappear. There is reason to believe that very few of them regard the instructions they have received, and consequently they get on the sick-list at almost every port where they may chance to stop. About half the number in our list are put down as "cured," but how long they remain so is a question.

It will be noticed that division of the stricture by any form of cutting instrument is rarely resorted to. The reliance is upon dilatation by graded instruments. The sizes are not recorded in the notes. After full relief has been obtained, the patient is directed to insert a bougie or sound himself every day or every other day for some time, and for a very long time afterwards to repeat the operation once or twice a week. This plan has given the greatest satisfaction, and, except in special cases, there is no reason to deviate from it. The hospital is not without its experience in most if not all of the various methods for treating stricture that have been and are recommended. From time to time these are tried upon appropriate cases, and we think a just estimate is placed upon their value, but nothing has yet arisen in the way of improvement as to induce us to abandon the present prevailing treatment.

The number of traumatic strictures and urethral injuries is extraordinary, there being thirteen of them in the fifty-five reported. This is by no means an index of the relation between the frequency of ordinary or gonorrhœal strictures and those arising from accident. Sometimes a case of the latter kind does not occur for a long while, and then they will follow each other rapidly. The writer remembers having at one time seven cases of rupture of the urethra under charge. In treating this severe accident, the surgeon considers both himself and the patient fortunate if immediate success follows the attempt to get a catheter into the bladder. If besides this there is infiltration, free incisions are at once made into the perineum.

While all are agreed that the catheter should be retained for a short time, say twenty-four or forty-eight hours, there is some difference of opinion among surgeons as to the necessity of keeping an instrument in the bladder during the whole progress to recovery. The fear of not being able to reintroduce the catheter no doubt has something to do with the continuous method.

Experience teaches us, however, that after the first two or three days have passed the instrument may be removed without much risk as to easy access afterwards, provided the manipulations are gentle, and the operation is performed by the same hand that originally found the way. The advantages of keeping the instrument out, and using it only at regular intervals for emptying the bladder, are so great, we believe, that the other course should only be followed in very exceptional cases. The evils of keeping the instrument in too long are more or less vesical irritation and cystitis, abscess along the urethral tract, and sometimes obstinate fistulæ. The original trouble is too apt to have this latter sequence, and we think it unwise to pursue a treatment that may add to the number of them. Again, through the curves of the instrument the secondary fistula is apt to occur at that most obstinate spot for remedial treatment, the peno-scrotal junction.

Strictures and Urethral Accidents.

No.	Age.	Sex.	Nature.	Cause.	Treatment.	Result.	Remarks.
1	30	M.	Stricture following gonorrhœa, 10 years since; was treated then *without injections.*	Gonorrhœa.	Symptoms were urgent on admission. Was etherized, and after much difficulty a catheter was introduced, and allowed to remain 5 days.	Relieved. Death.	Was discharged at own request after first relief. Two days after reapplied for admission, and died from peritonitis.
2	54	M.	Stricture of 25 years' standing.	Gonorrhœa.	Dilatation.	Relieved.	Bougies up to No. 13 passed easily. To continue treatment himself.
3	29	M.	Multiple stricture.	Gonorrhœa.	Dilatation.	Cured.	In hospital 15 days.
4	36	M.	Multiple stricture.	Gonorrhœa.	Dilatation.	Relieved. Eloped.	In hospital 4 days.
5	22	M.	In membranous portion.	Gonorrhœa.	Dilatation.	Cured.	In hospital 20 days.
6	26	M.	In membranous portion.	Gonorrhœa.	Dilatation.	Cured.	In hospital 48 days.
7	24	M.	Laceration of urethra.	Fell across a board, striking perinæum.	Catheter introduced with difficulty; retained all night; withdrawn, and used as required.	Cured.	In hospital 24 days.
8	26	M.	In membranous portion, 8 years' duration.	Gonorrhœa.	Dilatation.	Cured.	In hospital 35 days.
9	40	M.	Entered for retention of urine; stricture of 20 years' duration.	Gonorrhœa.	Catheter introduced with difficulty; opium suppositories, dilatation.	Cured.	In hospital 23 days.
10	54	M.	Admitted for retention; had old stricture.	Gonorrhœa.	Retention relieved by catheter.	Stationary as to stricture.	There was a urethral stone; removed 18 months before admission. In hospital 8 days.
11	52	M.	Inflammatory stricture.	From excess in drinking.	Catheter and belladonna suppositories.	Cured.	In hospital 1 week.
12	39	M.	In membranous portion, also near meatus.	Gonorrhœa.	Dilatation.	Cured.	In hospital 27 days.
13	21	M.	In spongy portion.	Gonorrhœa.	Dilatation.	Cured.	In hospital 26 days.
14	57	M.	In membranous portion.	Gonorrhœa.	Warm bath, rest, dilatation.	Improved.	In hospital 29 days.
15	53	M.	Traumatic stricture.	Fell across a beam, and struck perinæum violently; 3 weeks before admission.	There was great distention of bladder when admitted; relieved by flexible catheter, catheterism, baths, general treatment, etc.	Died.	No infiltration when admitted. Gradually abscesses formed, which were opened; one communicated with bladder. Patient first improved greatly, but gradually sank. In hospital 2 months, 14 days.
16	53	M.	Multiple stricture, many years' standing.	Gonorrhœa.	Dilatation.	Cured.	In hospital 3 months, 1 week.
17	24	M.	Multiple, with perineal abscess.	Gonorrhœa.	Clay dressing to perinæum.	Improved.	Refused operations for stricture.

Strictures and Urethral Accidents.—(Continued.)

No.	Age.	Sex.	Nature.	Cause.	Treatment.	Result.	Remarks.
18	73	M.	....	Accompanying enlarged prostate and cystitis.	Tonics, soda, buchu, dilatation, etc.	Improved.	In hospital 2 months.
19	25	M.	Rupture of urethra and fistula.	Five weeks before admission was caught between two cars, and severely injured about pelvic region.	Two operations were performed, the latter one plastic in character.	Improved.	In hospital 6 months.
20	32	M.	Rupture of urethra.	Fell astride of beam.	Free incisions, opium, etc.	Death from peritonitis.	There was a large swelling of groin from infiltration. In hospital 29 days.
21	25	M.	Irritable stricture, spasmodic (?).	....	While attempting to pass instrument patient urinated.	Relieved.	Admitted for a retention of 18 hours. A "nervous" case (?).
22	44	M.	Multiple stricture of urethra.	Gonorrhœa.	Dilatation.	Cured.	In hospital 65 days.
23	23	M.	Stricture in spongy portion.	Gonorrhœa.	Dilatation.	Cured.	In hospital 43 days.
24	18	M.	Traumatic (?).	From fall on deck of vessel.	Dilatation.	Improved.	In hospital 33 days.
25	36	M.	Traumatic (?).	From fall, while at sea.	Dilatation.	Improved.	In hospital 6 days.
26	26	M.	Inflammatory.	Had old stricture of 4 years' standing.	Bath, instrument, under ether.	Relieved.	In hospital 1 day.
27	59	M.	Traumatic.	Fall through a hatchway, struck perineum.	Catheterization, rest.	Cured.	In hospital 42 days.
28	27	M.	Traumatic.	Fell and struck bow of boat, then into water; 2 years before admission.	Catheter, first under ether, then dilatation.	Cured (?). To continue catheter.	In hospital 15 days. Patient had been several times in the hospital. There appears to be no difficulty if systematic catheterization is not neglected.
29	41	M.	....	Gonorrhœa.	Dilatation.	Cured.	In hospital 14 days.
30	56	M.	Traumatic.	Fell across plank; 9 years before admission.	Catheter and dilatation.	Improved.	In hospital 5 days.
31	38	M.	Inflammatory, on old existing stricture.	Gonorrhœa.	Catheter and dilatation.	Improved.	In hospital 5 days. Admitted for retention; 32 ounces of water drawn off at once.
32	45	M.	Chronic, 5 years' standing.	Gonorrhœa.	Dilatation.	Improved.	In hospital 8 days.
33	24	M.	In spongy portion.	Gonorrhœa.	Dilatation.	Cured.	In hospital 2 days.
34	27	M.	Acute inflammatory, with retention.	Gonorrhœa.	Catheter, etc., under rapid respiration.	Cured.	In hospital 10 days. Patient returned in 10 days for retention, and was again relieved.

35	28	M.	Multiple, of 12 years' standing.	Says was injured by horse falling on him; also has had gonorrhœa.	Dilatation, at first under ether.	Cured.	In hospital 13 days.
36	52	M.	Old stricture, with retention.	Gonorrhœa.	Retention first relieved, then catheterism continued twice daily.	Improved	In hospital 38 days.
37	37	M.	Old, with retention.	Gonorrhœa.	Dilatation.	Improved.	In hospital 52 days.
38	38	M.	Old.	Gonorrhœa.	Dilatation.	Improved	In hospital 48 days.
39	27	M.	Old.	Gonorrhœa	Dilatation.	Improved	In hospital 16 days.
40	50	M.	Old, traumatic	Fell upon perineum.	Bougies, rapid dilatation.	Cured	In hospital 15 days.
41	41	M.	Of rectum.	Abscess following accident, being caught between two horses.	Bougie, etc.	Improved.	In hospital 4 months, 21 days.
42	64	M.	Of urethra.	Gonorrhœa.	Rest, bougie.	Improved.	In hospital 20 days.
43	30	M.	Traumatic of urethra	Fell astraddle of an iron bar; 4½ months before admission.	Rest, dilatation.	Much improved.	In hospital 28 days. Patient had been operated on in Louisville and Cincinnati.
44	42	M.	Of urethra, old.	Gonorrhœa.	Dilatation.	Cured.	In hospital 6 days.
45	27	M.	Of urethra.	Gonorrhœa.	Dilatation.	Cured.	In hospital 19 days.
46	36	M.	Of urethra.	Gonorrhœa.	Warm bath, poultice, etc.	Relieved.	In hospital 1 day.
47	80	M.	Retention and incontinence.	No history	Catheter, etc.	Relieved.	In hospital 16 days.
48	23	M.	Retention.	From intoxication	Suppository of opium and rest.	Cured.	In hospital 1 day.
49	60	M.	Old stricture, with retention.	No history obtained	Catheter, rest, etc.	Died	In hospital 4 days. No autopsy obtained.
50	30	M.	Incontinence and stricture	Gonorrhœa.	Dilatation.	Relieved of incontinence. Stricture cured.	In hospital 3 months.
51	52	M.	Stricture of urethra.	Gonorrhœa.	Dilatation under ether.	Improved.	In hospital 14 days.
52	60	M.	Retention.	No history.	Catheter and palliatives.	Died	In hospital 9 days.
53	...	M.	Stricture of urethra and prolapse of rectum.	Gonorrhœa.	Catheter, prolapse reduced.	Cured	In hospital 12 days.
54	25	M.	Rupture of urethra.	Fell across a peg projecting from a log.	Catheterism, rest.	Cured.	In hospital 17 days.
55	28	F.	Laceration of urethra.	Fell from stepladder on a stick.	Rest, morphia, catheterism every three hours at first.	Cured.	In hospital 11 days.
56	24	F.	Of rectum.	Syphilis.	Dilatation and constitutional treatment. Operation by dividing stricture.	Died.	In hospital 8 months, 21 days.
57	32	M.	Of urethra. Came in for retention.	....	Rapid dilatation after relieving bladder.	Cured.	In hospital 2 days.

TETANUS.

THERE are 26 cases of this terrible disease recorded. All of these were traumatic in origin except one. All the patients were males except one. The far greater exposure of males to accident has something to do with this, but females are not without their exposures, and furnish many cases of lacerated wounds and burns to our surgical wards. If excess of nervous excitability has anything to do with the production of tetanus, women ought to have it more frequently than men in proportion to the number injured. It appears, however, by our list that they are remarkably exempt from the disease. 20 of the cases died and 6 recovered. The saying, therefore, "so many cases, so many deaths," does not hold good by any means. In fact, there is every reason to be encouraged by the results set forth in the table.

The tetanic symptoms appeared, on an average, on the tenth day after the injury was received. The fatal cases lived about four days after the disease began. When patients recover they are very slow in doing so, which is a well-known characteristic of tetanus. The average time in the hospital of five of those who got well was ninety-three days.

We have nothing new whatever to offer as to the pathology of the disease. Autopsies have revealed nothing to account for the extreme severity of the symptoms, or to explain the fatality of tetanus. As to treatment, however, there is much encouragement. Of the six cases which recovered, five were kept under chloral in large doses, and one got well under the use of conia. Five cases that were treated with chloral died. The other plans of treatment are in the table. Among the most essential elements to success are isolation, dark rooms, and silence. Nervous excitability is so great that the most trivial cause will serve to bring on violent spasms.

For this reason opening and shutting of doors and windows should be avoided as much as possible, and the patient should be kept out of all draughts of air. Nourishing food should be given at stated intervals, and in such a way as to cause the least possible disturbance.

Tetanus.

No.	Age.	Sex.	Nature.	Cause.	Treatment.	Result.	Remarks.
1	28	F.	Traumatic.	Followed operation for ovarian tumor.	Palliative.	Died.	(See Ovarian Tumor.)
2	24	M.	Traumatic.	Followed lacerated wounds of scalp and knee and contusions.	Morphia p. r. n., conium.	Died.	Tetanic symptoms on 8th day; died on 2d day after.
3	31	M.	Traumatic.	Followed having arm torn off at shoulder by belt.	Conia.	Died.	Tetanus on 6th day; lived 2 days after.
4	15	M.	Traumatic.	Followed railroad crush of right foot.	Morphia, chloral in large doses.	Died.	Tetanus on 11th day; lived 8 days after.
5	...	M.	Traumatic.	Followed burns. Case 45 of Burns.	Conia, atropia, quinine, etc.	Died.	Symptoms of tetanus on 17th day; patient lived 1 week afterwards.
6	19	M.	Traumatic.	Followed amputation for severe burn. Case 48 of Burns.	Atropia, etc.	Died.	Died on 3d day after tetanic symptoms set in; 11th day after amputation.
7	16	M.	Apparently idiopathic. Medical ward.	....	Chloral in large doses bromide of potassium, belladonna, stimulants, etc.	Cured.	In hospital 87 days. Patient bruised his knee slightly, but this was after he had already had some spasmodic pains, and did not appear to be the cause of the disease.
8	10	M.	Traumatic.	Followed fracture of femur and lacerations of forearm and thigh.	Atropia, etc.	Died.	Tetanus on the 16th day after injury; lived 6 days from that time.
9	15	M.	Traumatic.	Followed lacerated wound of hand.	Opium, Calabar bean, etc.	Died.	Tetanus first on 9th day after admission; died on 8th day from that time.
10	19	M.	Traumatic.	Followed lacerated gunshot wound of wrist.	Morphia, chloral continuously, in doses from 10 to 20 grains every two hours.	Cured.	Tetanus on the 16th day from injury. The symptoms were severe for about 3 weeks, and gradually subsided. In hospital 111 days.
11	17	M.	Traumatic.	Followed railroad crush of left thigh and laceration of right knee.	Opium, etc.	Died.	Tetanus on 6th day; died on 2d day of disease.
12	5	M.	Traumatic.	Burn.	Chloral, etc.	Died.	Case 65 of Burns. Tetanus on 10th day. Died 4 days after.
13	7	M.	Traumatic.	Followed compound fracture of arm and laceration of thigh.	Bromide of potassium, stimulants, etc.	Died.	Died on 4th day from injury.

Tetanus.—(Continued.)

No.	Age.	Sex.	Nature.	Cause.	Treatment.	Result.	Remarks.
14	35	M.	Traumatic.	Followed compound fracture of elbow and dislocation of left ankle.	Chloral, etc.	Died.	Elbow-joint incised on 5th day after admission. Too great shock for immediate operation. Tetanus appeared 30 hours after operation, and death occurred on 2d day following.
15	12	M.	Traumatic.	Followed crush of right foot, from being caught between elevator and floor.	Pirogoff's amputation performed at once. Chloral, etc., morphia, for tetanus.	Died.	Tetanus on 9th day; death 3 days after.
16	45	M.	Traumatic.	Followed a crush of second finger of right hand; 10 days before admission.	Chloral, gr. x, every 2 hours; finger dressed with ext. belladonna. Chloral reduced as symptoms improved; then increased to xv grs. every 3 hours; then p. r. n.	Cured.	Patient was admitted for the tetanus. In hospital 57 days.
17	27	M.	Traumatic.	Followed leg being caught between bumpers of cars; contusions; no fracture; gangrene.	Opium, hypodermics of atropia, stimulants, etc.	Died.	In hospital 28 days. Tetanus began on 27th day; died next day.
18	11	M.	Traumatic.	Followed crush of feet, from being run over on railway.	Opium, atropia, etc.	Died.	In hospital 6 days. Tetanus began on 2d day.
19	22	M.	Traumatic.	Followed amputation of leg for necrosis of ankle-joint.	Atropia, etc.	Died.	In hospital 11 days. Tetanus began on 7th day after operation; died 3d day after.
20	21	M.	Traumatic.	From burns.	Chloral, etc.	Died.	Tetanus on 19th day; died next day. Case 118 of Burns.
21	20	M.	Traumatic.	Followed gunshot wound of axilla.	Chloral, etc.	Cured.	In hospital 52 days. Tetanus on 16th day; under chloral treatment 18 days.
22	4	M.	Traumatic.	Followed railroad crush of foot and thigh.	Opiates, etc.	Died.	In hospital 2 days. Tetanus on 2d day; died in few hours.
23	21	M.	Traumatic.	Followed laceration of foot.	Chloral, etc.	Cured.	In hospital 6 months.
24	13	M.	Traumatic.	Followed gunshot wound in palm of left hand; 24 hours before admission.	Hypodermics of morphia, etc.	Died.	Tetanus on 10th day; died 4 days after. In hospital 14 days.
25		M.	Traumatic.	Followed being caught and twisted by the spokes of a wheel in motion. Great injury to knee-joint.	Conia, gr. 1/4, every two hours.	Cured.	
26	35	M.	Traumatic.	Followed excision of elbow-joint.	Morphia and chloral.	Died.	Tetanus on 5th day; died on 8th day from injury.

WILLIAM HUNT.

FRACTURES.

1.—*The Treatment of Fractures, with a Description of the Forms of Apparatus employed and the Methods of their Application.*
2.—*Statistics of Fractures.*

1.—THE TREATMENT OF FRACTURES.

WHEN a patient is received into the accident ward, suffering with a recent injury involving the bone, his treatment is guided by certain principles, the importance of which is well understood by the resident surgeons and the attendants. In the first place, we insist that the limb shall never be roughly handled; all manipulations must be made with extreme care and gentleness.

SHOCK.—If the fracture be of sufficient gravity to cause surgical shock, the first efforts are directed to the relief of this condition. Hot-water cans and blankets are applied, the foot of the bed is elevated, and beef-tea, quinine, coffee, and stimulants are administered internally. Great care is taken to avoid overdosing the patient with alcohol, especially as many of those admitted have already been given some stimulant of this kind at the place where the accident has occurred. Again, many patients have received the injury while in an intoxicated condition, and are perhaps suffering the depression caused by the overuse of alcohol. Hence small repeated doses of nutriment, combined with quinine and coffee, with but little alcohol, are looked upon as being the preferable means of producing reaction. Recently Dr. Levis has suggested the use of hot-water enemata in cases of shock. If there be any hemorrhage, it is controlled at once by ligatures or pressure.

EXAMINATION.—As soon as it is deemed proper to disturb the patient a thorough examination of the part is made. In order to spare the patient the pain and nervous exhaustion which might otherwise ensue, as well as to overcome muscular spasm, it is the custom in severe cases to administer ether during the primary examination, as well as during the subsequent setting of the fracture. In this manner we escape the embarrassment of resistance by the patient, whose struggles and complaints would greatly interfere with the necessary manipulations; and, moreover, the muscular contraction having been overcome by the anæsthetic, the surgeon is enabled to arrive at a clear and definite diagnosis that would otherwise be impossible in obscure cases. This expedient is particularly available in the case of children. Should the injury be in the neighborhood of a large joint, such as the hip, knee, shoulder, or elbow, an anæsthetic is often of the utmost assistance; in fact, it is sometimes absolutely necessary in order to arrive at anything like a correct understanding of the case.

Ether is the anæsthetic used, while chloroform is never employed, since the surgeons look upon its administration as fraught with too much danger. Occasionally nitrous oxide has been given, but it is now seldom used, as it is less convenient than ether. Recently bromide of ethyl has been used as an anæsthetic by Dr. Levis with favorable results, and further experiments are being made to determine its comparative value.

When examining the injured part, there should be no hasty or careless handling; each step of the inquiry should be conducted with care, system, and deliberation. Rough manipulations may not only injure nerves and blood-vessels, but may also provoke cellular inflammation. When a patient is brought into the hospital in a dirty condition, either from the accident or from the ordinary occupation of the individual, he is given a bath, or carefully cleansed by the nurse, *before* the fracture is dressed, so that the patient shall be disturbed as little as possible after the surgeon has examined and set the fractured bone.

In the after-treatment of these injuries, we insist upon two principles as essential in order to obtain the best results: first, the great importance of frequent inspection of the injured part; and, secondly, simplicity of apparatus. It is scarcely necessary, at the present day, to do more than to barely allude to the value of a daily inspection of a fracture in order to promptly correct any tendency to deformity, or to early detect the occurrence of phlegmonous inflammation, on the one hand, or defective circulation upon the other, which might be produced by a badly-fitting apparatus, or other and less obvious causes. There appear to be a few practitioners, however, who believe that all a fractured bone requires is to be set in position, and placed in a permanent apparatus; and at the end of a certain time good union and a perfect result may be looked for. On the contrary, experience has demonstrated in the cases under our observation that the old-established rule of frequent inspection, occasional washings, and perhaps carefully rubbing the limb with bathing whiskey, soap liniment, or some form of stimulating embrocation, add materially to the chances of rapid and perfect consolidation of the bone, and the absorption of the products of inflammation and the provisional callus. We ask, "What is more likely to favor non-union of a fracture than an enfeebled circulation?" But this is exactly what may be expected from the use of fixed dressings, under which the skin becomes dry and inactive, the capillary circulation sluggish, and the general nutrition of the part decidedly reduced. We can only point to the remarkable immunity from the occurrence of ununited fracture in our wards which we have enjoyed, more remarkable still if we consider the great number of fractures that have been under treatment. During the sixteen years of Dr. Morton's connection with the Pennsylvania Hospital no less than 6500 cases of recent fractures have been received, among which there have been but few cases of ununited fracture; and only occasionally has delay in union occurred.

TIME OF DRESSING.—As a rule, fractures are dressed immediately, but when the injury to soft parts is very great it is sometimes deemed advisable to wait a few days, until swelling and other inflammatory symptoms have subsided, before attempting accurate adjustment with bandages and splints. When the great amount of swelling endangers the safety of the limb by arresting circulation, free incisions are made through the tense integument to permit the fluids to drain away, and especially to relieve the pressure upon vessels and nerves. Much stress is laid upon this measure, as threatening gangrene has been averted by these free cutaneous incisions.

COMPOUND FRACTURES.—In these injuries we endeavor to convert the fracture into a simple one by closing the wound at once. If there is protrusion, the ends of the fragments are reduced by manipulation and extension, by enlarging the wound, or by tenotomy. When suppuration occurs, the limb is usually placed in some form of fracture-box, and surrounded with bran, which may be rendered disinfectant by the admixture of some carbolized powder. Occasionally resection of the fragments has been done when it has been found impossible to replace the projecting pieces of bone. The contiguous ends are sometimes then united by wire.

APPARATUS.—In regard to apparatus, simplicity is the first requirement. A good surgeon can obtain satisfactory results from improvised splints, while a poor one would fail even with a costly apparatus; in other words, ingenious and complicated appliances can only to a limited extent make up for the deficiencies of the attendant. A good surgeon does not need them; a poor one cannot use them so as to obtain a perfect result. After the diagnosis of the fracture is established,—and this is all-important,—the treatment is usually simple, and any surgeon ought to be able to improvise suitable splints in an emergency; indeed, as we shall subsequently point out, very few splints are really required, and these need only be of a simple and inexpensive character.

DRESSING.—In dressing fractures, therefore, all complicated forms of apparatus are rejected. Indeed, of late there has been a decided tendency to the greatest simplicity. In many instances of fracture of the upper extremity, splints, properly so called, have been discarded entirely, and the postural method insisted upon. In other words, the limb is put in a position which relaxes all displacing muscles, and is retained there by roller bandages or adhesive strips. For fractures of the femur, continuous extension, and for those of the tibia and fibula, the hinged fracture-box, are almost invariably the means adopted. Ordinary picked oakum is always used for padding.

The well-known fracture-bed, with a movable centre-piece in the mattress to allow the introduction of a vessel to receive the alvine discharges, is often used, but is not believed essential. Its use is objectionable, because it is found difficult to get the attendants to keep the receptacles always clean. The bed-pan can be employed without any risk of interfering with repair even in fractures of the thigh. When the wards are cleaned, or the patients taken to the clinical lecture-room, the bed-carriage devised by Dr. Morton in 1872 is brought into requisition. This apparatus can be pushed under the bed, which is then raised from the floor by a few turns of a crank. By this means patients can be transported from place to place without jarring or injury.

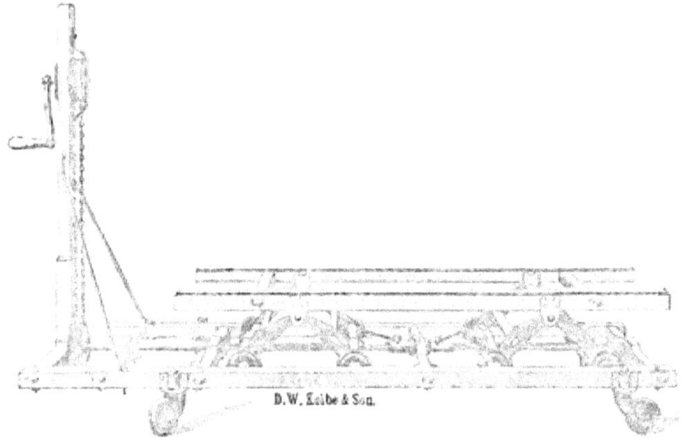

The carriage consists of a double truck, the upper one elevated by a series of cams which run upon a narrow iron track; a long right and left screw, worked by an endless chain upon a crank at one end, raises and depresses the cams. It is made of oak, and the hinges of brass; the wheels are supported by steel pins.

When the diagnosis of the fracture has been satisfactorily made, and the limb has been properly cleaned, an effort should be made to place the broken bone in as good a position as possible; in other words, to "set the fracture," the fragments being adjusted as nearly as possible in their normal condition. When the line of the fracture is transverse, this can be easily done, but most frequently the line is oblique, when it is difficult, and often impossible. We may, in the latter case, under the relaxing influence of the anæsthetic, completely reduce the deformity, but when the muscles again resume their functions the overlapping will recur, so that the setting of a fracture may require many days. But with careful manipulation and constant attention, the muscles being kept at rest by proper restraint, such as we get by the use of appropriate splints, or by extension, the tendency to deformity will eventually be overcome and a good position maintained. A good splint and bandage are of great assistance in keeping the muscles quiet and in a relaxed condition, so that in a few days they may be considered as partly paralyzed, and then offer much less resistance to any necessary manipulations.

Immediately after the occurrence of a fracture more or less swelling may be expected, produced either by the injury, or by the ends of the bone lacerating the soft parts, or else by the contusion from direct injury to the tissues received at the time of the accident. The swelling is most marked when a joint is involved in the fracture, and in this case it rapidly occurs, making satisfactory examination impossible. We are, therefore, content to let the patient rest for a short time, and in case of thigh, leg, patella, ankle, and foot fractures we simply apply, for the first few days, an evaporating and anodyne lotion (lead-water and laudanum generally), using for the thigh lateral supports by sand-bags, without extension, for the patella and knee a posterior straight splint, and for the leg, ankle, and foot the fracture-box. In some cases, however, the extension should be applied to fractured thighs immediately, because the overlapping fragments and the spasmodically contracting muscles are causes of pain and increased inflammatory action. These conditions can be overcome more quickly by powerful extension than by any other means, hence a greater amount of weight should be applied at first than at any subsequent period. In the course of a few days the swelling largely disappears, and a satisfactory diagnosis can be made. The details of treatment will be considered later. We never apply an encircling or tight roller bandage upon a fracture until the swelling and œdema begin to subside. When this has markedly commenced, the absorption of the remainder of the effusion may be materially assisted by the application of the uniform pressure of a bandage. Nothing is gained, but much risk is incurred, by attempting to keep a fracture in rigid apposition immediately after the injury by means of roller bandages. It is then impossible to know the amount of the swelling, and a tight bandage of itself often produces irritation, and thus adds serious complications to what might otherwise have been a trivial injury.

We well remember the history of a case of simple fracture of the leg, produced by the fall of a mass of coal, where the soft parts were apparently not much involved. In this case a firm roller bandage had been applied, and lateral splints firmly bound upon the leg. In this condition the patient was put upon the cars to be sent to this hospital from the coal regions. When he arrived, on the following day, the limb was already gangrenous, from the interruption to the circulation caused by the swelling and the firm bandage. Fortunately such cases are not common, yet they are sufficiently frequent in their occurrence to warrant these words of caution. Should it be deemed desirable to apply a retaining bandage around the limb immediately after the fracture, it should be done with great caution; and, if there should occur the slightest numbness or coldness of the limb, or blueness of the parts below, the bandage must be cut. When it is remembered that there is no necessity for immediate reduction of a fracture, since union does not begin for some time, during which the bone may be set, we see that there is no need for undue haste in adjusting the fragments. At the same time, we do not wish to be understood as implying that all proper means should not be taken to reduce a fracture at once, with the aid of ether, where necessary; but we would insist that only moderate extension and compression should be exerted in keeping a fracture in position during the first few days.

FIXED DRESSINGS.—The application of dressings of plaster, silicate of sodium, and such materials is never resorted to in the early stages of the treatment, but is frequently employed later in fractures of the leg and thigh. We are opposed to what are called "fixed dressings," but often use hardening materials to make circular dressings, which are subsequently split open and reapplied. The immovable or fixed dressing should seldom be employed, except in the late stage of fractures, while the patient is still under strict surgical observation, or in special cases of delayed union. Their use in fractures of the upper extremity is exceptional, because the patient can be on his feet and take exercise with the ordinary dressings for such injuries. In cases of broken femur or tibia, however, the patient need not be kept in bed more than five or six weeks, if some form of permanent dressing is then applied. Silicate of sodium is the hardening agent generally employed. It is cleanly, dries with moderate rapidity, and is always available at once. Plaster of Paris is rather dirty in its application, but is considered desirable where rapid solidification is required; hence in cases of delirium tremens occurring in the course of the treatment, it has been used, and is considered of value.

At one time many patients' limbs were encased in a fixed dressing made of glue, to which powdered oxide of zinc had been added to hasten desiccation. The method of application as employed by Dr. Levis, who has adopted this dressing in many instances, is as follows:

Several pieces of flannel—old blankets or worn-out underclothing answering the purpose admirably—are selected and cut the requisite size. One of these is laid around the limb, and the two edges are tightly stitched together along the anterior surface, allowing the edge to project above the seam; then the melted glue, mixed with oxide of zinc, is painted on with a brush. The dressing may be strengthened by an additional layer of flannel or blanket, saturated with the glue and oxide of zinc, and made to adhere to the underlying layer. A third or even a fourth layer may be thus applied, if it is deemed necessary, and the limb supported until the dressing dries, which requires from four to eight hours.

The stitches of the seam on the front of the limb having been cut with scissors, the edges of this elastic case are sprung apart, and the dressing removed. The edges are then trimmed smooth, and a number of eyelets inserted, in order that the case may be laced like a shoe, and the degree of pressure regulated.

This fixed fracture-apparatus is exceedingly light, is made from materials almost everywhere obtainable, and is much cheaper than the silicate dressing. There are at all times pieces of waste flannel or cloth about a large hospital which can be appropriated, while the glue and zinc could probably be supplied at a very low price per pound,—a quantity sufficient for the manufacture of many such splints. Another advantage is its elasticity, which permits its removal without endangering the splint, for it can be pulled apart, and immediately springs into place around the limb to which it has been moulded.

By a little care and dexterity in stitching on the layers of flannel, the surgeon can readily shape the dressing so that both the leg and the foot are completely encased. Old stockings make an excellent basis for such splints.

This dressing, though still recommended, is not employed as much as the silicate, probably because the latter is always ready for use as it comes from the chemist's, and requires no heating.

In any of these dressings a hole can be cut opposite the wound in instances of compound fracture, and frequent change of dressing thus be obviated.

Passive Motion.—As a rule passive motion is not insisted upon at as early a period as frequently recommended. If there is no direct inflammatory involvement of the contiguous joints early passive motion will not be required, as stiffness can hardly occur within a period of three or four weeks. On the other hand, if the joint is implicated in the fracture, some degree of anchylosis will probably occur, but it will be rather increased than not by passive motion in the early stages. This serves to enhance inflammatory processes at a time when every effort must be to reduce the activity of the pathological phenomena. It is better to keep the joint at rest at this time, and trust to strong passive manipulation at a later period, when there is less likelihood of awakening severe inflammation.

Thus much for the general subject of treating fractures. It will now be proper to take up in turn the different regions of the body, and discuss the methods of dealing with fractures in each locality. The injuries of this nature that occur in the bones of the trunk will first be considered, then those of the head and face, afterwards those of the upper, and finally those of the lower extremity. The records of the hospital will furnish illustrative cases when necessary.

Fractures of the Spinal Column.—Fractures of the vertebral column are the most unsatisfactory fractures treated in the hospital, for in many instances it is impossible to make a correct diagnosis, and, even when the nature of the injury is patent, the treatment is almost uniformly unsuccessful. As a rule the paralyzed patient dies exhausted by numerous bed-sores, though occasionally a man has survived and has been discharged, after the lapse of many months, with incurable paralysis of the lower extremities.

The number of cases of fracture of the spine treated appears from the records to be small, but it is to be remembered that many injuries which appear as concussion of the spine, etc., would probably show fracture of the vertebræ if an autopsy was obtained. The following is an example:

Fracture and dislocation of the spinal column, with injury to the cord.—A German was admitted October 18, 1875, who, two days previously, had fallen backwards from a fence, on which he had been sitting at a distance of three or four feet from the ground. On admission there was complete paralysis of motion in the legs; the right arm could be moved a little, but the left was powerless. The head could be turned only slightly to the right. There was no sensation below the level of the nipples, and none of the arms below the deltoid region. The patient was perfectly conscious and rational, and suffered with retention of urine and priapism. The diaphragm moved freely. Gradually difficulty of respiration appeared, and he died on the third day. No fracture had been determined by any marked deformity, and the fall which he had sustained hardly seemed sufficient to cause so severe an injury, yet the symptoms pointed in that direction, and the autopsy proved the existence of grave structural changes.

The fifth cervical vertebra was separated from the cartilage on its under surface, and had been displaced forwards on the sixth vertebra half the width of its body. All the ligaments were torn except a part of the anterior common ligament and those of the transverse processes, while the vertebræ themselves appeared uninjured at this point. The spinous processes of the first and second dorsal vertebræ were fractured at a point about half-way in their length.

The condition of the cord was thus described by the pathologist of the hospital: "The spinal cord makes a bend, and is compressed at the point of dislocation. The dura mater, as well as the ruptured cartilage and ligaments of the vertebra, is stained red. . . . The dura mater is reddened also at the part situated in the body of the second dorsal. At both of these points the dura was slightly distended. At the upper point it is somewhat broadened and flattened by pressure, whilst at the lower one it is rounded. . . . No adhesions were found between it [dura mater] and the arachnoid. The arachnoid was normal and transparent, showing no evidence of hemorrhage and no adhesion to the pia mater. At the seat of the injuries the pia mater and cord show a darker color, and are slightly enlarged. The cord is very much softened, giving a feeling of fluctuation, but on incision it is not found diffluent, although the nervous tissue wells up through the cut surface of the membrane. The softening is more marked at the upper injury. The length of the cord between these two places is decidedly less firm than normal, but otherwise shows no marked change. Microscopically, the nervous tissue taken from the softened cord shows abundance of blood-coloring matter, with numerous broken tubules, quantities of granular detritus, and a very few large granular corpuscles." *

This history is of interest, for the reason that the injury was the result of a slight fall, and that dislocation took place without the occurrence of fracture at the same point. In addition, the fracture of the dorsal spinous processes did not present any deformity recognized before the autopsy. All of these facts render the case an instructive one.

The following case attracted considerable attention during the progress of the treatment, on account of the difficulty of locating the fracture and the very excellent recovery which occurred:

Fracture in the cervical region, with almost perfect recovery.—James G., aged 34 years, fell from a scaffold, striking his head and shoulders. On admission, June 14, 1876, into Dr. Morton's ward, there were evidences of severe shock, and there

* *Trans. Pathological Society*, vol. viii. p. 3.

was found complete paralysis of all the extremities. It was expected that death would necessarily ensue shortly. The diagnosis, made after examination, was fracture of the third cervical vertebra. He was treated in accordance with the general rules of the hospital, by dry cups, tonics, etc. For a time it was necessary to withdraw the urine with the catheter, but in a few days he was able to urinate voluntarily. In the course of three or four weeks he had regained considerable power in his hands and arms. After remaining in the wards until November 9, and having been treated by electricity, he was discharged to the Orthopædic Hospital. At this time the notes say there was some rigidity of the muscles of the forearm,—that they obeyed his will slowly and in a labored manner. Some loss of co-ordination was also believed to exist.

At this hospital he was treated with iodide of potassium, electricity, and massage, having still at this time partial paralysis of both legs and arms. The muscles responded very well to the current, and there was no anæsthesia or hyperæsthesia. The injury was here regarded as having involved the spinal column in the region of the fourth to sixth cervical vertebræ. In May, 1877, he was readmitted to the Pennsylvania Hospital, and placed in the medical ward.

Here he was treated with tonics and strychnia, and by November he had materially improved. The neck could be more freely moved, the arms were less rigid, and he went up and down stairs with greater facility. The grasp of the right hand was feebler than the left, and he raised his right leg with difficulty. The case was regarded at this time as a hemiplegia rather than a paraplegia, and ascribed to pressure on the cord in the cervical region.

About this time iodide of potassium was given in doses of twenty grains three times daily, and blisters applied locally. The improvement continued, so that he obtained pretty good use of his arms, had a good grasp, and was able to write with a pencil. He was able to stand on either leg. He was discharged in February, 1878, and the record made that he was cured of the paralysis for which he had been admitted into the medical ward.

In April, 1878, he was seen almost entirely well of the paralytic condition, though the neck was a little stiff. There still remained a slight impairment of power in the fingers. In January, 1880, he was working as porter in a store.

Another case of fracture of the spine deserves special mention, because of the circumstance that it occurred in a spinal column that had been the seat of previous inflammatory changes, causing anchylosis. The patient's friends knew nothing of this previous affection, and he himself was not in a condition to give any information about it:

Fracture of the spine in a case of previous disease of the vertebræ: separation of the fragments the probable cause of death.—A man, aged 50, colored, admitted November 2, 1873, had fallen down-stairs at 3 A.M. of that day. He lay some hours, unable to move or call for help; but was brought to the ward at 6 A.M., unable to articulate, and with arms and legs paralyzed. The neck was prominent in front, and a depression was felt at about the seventh cervical spine. In the afternoon he could talk without difficulty; there was paralysis of motion and sensation of the left hand and forearm, but he had regained motion and sensation over the rest of the body. He could place either hand on his head, but moved them hesitatingly. The head moved a little to the left side, but not to the right. Complained of discomfort, but no severe pain. He required the use of the catheter. He was found dead at midnight, having continued in the same condition until that time. There was some evidence which led to the belief that a sudden movement of the body caused a dislocation of the fracture. There was no history obtainable of previous disease affecting the neck or throat, and no evidence of syphilitic infection.

The specimen was examined by Dr. Longstreth, the pathologist, and shows disease of nearly all the cervical vertebræ, most markedly on their bodies and articular processes. The axis and the third vertebra are fused into one mass, showing their parts, however, distinctly. The anchylosis affects not only the bodies at the intervertebral disks, and the articulation at the processes, but also the back parts of their arches at the base of the spinous processes.

The lower part of the third vertebra has the lower part of its body, which when normal projects downwards and forwards to meet the fourth, very much spread out, and shelved forwards more than a half inch, and this part is very uneven and jagged with small spiculæ of bone. It would seem as though part of its body had been destroyed by caries. The junction of the third and fourth vertebræ gives the appearance on their anterior surface of a backward dislocation of the former, but the interior of the spinal canal disproves the occurrence of this condition.

The body of the fourth vertebra is increased in size in all directions.

The fifth and sixth vertebræ are firmly anchylosed throughout, showing as the only openings between them the two intervertebral foramina, and a small fissure at the base of their spinous processes.

The seventh vertebra has its body distorted, its anterior surface roughened and irregular, the edges projecting and jagged. The intervertebral cartilage between it and the first dorsal is normal, but the first dorsal has suffered to the same degree as the seventh cervical; both of them except in their bodies appear nearly normal.

The spinous processes of the sixth and seventh cervical vertebræ were widely separated by the fracture, so that the membranes of the cord were exposed at the bottom of the fissure.

The examination of the spinal cord was very imperfectly carried out, it being impossible to remove it except in fragments, without destroying the bones as a specimen, owing to the firmer connection of the dura mater in the spinal canal than usual.

The dura mater appeared, as seen through the opening between the separated and the fractured vertebræ, uninjured, but having a red color from blood-staining. The cord was softened, and there was considerable hemorrhage at the seat of injury.

The fracture and the probably subsequent dislocation were, as stated above, at the junction of the sixth and seventh cervical vertebræ.

These three cases show the difficulty surrounding cases of supposed spinal fracture in regard to diagnosis; hence, as may be supposed, the treatment is always expectant.

The patients are occasionally subjected to cupping, either as a counter-irritant or as a depletant, when first admitted. If any deformity is detected, an attempt at replacement is made, but no severe manipulation or operation, such as trephining, has been attempted, as far as we are cognizant of the facts. The great liability to bed-sores renders it imperative to place the patient upon an air- or water-bed. The former is preferable because less heavy, and not so troublesome when it begins to wear out and to leak. The wards have, on one or two occasions, been almost deluged by the rupturing of a water-bed. All points of the patient's body or limbs that sustain pressure are washed, and carefully bathed with whiskey at frequent intervals. When bed-sores have actually appeared they are treated on general principles. The urine is withdrawn by means of the catheter during the period of retention; and even when incontinence occurs, it is well to introduce the instrument occasionally to give exit to any residuary urine that may not be expelled. Sometimes the bladder is irrigated.

If the injured man survive the primary results of the fracture, iodide of potassium and mercury are administered to prevent or relieve the paralysis due to mischief done to the spinal marrow either primarily or secondarily. Electricity has at times been employed, but without giving any very satisfactory results.

Instances of recognized fracture, unless, perhaps, of a spinous process, have seldom been discharged cured. The list of cases terminates monotonously with the words "died," "incurable," "removed by friends." It must not be inferred that sufficient time is not given for cure to be effected, because patients have repeatedly been kept during a period of many months.

During a period of forty-four years, according to the statistics given by Dr. Agnew, seventy-three instances of fractured spine were treated; of these, forty-seven are known to have died. Many others doubtless succumbed after being removed by their friends.

Fractures of the Pelvis.—Fractures of this portion of the skeleton are not very common, and even when seen present very few points of interest as far as treatment is concerned. If the patient has been crushed in such a way as to sustain a bad fracture of the pelvic bones, the viscera are probably injured, and he soon dies of shock or peritonitis. If the injury consists of a mere splitting off of the crest of the ilium or of one of the spinous processes, very little treatment is demanded.

Occasionally the acetabulum suffers fracture of the rim or of the bottom of its cup-shaped cavity from the femur being forcibly driven against the socket. Such cases are treated by extension. Quite recently the following interesting example came under the care of Dr. Morton:[a]

Fracture of the lower rim of the right acetabulum; subsequent spontaneous luxation of the head of the femur into the sciatic notch; reduction; recovery.—John G., aged 46, a miner, was admitted September 5, 1879. While loading a coal-car in a mine near Hazleton, Pennsylvania, a mass of slate, weighing about 3 tons, broke away and fell upon his right side, throwing him down upon the left side, and producing serious contusions and wounds.

On September 6 he was brought to the hospital. The thigh and leg were flexed, and the knee presented to the middle of the left thigh; there was some fulness of the buttock, and great pain with every motion. The appearance of luxation on the dorsum ilii was marked, but with considerable effort the man was able to straighten the limb, finally to stand and bear his entire weight upon it, and at the same time to elevate the other, showing that the deformity was purely muscular; a few steps were also made, but this gave great pain. On examination, in a horizontal position, under ether, it was found that when the limb was extended the head of the femur could be drawn downwards to an unusual extent, showing that the ligamentum teres was certainly ruptured, while the head of the femur could be drawn on to the broken rim of the acetabulum. Patient was simply confined to his bed, with anodyne embrocations to the hip and limb.

On the 22d, seventeen days after the injury, and during the night, while turning in bed, he experienced a sudden and severe pain, with the sensation of something giving way in the hip. On examination, next morning, the femur was found luxated in the sciatic notch; pain was very great. Under ether the diagnosis was confirmed, and the luxation was readily reduced. A broad band was then passed around the pelvis and made firm; then extension to a moderate degree, with sand-bags, was applied to keep the limb in position.

[a] Paper read before the Philadelphia Academy of Surgery, November, 1879.

December 1.—Has been up and about the hospital wards for some days, and leaves for his home soon; has fair use of the limb.

December 28.—Discharged, cured, with fair use of the limb.

Fractures of the Ribs.—The ribs are not unfrequently subjected to fracture from falls against curb-stones, and from carriage-wheels passing across the thorax. When a man is struck by a locomotive there are often a number of the ribs broken, and great injury inflicted upon the intra-thoracic viscera.

In the majority of instances the injury is simple, though occasionally it is compound, and sometimes there occurs perforation of the costal pleura or puncture of the lung itself. In the 221 cases of fracture of the ribs admitted into the hospital between January, 1850, and January, 1874, there were, according to Dr. Agnew, 212 simple fractures. Emphysema of the superficial cellular tissue often shows the existence of perforation of the lung, but these cases are not at all similar to compound fractures.

There has been found as a rule little actual separation and but slight tendency to displacement. This at times renders the diagnosis obscure. A point of diagnostic value insisted upon by Dr. Levis is this: if strong pressure be made upon the sternum and anterior part of the chest, while the patient lies upon his back, pain will be referred to the point of fracture upon the lateral aspect of the thorax, if such fracture exists. The reason for this production of pain at the seat of fracture is readily appreciated by recollecting that any springing outwards or displacement of the fragments gives rise to pain. If no fracture is present, pressure anteriorly cannot give rise to pain at the side or back of the chest.

The method of treating fractured ribs is uniform. Motion of the affected side is prevented by encasing it in adhesive plaster, which may be applied in one broad piece, or in strips of two or three inches width, overlapping each other. By this means costal respiratory motion is checked and pain avoided. It is better, perhaps, to apply the plaster tightly during expiration, and to enclose only the affected side, if completely encircling the chest should interfere too much with respiration and the patient's comfort. Dr. Morton has recently introduced an apparatus for the treatment of fracture of the ribs. It consists (see figure) of a broad band of stout linen, which encircles the chest, and can be adapted to a large or small patient; by lacing it in front any degree of tightness can be secured. This form of apparatus is especially applicable when treating women, or men who may have a considerable amount of hair on the chest or back. The adhesive plaster in such cases is often exceedingly annoying. If much pain exists, if emphysema is present, or if several bones are broken, the patient is kept in bed for a few days, otherwise he is allowed to go at once to the out-patient department for further treatment.

Local pneumonia or pleurisy occurring as a complication is treated upon general principles. Cellular emphysema receives no special attention, but serves to direct the attendant's mind to the possibility of intra-thoracic inflammation.

Cure is always expected unless the severity of the injury, such as occurs from railroad accidents, renders death from shock probable. A case of this nature was admitted where laceration of the lung and rupture of the liver and of the kidney occurred in addition to fracture of six ribs of the right side. Death occurred in one and one-quarter hours. In 14 cases observed during a short period there were 9 deaths. In all of these cases, except two, some severe complication existed, such as rupture of intestine, laceration of lung, or fracture of extremities.

Fractures of the Sternum.—The sternum, supported upon the costal cartilages as upon springs, is seldom fractured; hence there is little to be said about the management of this injury in the hospital wards. In forty-four years there were recorded but 13 examples of fracture of this bone. From April, 1873, to April, 1878, there was only one case of fracture of the sternum admitted. This was the result of a crushing injury, which fractured the ribs and lacerated the lungs. Death occurred shortly after the patient's admission.

FRACTURES OF THE HEAD.

Fractures of the Cranium.—The frequency of occurrence of these fractures gives an interest and importance to the subject that can hardly be overestimated. The organ contained within the cranium is so liable to be damaged by the force that fractures the skull, or by the inflammatory processes occurring subsequently, that death is to be expected in a large proportion of cases. In 305 cases treated in the hospital during the period of forty-four years, already mentioned, a fatal issue occurred in 146 cases.

The treatment adopted in fractures of the skull, when no operative procedure is required, is as follows: the patient is placed in bed, with the head low, and moderately stimulated until brain-shock or concussion is overcome. Afterwards the head is elevated, and cold affusions employed, while purgatives, and bromide of potassium in large doses, are administered internally. These measures are adopted to prevent cerebral congestion and inflammation. The amount of bromide given is usually two to four drachms during the twenty-four hours. Occasionally opium in small amounts and mercurials are resorted to for a similar reason. A study of the hospital notes, however, shows that more general reliance is placed upon bromide of potassium and purgation. Cupping has at times been employed, but phlebotomy seems to have been almost ignored as a remedial agent.

Compound comminuted fractures are usually managed by removing the loose fragments and by elevating depressed portions of bone. The trephine is frequently employed to attain the latter object, and, as it is only required to get an opening by which to introduce the elevator, the smallest trephine is recommended.

When there are serious symptoms in a simple fracture of the skull it is certainly good surgery to make a free exploratory incision in the scalp, and thus obtain a complete understanding of the nature of the injury. The fact that the fracture is thus made compound argues little against the practice, when it is remembered that comminuted fragments and depressed bone may be unrecognized through the overlying scalp. Cases have doubtless died of secondary inflammation, or of compression of the brain, that might have been given a good chance for recovery by this operative exploration. It is felt that too great stress can hardly be laid upon the importance of this exploratory incision in severe injuries of the head with suspected fracture of the cranium.

A remarkable case of brain injury, due to fracture of the skull, was treated in Dr. Morton's ward in 1873, and is deserving of special record:

Case of compound comminuted fracture of the skull, with paralysis of the first, second, third, fourth, fifth, sixth, and seventh cranial nerves; death three months afterwards: autopsy (see figure). —The injury, of which the history was obscure, was evidently due to a railroad casualty. The man, aged 47 years, was

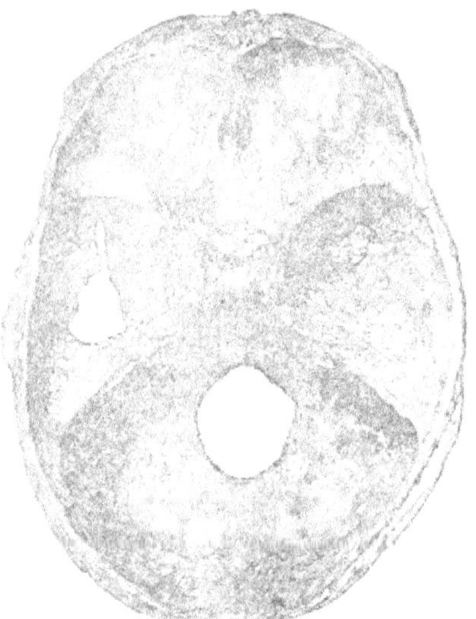

Fracture of the base of the skull.

admitted July 24, 1873. The left temporal bone was comminuted, presenting an opening one and one-half inches in diameter. The dura mater was extensively torn, and numerous fragments of bone were removed from the brain. The base of the skull appeared to be involved in the fracture. Profound coma was present during the succeeding ten days, when a gradual improvement was observed, and paralysis was found on the left side of the face.

On August 30 the wound had almost closed, and there was complete facial paralysis of the left side. The patient walked about the hospital, but had some difficulty in masticating and swallowing his food. His intellect was stated to be "probably as clear as before the accident." A few days subsequently the condition of the nerves was examined with care. The cranial nerve-distribution on the right side was normal. On the left side the first, second, third, fourth, fifth, sixth, seventh, and possibly the ninth nerves, were palsied.[1]

On October 22, 1873, he died, after having suffered several weeks with headache and prostration, which terminated with coma.

The autopsy showed the fracture to involve the petrous and squamous portions of the temporal bone, the sphenoid, palate, and malar bones. All the portion of bone included in the area of fracture, viz., the base of the skull, together with the styloid process of the temporal bone, etc., was entirely displaced from its original location, and was driven downwards and inwards, and firmly united angularly one inch below its normal position. The anterior portion of the middle lobe of the brain was softened and wasted, and by incision a large abscess was found involving the entire extent of the left middle lobe, the fissure of Sylvius, and, to a considerable extent, the anterior lobe. The olfactory nerve of the left side, and all the nerves passing through the optic, sphenoidal, round, oval, and stylo-mastoid foramina of the same side, were involved.

This remarkable case is worthy of consideration, because the patient was under observation for a long time and a careful autopsy was made.

Fractures of the Face.—Having discussed fractures of the cranium, it becomes necessary to consider fractures of the bones of the face. Little or nothing is to be said of the nasal bones, because fracture of these bones is usually treated at the out-patient department, on account of the insignificant character of the injury. Occasionally a man may receive a powerful blow upon the nose, causing fracture, which may be complicated with fracture of the ethmoid bone or concussion of the brain. These cases are of course rare. There was admitted some years ago a man who had fallen from a height, sustaining a simple fracture of the nose, with considerable laceration of the face and scalp, and who was insensible when brought to the hospital. He recovered, and was discharged, cured, after being forty-seven days in the wards. There is very little special treatment necessary for fracture of the nasal bones. The parts, if displaced, are put in position, and union soon occurs. Separation of the lateral cartilage from the nasal bone appears to be quite common.

Fractures of the malar or superior maxillary bones are rare injuries, because of the position of these components of the face. It is unusual to see either bone broken, except in instances of severe injury to other parts, as in the case of a man who was treated for simple fracture of the upper jaw, fracture of the radius, and fracture of one or more ribs, showing the violent character of the force by which he was hurt.

Fractures of the Lower Jaw.—The fracture of the face which, from its frequency and importance, deserves great consideration, is fracture of the lower jaw, and yet, from the notes of the patients treated in the wards, it

[1] *Phila. Med. Times*, October 3, 1874, p. 1.

would seem to be a rather unusual injury. This again is due to the fact that many such cases are treated as out-patients, and never enter the wards.

Although nearly all fractures of the inferior maxilla are complicated with rupture of the mucous membrane covering the alveolar process, and are consequently compound, it seems that they behave, as a rule, like simple fractures.

The usual method of treatment adopted is the application of a pasteboard cap to the chin, which is retained with the figure-of-eight bandage of the jaw and occiput. It is very rarely that wiring of the teeth or the use of the interdental splint is required. Necrosis and delayed union of the bone appears to have been exceedingly rare. Food is introduced in a liquid state between the crevices of the teeth, and but little difficulty has been experienced in this regard. In the majority of cases the patient is never obliged to keep his bed, and, as stated above, very many instances of this fracture are at once sent to the out-patient department, while others are retained in the wards for a few days, and then discharged to the dispensary for a continuance of treatment.

Occasionally multiple fractures are seen, as in a case where there were four fractures near the symphysis and along the body. This patient was discharged, cured, in fifty-seven days. In some instances the line of fracture, and the tendency to displacement, have been such that deformity could not be prevented by the most careful treatment. This is believed to be in accord with the experience of other hospitals.

Fractures of the Scapula.—This bone does not often sustain fracture, which is due probably to its attachment to the trunk being mostly muscular. The acromion has been the portion most frequently broken; and a rare fracture, that of the coracoid process, has been recorded at least once. It is possible that the diagnosis was not fully established in this instance.

Fracture of the acromion is treated by bandaging the arm to the chest, and forcing the humerus up against the acromion process as much as possible. A similar injury to the body of the bone would be dressed with compresses and bandages to prevent motion.

Inspection of the tables at the end of this article will give an idea of the unfrequency of fractured scapula.

Fractures of the Clavicle.—The multiplicity of methods and apparatus in general use would seem to show that there is no one recognized plan for the treatment of this fracture. Dr. George "Fox's apparatus," which is known all the world over, was introduced into the Pennsylvania Hospital by that surgeon in 1828. It forms a very effective dressing, and at once fulfils all the indications in the treatment of this fracture by keeping the arm upward, outward, and backward. The apparatus, as is well known, consists of a wedge-shaped pad, which is placed well up in the axilla. A ring or collar encircles the opposite shoulder, to which the axillary pad is fastened by strings across the chest, back and front. The arm is supported by a strong muslin sling. There is no apparatus better than this, we think, for general use. Nevertheless, occasionally some deformity may result from

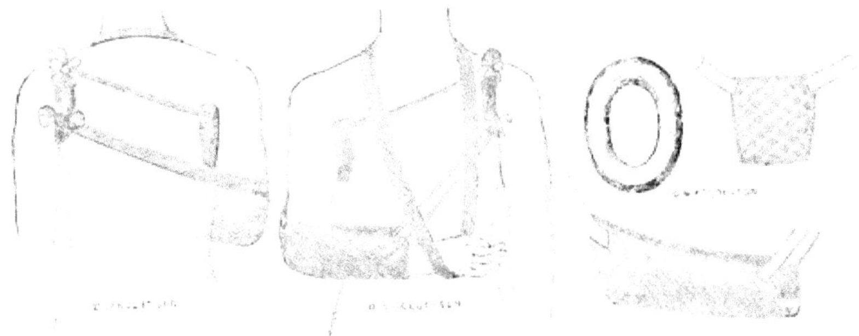

Fox's apparatus for fracture of the clavicle.

the fact that the patient in moving about produces a jarring of the limb, and the bone may become overlapped. The apparatus known and figured in systematic works on surgery as Levis's dressing for fractured clavicle is sometimes employed; but even Dr. Levis himself prefers, as a rule, the dressing

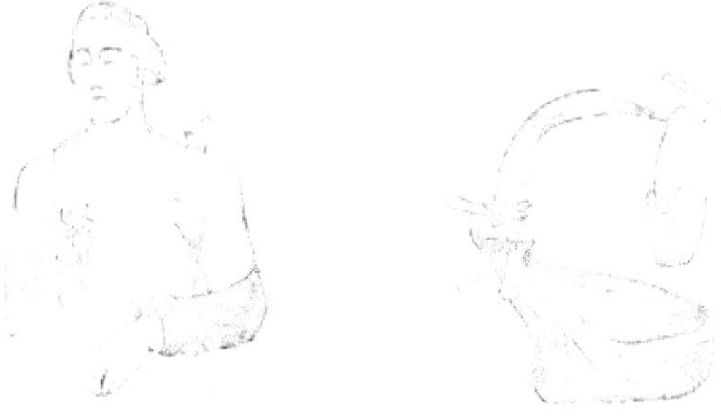

Levis's apparatus for fracture of the clavicle.

of adhesive plaster, applied somewhat in the manner of Velpeau's bandage. This is used for walking cases, but the supine position, with the head slightly elevated to relax the sterno-mastoid muscles, and with perhaps a bag of shot upon the shoulder, is the best treatment when it can be enforced. In treating a broken clavicle in the female it is often important that there shall not be the slightest deformity. In these cases we adopt this "postural treatment," placing the patient upon a moderately firm mattress, with the shoulders kept horizontal, the head resting on a very thin pillow, with the hand and forearm upon the chest. We treat many of our clavicle fractures by this method, and we generally have no deformity. It is not necessary to keep the patient in bed more than a fortnight or so, since the tendency to overlapping is overcome as soon as partial solidification occurs.

Fractures of the Humerus.—Fractures of the shaft of the humerus (in its upper portion), including the head and neck of the bone, are treated either with the anterior or internal angular splint. If with the latter, we use a moulded binders' board humeral splint, but we more commonly bring the humerus to the side, close to the ribs, which serve admirably as an internal splint. For the first day or two after fracture of the humerus the patient is confined to his bed, and the figure-of-eight bandage applied, thus keeping the parts at rest, which should be covered with lead-water and laudanum, if painful or swollen. Then, when the acute symptoms have commenced to subside, a roller bandage may be applied from the hand to the shoulder, and either an internal or anterior angular splint may be adjusted, with a moulded shoulder and humeral support of binders' board, so cut as to cover the acromion and humerus to about its middle. In warm weather a light towel or napkin should be placed in the axilla, which will absorb the perspiration and prevent irritation and excoriation. In fractures of the condyles we use either the anterior or internal angular splints. And in order to avoid, if possible, anchylosis we vary the angle of the splints very often. When repair begins we constantly use passive motion. In fractures near the head of the humerus, where the upper fragment is tilted outwards, it may be necessary to abduct the arm to a considerable degree in order to get good apposition. Continuous extension even may at times be required.

There is no question, it would seem, as to the propriety of removing all splints early, and allowing the patient to gradually get in the way of flexing and extending the joint. It is seldom that elbow-joint fractures recover perfectly, but the too long use of rigid splints only tends to increase this condition.

Fractures of the Forearm.—In fractures of the shaft of one or both bones of the forearm we use two straight splints: a palmar one, extending from the bend of the elbow to the finger-ends, and one on the opposite aspect of the arm, from the elbow to the wrist only. Great care is taken to exclude the epicondyle of the humerus from pressure by the internal splint. In applying these splints we have the palm of the hand presenting inward, and consequently the thumb upward and slightly outward. In this position the radius and ulna are most widely separated.

Fracture of the olecranon process is treated by extreme extension, the arm and forearm being placed on a straight splint.

We commonly recognize two fractures involving the lower end of the radius; the one "Colles's fracture," the other "Barton's fracture," the essential character of which is that the fracture extends obliquely from the articulation backwards and upwards. In this the posterior articulating surface is more or less separated. We use commonly the old splint devised by the late Dr. Henry Bond, of Philadelphia, with excellent results, not only in the Barton, but in the Colles fracture. Within a comparatively short time Dr. Levis has introduced a splint which he believes fills the indications in fracture of the lower end of the radius. He desires, however, to lay stress upon the principles involved rather than upon the particular form of splint.

This splint, with the fracture it is intended for, has been so ably described by Dr. Levis, that his account from the *Transactions of the Medical Society of Pennsylvania* is deemed worthy of a full reproduction. It will be seen that he denies the distinctive character of the Barton's fracture.

"The correct nature and mechanism of the ordinary form of fracture of the lower end of the radius are now, after much controversy, generally admitted and properly comprehended. With this proper understanding the indications of treatment become rational and decisive.

"In the usual and very characteristic fracture of the carpal end of the radius the primary line of fracture is, with little tendency to deviation, *transverse* in direction. Associated lines of fracture are generally those of comminution of the lower fragment, and are caused by the upper fragment being driven vertically into it and splitting it, usually in directions towards its articular surface.

"The displacement of the lower fragment is towards the dorsal aspect of the forearm, and its articular surface is inclined in the same direction, abnormally presenting backwards and upwards.

"The mechanism of the fracture is its production by falls upon the palm of the hand, which, with the carpus, undergoes extreme extension, and the fracture is caused by an *act of leverage* or *transverse strain*. This direction of force has also been called *cross-breaking strain*.

"In this fracture actual displacement of the lower fragment may not exist at all, or it may be to the extent of complete separation from contact of the broken surfaces, varying with the amount of force applied and with the retaining influence of the surrounding dense structures.

"The first essential of the treatment of fracture of the lower end of the radius is *the complete reduction of the displacement*. The action of replacement must be directed to the lower fragment itself. The reduction of the fracture can usually be thoroughly effected, under anæsthesia, by *strong extension applied to the hand, associated with forced flexion of the wrist, and with pressure applied directly on the dorsal surface of the lower fragment*. Unless vertical splitting or comminution of the lower fragment exists, the maintaining of partial flexion of the wrist, with pressure of a pad on the dorsal surface of the fragment, will prevent return of deformity.

"With the object of retaining the apposition of the fractured surfaces, by overcoming displacing forces, I have practised for many years on the principles involved in the splint here illustrated, the application of which will not require much description.

"In the treatment of fracture of the lower end of the radius it is essential that proper allowance be made for the curvature of the anterior or palmar surface of this part of the bone. This is insured in the splint which I have devised, which follows correctly the radial curvature; and the fixing of the

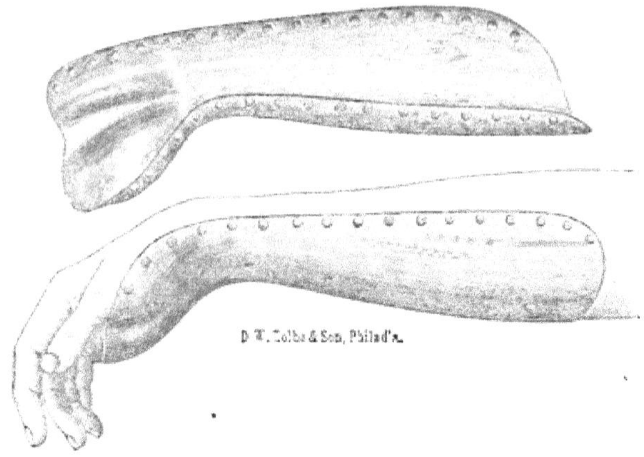

thenar and hypothenar eminences of the hand in their moulded beds maintains the splint immovably in its correct position with reference to the radial curve.

"To neglect of complete primary reduction of the displacement of the lower fragment, and to inefficient restoration and retention of the normal radial curve, are due the frequent unfortunate sequences of this fracture.

"The splint is made of copper, so as to be readily conformable by bending to suit the peculiarities of size and form of forearms. The series of little pointed elevations along the edge is for the purpose of keeping the bandage from slipping. It is tinned to prevent oxidation.

"The splint will usually fit the forearm so accurately that but little padding will be required, and a piece of woven lint or of cotton or woollen flannel is all that is necessary for its lining. No dorsal splint is needed, but, as before referred to, a small pad will, in most cases, be required over the dorsal surface of the lower fragment. For retention of the splint, an ordinary bandage, two inches and a half to three inches wide, is all that is necessary."

Fractures of the Fingers.—Fractures of the fingers are often compound and comminuted; but we never lose sight of the fact that any portion of a finger saved is of immense value, especially to the artisan. We therefore adopt the rule to *never* make a primary amputation of a finger, be it never so seriously injured, but wait until nature gives the indication for a secondary operation. In the article on Amputations this subject has been thoroughly discussed, and does not require further mention here.

Fractures of the Femur.—In the treatment of fractures of the femur we employ extension either by adhesive plaster or by the apparatus devised by Dr. Morton some years since. In fractures of the neck of the femur, especially in old subjects, some of the staff are quite partial to the anterior splint of Prof. N. R. Smith, with suspension. This method is very grateful to the patient, and there is less risk of excoriation and bed-sores; for the back can be examined at any time, friction can be made, and the patient can alter his position without interfering with the fracture. We make no difference in the manner of our treatment of femur fractures, be the break at or involving the condyles, or at any part of the shaft or neck of the bone, save in treating very old people, to which allusion has already been made.

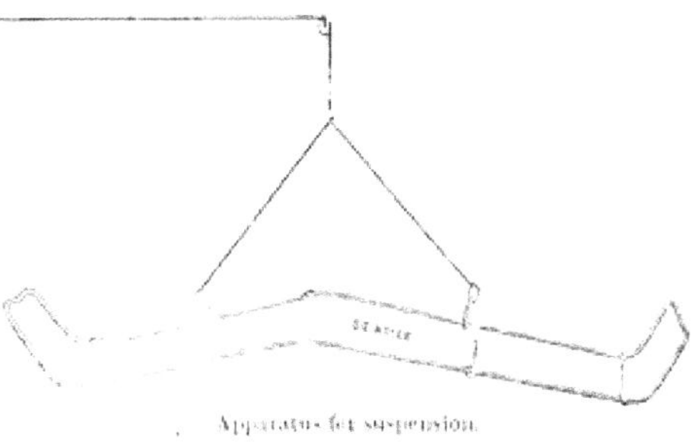

Apparatus for suspension.

The practice of making extension by lateral adhesive strips was early employed in the Pennsylvania Hospital. The apparatus is readily made and quickly applied. (See figure.) It is simply a two and a half or three inch adhesive strap, long enough to reach from just above the knee to the sole of the foot, and up the other side to the same level, always taking care that the strips do not overlap the broken extremity. In the middle of this strap a piece of wood, two and a half by four inches, is placed, which serves to keep the lateral straps from pressing on the malleoli, and to it the weights for extension are attached. Strips of plaster are also made to gird the limb above and below the knee and above the ankle in order to prevent the apparatus

slipping. A roller bandage is applied outside the dressing from the foot to the hip, and sand-bags are placed one on each side of the limb to support it in good position. The lateral strips should go just above the knee and below the point of fracture. If the adhesive strips are applied tightly at once after the injury, or before the swelling has commenced to subside, they often give rise to considerable discomfort, and when the swelling subsides there will be wrinkles and folds, which may cause ulceration. The adhesive plaster sometimes slips during warm weather, and requires re-application; at times a rubber plaster is used. In all cases the limb should be shaved before the plaster is applied.

We make extension by means of pound weights, which are placed in a small wire crib, as suggested by Dr. Morton, with a base of the same size as the weights employed. Formerly we used bags of shot or sand, and more commonly, bricks; but neither of these methods are as convenient or reliable as that first described.

The form of apparatus, devised by Dr. Morton, used in the hospital is the same in principle as the one already described, and it is made as follows: The cir-

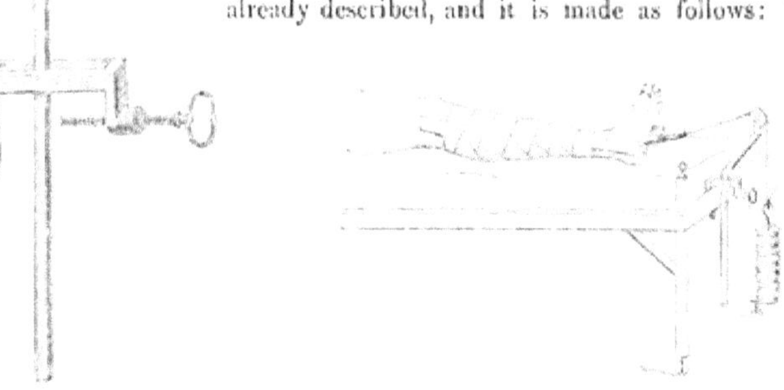

Pulley and shaft devised by Dr. Levis, which is screwed to the end of the bed; over this the cord passes which is attached to the extension apparatus.

cumference of the thigh and leg is taken,—the former four or five inches above the knee, and the latter four inches below the joint; leather, buckskin, or any stout material is cut according to the measurements to encircle the limb between these two points, with an opening left for the patella; the portions of the apparatus above and below the knee should buckle or lace up above separately. On the outside of this a band of any stout material is buckled or sewed, which should be long enough to extend to the sole of the foot and up on the inner side of the limb to the knee, where it is fastened in like manner. Just above the ankle a band encircles the limb, which keeps the extension strips in position. As the band passes around the sole it is attached to a piece of wood, the same as is used in the adhesive plaster extension, and sand-bags complete the dressing. (See figure.) This apparatus can be readily made in a short time by any one, or may be obtained from an instrument-maker by sending the girth of the limb above and below the knee and its length. It is easily applied, and can be removed as often as desired to wash or rub the limb; it does not interfere with circulation, and is very comfortable to the patient. The apparatus can be used on either limb.

After applying either form of apparatus described and the extending weight, the foot of the bed should be elevated about four inches. The muscles,

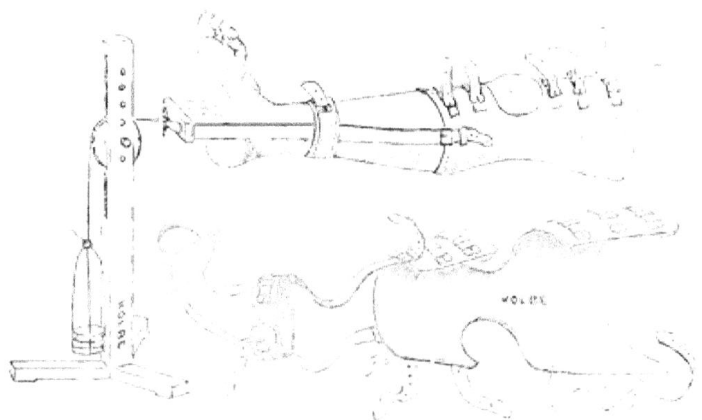

Dr. Morton's apparatus for making extension in fracture of the femur.

shortly after a fracture, often contract spasmodically, producing pain, and at the same time disturbing the fragments; but in the course of a few days, by the lateral pressure of the sand-bags and the steady extension, this condition is gradually overcome, and in the course of three or four days the fractured ends of the bone can be brought in absolute line, and in perfect apposition.

Measurement of the limbs in cases of suspected fracture of the femur, or as a means of determining results of treatment, is considered of value to a very limited degree. Several years ago Dr. William C. Cox, one of the resident surgeons, made a series of measurements of uninjured legs of patients in the hospital, and proved that it was no unusual thing for one extremity to be longer than the other.* This hospital was therefore the first in which attention was called to the now well-known fact that asymmetry of the lower extremities is not at all exceptional. Subsequently Dr. J. B. Roberts, formerly connected with the hospital, measured a number of skeletons, and found that the difference existed in the bones themselves,† and that therefore the former results were not due to inaccurate observation, as seemed to be believed by some. The subject has been investigated still further by other observers, especially by Dr. Wight, the late Mr. Callender, of London, and by Dr. Garson. The latter measured the skeletons in the Museum of the Royal College of Surgeons of England, and found asymmetry in the large majority of cases; and, finally, Dr. Morton has measured a very large number of youths, the results of which will be found in the paper on Asymmetry of Limbs.

* *Amer. Jour. of Med. Sciences*, April, 1875. † *Phila. Med. Times*, Aug. 3, 1878.

For this reason measurement of the limbs has been almost discontinued, for there is no possible method of insuring accuracy, when the injured leg may have been a half-inch longer or shorter than the other before the receipt of fracture. Great attention is therefore to be given to the appearance of the thigh, and deformity avoided by accurate inspection.

Fractures of the Patella.—We believe simple fracture of the patella to be almost invariably produced by muscular contraction. The forms of apparatus devised for fracture of this bone may be said to be legion. A great

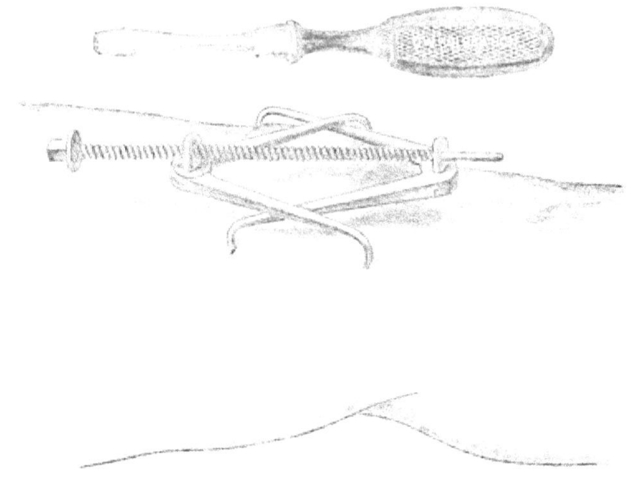

Dr. Morton's hooks for fracture of the patella.

many methods, too numerous to mention, have been recommended from time to time. In order to have bony union we must keep the fragments in close apposition, and this without cutting off the supply of blood to the parts, and thus interfering with repair. Much of the apparatus invented has the serious fault of interfering with the circulation of blood to the patella, by pressing upon the bone upon all sides, and allowing little or no blood to reach it. Want of union is, therefore, a frequent result.

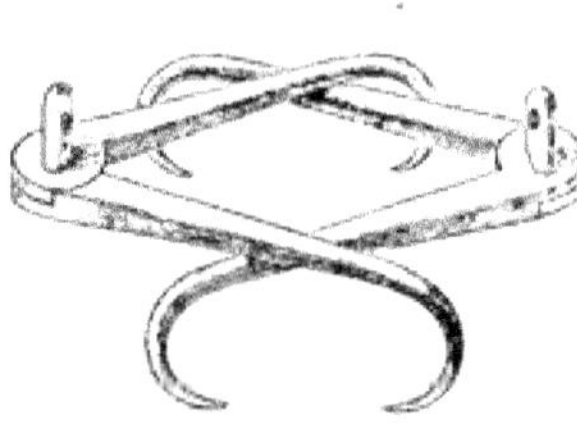

Dr. Agnew's apparatus consists of a straight posterior splint, having cross-pieces or pegs above and below the knee. Strips of plaster are applied below the lower fragment, and then attached to the two upper pegs. The upper fragment is drawn down by a similar application of plaster. By rotating the pegs the tension can be regulated.

Malgaigne's hooks, or a modification of them, have been used by Dr. Morton and Dr. Levis for years past in all cases of fractured patella which have been admitted into their wards, and are believed by these surgeons to answer

Malgaigne's hooks for fracture of the patella, separated by Dr. Levis.

the purpose most perfectly. Dr. Levis made a modification of these hooks a few years ago by separating the set into two pairs, which act independently. (See figure.) Another arrangement for the same purpose that Dr. Morton uses, consists of a set of four movable hooks (see figure), two for the upper fragment and two for the lower, which, like the preceding, are brought together by a long screw. Either of these forms of hooks give excellent results.

The first case in which the Malgaigne's hooks were recently used (in 1873) was that of Bridget S., who recovered with perfect osseous union. She was admitted into the hospital with a transverse fracture of the patella; an interval of two inches existed between the fragments. A month after her discharge Dr. Morton presented her for examination before a number of surgeons, who were unable to detect which patella had been injured except by the little scars left by the hooks. The extension and flexion were perfect.

Fracture of the patella treated with a modification of Malgaigne's hooks.—Mary J. was admitted with transverse fracture of the left patella, the fragments of which were separated by a distance of three-quarters of an inch. After elevation and anodyne lotions had been employed for three days, she was etherized, and Dr. Levis's modification of Malgaigne's hooks applied to bring the pieces of bone into accurate approximation. The modification consists simply in having two pairs of single hooks instead of the pair of double hooks usually seen. By this change the bone is more firmly and accurately held, because on account of the irregular surface of the patella it is difficult to make the double hooks sit accurately; they are like a four-legged stool on an uneven floor. After the hooks had been adjusted and firmly screwed together, carbolized oil was applied, and the leg placed on an inclined plane.

The hooks gave accurate apposition of the fragments, and were kept in position a little over five weeks. After their removal the same dressing was continued to the small ulcerated points, and the limb kept extended. The two patellæ were accurately measured, and found to be of the same length. She was soon allowed to walk with a posterior straight splint, and passive motion was instituted when this was discarded.

Four months after the receipt of injury the leg could be flexed to a right angle, and the patient, though she still walked somewhat lame, was rapidly gaining perfect control of the limb. With the finger there could be felt a slight ridge at the point of fracture, but the two patellæ seemed to be of the same length. Subsequently she re-fractured the bone, but was treated in another hospital.

In comminuted or stellate fracture of the patella union is often remarkably good, because it would seem that the greater injury leads to more irritation, and consequently a large supply of blood, and, as a result, more callus is deposited. In simple fracture of the patella there often appears insufficient irritation to produce bony union, and as apposition is difficult the hooks act very beneficially. The hooks should never be applied until all inflammation, as a result of the injury, has subsided; this is usually between the fifth and seventh day. In the application an anæsthetic should be employed.

We occasionally find immense swelling and effusion consequent upon a fracture of the patella, and this often continues notwithstanding our topical applications. At the end of a week, if this has not subsided, the fluid may be evacuated by the aspirator, and the hooks then applied. In one case lately, in Dr. Morton's ward, three tappings were made before the apparatus could be employed. The case ultimately did remarkably well. In another case, on account of the effusion and inflammation the hooks could not be introduced until the twentieth day, but so much effusion reappeared, with redness of the skin at the points of the entrance of the hooks, that Dr. Morton abandoned the treatment by the hooks, and the fractured bone remained in excellent contact for nearly one day, then an inch separation occurred. No other form of treatment could have been adopted, for the parts would not have borne any pressure. This last case referred to had been a patient in the hospital several years previously with fractured patella of the right side, but even with two inches separation of fragments has remarkably good use of the limb. It would seem that in some cases the separation of the fragments, even to the extent of two or more inches, is not incompatible with almost entire usefulness.

Fractures of the Tibia and Fibula.—In order that a patient with a broken leg may have the greatest freedom, and be privileged to sit up in bed, we sometimes use the suspended anterior splint. But the treatment of fracture of the leg for which this institution is best known is that by means of the fracture-box (see figures), made the length of the leg, with a permanent foot-piece and falling sides. A patient with a recent fracture, when admitted, has his leg enveloped in an evaporating anodyne lotion (generally lead-water and laudanum), a small pillow is placed in the fracture-box, and on this the limb is placed. The foot is secured to the foot-piece, and two pieces of bandage are tied around the box to secure the sides from falling. This treatment favors frequent inspection, and the dressings can be changed as often as desired. When union has occurred lateral splints of binders' board, starch, silicate of sodium, or any firm material is then applied. In compound frac-

tures the box is often filled with bran with more or less carbolized powder. If desirable the apparatus can then be swung in the same manner as the anterior splint.

The fracture-box has universally given excellent results, and is in constant use. It is, we think, applicable to all fractures of the leg and foot, with perhaps one exception. We refer to the fracture with luxation of the ankle when one or both malleoli have been broken. The tibia is dislocated backwards, it being, in fact, the backward tibio-tarsal luxation of Maigaigne. If the fracture-box is used in such a case the heel is apt to be pressed up, and the deformity, which can hardly be reduced without ether, is likely to recur.

A very interesting example of this form of fracture with luxation has lately been under treatment in the hospital, and was successfully treated after the luxated tibia was reduced by the anterior splint and suspension. In fracture at the ankle with great deformity, section of the tendo Achillis has repeatedly been required to allow perfect apposition.

Fractures of the lower end of the fibula are commonly treated in the fracture-box, or sometimes more readily by Dupuytren's splint. Starch, silicate, or pasteboard dressings may also be very judiciously used in proper cases. Fractures of the tarsus and metatarsus are treated, if severe, in the ordinary fracture-box, or are allowed to rest upon a pillow.

Simple fractures of the toes are not often seen. Usually the injury is complicated with dislocation, comminution, or great laceration. These conditions are treated by amputation, or by sutures and ordinary dressings. Conservatism is not carried to the same extent as in fractures of the fingers, because the loss of a portion of a toe is not as serious a matter to an artisan as that of a corresponding part of a finger. Again, a workman can hardly give up his occupation for the long period required for conservative surgery, when he knows he will be unable to walk during the time of treatment. Hence partial amputations, giving rapid recoveries, are more often resorted to in injuries of the toes than of the fingers.

2.—CONDENSED TABLE OF FRACTURES

Treated in the Wards of the Hospital from April, 1873, to April, 1878. Compiled by Drs. John B. Roberts and William C. Cox, from the Hospital Notes, vols. i. to xxi.

Region or Bone.		Nature.		Locality.		Age.		Result.	
Spine	16	Simple	16	Cervical	1	Under 20 years	1		
"				Dorsal	11	From 20 to 40 years	11	Relieved	6
"				Lumbar	2	From 40 to 60 years	4	Died	10
"				Not given	2				
		Total	16		16		16		16
Pelvis	28	Simple	24	Pubes	2	Under 20 years	3	Cured	13
"		Compound	2	Ilium	11	From 20 to 40 years	9	Relieved	3
"		Not given	2	Ischium	1	From 40 to 60 years	12	Died	12
"				Sacrum	3	From 60 upwards	2		
"				Acetabulum	3	Not given	2		
"				Not given	8				
		Total	28		28		28		28
Ribs	135	Simple	129	From 1st to 3d rib	4	Under 20 years	6	Cured	75
"				From 4th to 7th rib	55	From 20 to 40 years	45	Relieved	21
"		Not given	6	From 8th to 12th rib	36	From 40 to 60 years	60	To out-patient department	14
"				Not given	40	From 60 upwards	20	Died	25
"						Not given	4		
		Total	135		135		135		135
Sternum	1	Simple	1	Not given	1	From 20 to 40 years	1	Died	1
Cranium	79	Depressed	26	Orbit	1	Under 20 years	22	Cured	34
"		Simple	27	Frontal	21	From 20 to 40 years	28	Relieved	7
"		Compound	22	Parietal	23	From 40 to 60 years	19	To out-patient department	1
"				Temporal	9	From 60 upwards	3	Died	36
"		Not given	4	Occipital	4	Not given	7	Unimproved	1
"				Base	15				
"				Not given	6				
		Total	79		79		79		79

Nose	8	Simple	5			Under 20 years	1	Cured	3
"		Compound	2			From 20 to 40 years	6	To out-patient department	4
"		Not given	1			From 40 to 60 years	1	Died (ribs, etc.)	1
		Total	8				8		8
Upper jaw	2	Simple	2	Not given	2	From 20 to 40 years	2	Cured	2
Lower jaw	35	Simple	25	Near symphysis	9	Under 20 years	7	Cured	24
"		Compound	4	Near angle	7	From 20 to 40 years	18	Relieved	3
"		Compound and comminuted	1	Between symphysis and angle	6	From 40 to 60 years	9	To out-patient department	7
"		Not given	5	Double	4	From 60 upwards	1	Died	1
"				Four places	1				
"				Not given	8				
		Total	35		35		35		35
Clavicle	78	Simple	73	Sternal end	4	Under 20 years	12	Cured	55
"		Compound	1	Acromial end	25	From 20 to 40 years	31	Relieved	8
"		Not given	4	Middle	25	From 40 to 60 years	25	To out-patient department	8
"				Not given	24	From 60 upwards	8	Died	7
"						Not given	2		
		Total	78		78		78		78
Scapula	19	Simple	16	Neck	2	From 20 to 40 years	12	Cured	13
"		Not given	3	Acromion	7	From 40 to 60 years	4	Relieved	1
"				Coracoid	1	From 60 upwards	3	To out-patient department	1
"				Body	4			Died	4
"				Glenoid cavity	1				
"				Not given	1				
		Total	19		19		19		19
Humerus	191	Simple	144	Upper third	64	Under 20 years	58	Cured	124
"		Compound	15	Middle third	31	From 20 to 40 years	62	Relieved	20
"		Compound and comminuted	31	Lower third	69	From 40 to 60 years	46	To out-patient department	21
"		Not given	1	Not given	27	From 60 upwards	22	Died	26
"						Not given	3		
		Total	191		191		191		191

Condensed Table of Fractures.—(Continued.)

Region of Bone.		Nature.		Locality.		Age.		Result.		Bone Fractured.	
Forearm	185	Simple	145	Upper third	25	Under 20 years	40	Cured	111	Radius	91
"		Compound	20	Middle third	21	From 20 to 40 years	60	Relieved	31	Ulna	27
"		Compound and comminuted	13	Lower third	95	From 40 to 60 years	64	To out-patient department	27	Both	62
"		Not given	7	Not given	44	From 60 upwards	18	Died	15	Not given	5
"						Not given	3	Not given	1		
		Total	185		185		185		185		185
Hand	68	Simple	14	Carpus	6	Under 20 years	25	Cured	18		
"		Compound	40	Metacarpus	24	From 20 to 40 years	30	Relieved	7		
"		Compound and comminuted	12	Phalanges	34	From 40 to 60 years	11	To out-patient department	40		
"		Not given	2	Not given	4	From 60 upwards	1	Died	3		
"						Not given	1				
		Total	68		68		68		68		
Femur	256	Simple	218	Upper third	94	Under 20 years	80	Cured	179		
"		Compound	6	Middle third	96	From 20 to 40 years	47	Relieved	29		
"		Compound and comminuted	25	Lower third	24	From 40 to 60 years	71	Died	48		
"		Not given	7	Not given	42	From 60 upwards	50				
"						Not given	8				
		Total	256		256		256		256		
Patella	39	Simple	36	Transverse	27	Under 20 years	1	Cured	36		
"		Compound	2	Oblique	2	From 20 to 40 years	26	Relieved	1		
"		Compound and comminuted	1	Not given	10	From 40 to 60 years	9	Died	1		
"						From 60 upwards	2	Unimproved	1	(Admitted for separation of old fracture.)	
"						Not given	1				
		Total	39		39		39		39		

Leg	507	Simple	352	Upper third	21	Under 20 years	88	Cured	395	Tibia	81
"		Compound	62	Middle third	[illegible]	From 20 to 40 years	227	Relieved	25	Fibula	95
"		Compound and comminuted	77	Lower third	[illegible]	From 40 to 60 years	149	To out-patient department	5	Both	329
"		Not given	16	Not given	117	From 60 upwards	32	Died	82	Not given	2
"						Not given	11				
		Total	507		507		507		507		507
Foot	65	Simple	25	Tarsus	18	Under 20 years	14	Cured	43		
"		Compound	12	Metatarsus	17	From 20 to 40 years	25	Relieved	9		
"		Compound and comminuted	26	Phalanges	21	From 40 to 60 years	22	To out-patient department	4		
"		Not given	2	Not given	9	From 60 upwards	3	Died	9		
"						Not given	1				
		Total	65		65		65		65		
Larynx	1	Simple	1	Thyroid cartilage	1	From 20 to 40 years	1	Relieved	1		

Total Number of Fractures.

Spine	16	Forearm	185
Pelvis	28	Hand	68
Ribs	125	Femur	256
Sternum	4	Patella	39
Cranium	70	Leg	507
Nose	8	Foot	65
Upper jaw	2		
Lower jaw	35		1712
Clavicle	78	Larynx	1
Scapula	49		
Humerus	191		1713

This summary of the cases of fracture treated in the hospital during a period of five years was compiled in the following manner. The bound volumes of notes, written by the resident surgeons on duty at the time the injury was treated, were looked over, and every case of fracture abstracted and placed under its appropriate heading. These voluminous tables were then gone over, and the facts condensed into the form here presented. In such an amount of work errors are liable to creep in, and cannot be avoided. For example, a note written may have been lost before the time arrived for binding the volume; there may have been a mistake in diagnosis, which was revised at the time of treatment, but, through forgetfulness, not entered upon the notes. Another source of error may be the fact that the exact nature and locality of the injury is not accurately given; and, finally, those copying the original notes may not make the abstracts by exactly the same method. Notwithstanding these tendencies to inaccuracy, the table has been made from such a large number of cases, and with such a watchful eye, that probably nothing of the kind vitiates the general proportions of the summary, since one region would be as liable to suffer as another.

A glance will show that comminuted fractures are not separately classified. This is owing to the fact that the notes showed that little attention had been paid to recording this complication, unless the injury was in addition compound. Hence comminuted simple fractures would have appeared exceedingly and disproportionately rare. Therefore it was deemed advisable to place all of them under the heading "simple fractures," or "compound comminuted fractures," as the case might be. In tabulating fractures of the cranium the most prominent characteristic was taken as the basis of classification. If an injury was simple and depressed it was placed under "depressed fracture," and so in other cases.

In determining the locality when there were double fractures, or when two parallel bones were broken, as tibia and fibula, the point showing most serious injury was adopted as the locality of the fracture.

It must also be recollected that this is a table of fractures and not of patients. Therefore a death or a cure, following fracture of several bones in one individual, appears as two or three deaths or cures, as the case may be. It was found impossible to give an idea of the comparative frequency, locality, etc., of special fractures in any other way than that adopted. Again, a patient who suffered a primary or secondary amputation for fracture and recovered, was entered upon the books as cured. This table, therefore, will have to be considered in connection with the statistics of amputation to give an accurate idea of the number of cases where this method of treatment was deemed advisable. At times, where amputation has been done, death, of course, has occurred within short intervals.

Fractures of the Spine.—The most noteworthy fact respecting injury in this region is that the fracture occurred in the dorsal region in 11 cases out of 16, though it is generally believed that the more mobile cervical and lumbar vertebræ are generally broken in injuries affecting the spinal column. The unfavorable prognosis is exhibited by the record of no cures, and only 6 relieved, out of a total of 16 cases.

Fractures of the Pelvis.—The proportion of cured patients here is probably larger than would be anticipated, considering the liability of involvement of the pelvic viscera.

Fractures of the Ribs and Sternum.—The number of patients marked cured after rib fractures is reduced by the fact that many of them are soon discharged to be treated at the out-patient department, or are sent to their own homes before the cure is complete, and hence are stated as being relieved. The unfrequency of fractures of the sternum is evident, since only one case occurred in a period of five years.

Fractures of the Cranium.—The number of cases successfully treated is very satisfactory, when the large proportion of depressed and compound fractures is considered. The base of the skull was the seat of injury in a rather large number of instances; the result in such cases is almost uniformly fatal.

Fractures of the Face are few in number, because so often treated at the out-patient department, and never admitted to the wards. The ramus of the lower jaw was broken in no instance, except where there were multiple fractures of the bone.

Fractures of the Clavicle and Scapula.—A number of deaths occur in these tables, which must be attributed to the patients having suffered in many, if not in all, instances from some other more serious injury.

Fractures of the **Arm, Forearm, and Hand.**—In these lists there is seen again the effect of the out-patient department, which treats a large proportion of the fractures of the forearm and hand, and some of the humerus. The frequency of fractures of the radius at its lower extremity is made apparent by the number of cases reported as fracture of radius alone, and by the number of injuries described as being in the lower third of the forearm. The frequency of compound comminuted fractures of the hand is worthy of remark. These are usually crushes in printing-presses and carding-machines, and appear rather more frequently because the simple fractures are not so apt to be treated as ward patients.

Fractures of the Femur.—Here the preponderance of fractures in the upper and middle thirds is very great. The number marked "relieved" is much increased by the frequency of non-union in intra-capsular fractures of the neck.

Fractures of the Patella.—This fracture appears to have been generally transverse, and doubtless usually produced by muscular contraction.

Fractures of the Leg and Foot.—The great number of cases of leg fracture involving both tibia and fibula, corresponds with general experience in all hospitals. The results after fracture of the foot are about the same as in similar injuries to the hand.

Fracture of the Larynx.—One case of fracture of the thyroid cartilage was admitted, but was discharged by request, having remained in the hospital but one day.

General Résumé of Fractures, Simple, Compound, Ununited, etc., from 1751 to May, 1878.

Unclassified fractures	2,777
Simple fractures	8,160
Compound fractures	2,975
Ununited fractures and delayed union	100
Total	14,012

General Classified Résumé of Fractures.

	Number.	Remain.	Total.
Simple, unclassified	2,460	317	2,777
" cranium and face	464	22	486
" trunk and pelvis	812	57	869
" superior extremities	2,910	211	3,121
" inferior extremities	3,170	514	3,684
Compound, unclassified	240	48	288
" cranium and face	299	20	319
" trunk and pelvis	35	......	35
" superior extremities	947	102	1,049
" inferior extremities	1,122	162	1,284
Ununited fractures and delayed union	85	15	100
Total	12,544	1468	14,012

THOMAS G. MORTON.
JOHN B. ROBERTS.

FISTULA.

Of the 71 cases of fistula recorded, there were 49 anal, 13 perineal, 1 urethral, 5 vesico-vaginal, 2 recto-vaginal, and 1 recto-vesical.

The vesico-vaginal and recto-vaginal fistulæ are described under another head.

Anal Fistulæ.—Of these, 26 were cured, 17 relieved, 3 were unimproved, 2 eloped, and 1 died from phthisis. The average age of the patients was 35$\frac{5}{8}$ years.

The relieved and unimproved cases include some who left the house before they could be called cured, and who were not afterwards traced, and others who sought temporary relief only.

They were cases often combined with phthisis. Operations on this class are, as a rule, discouraged; not so much, if at all, from the idea that the march of the constitutional affection is accelerated by closing up diverting drains, but from the fact that the closing up is easier to talk about and to try, than it is to do. Very frequently the efforts are entirely abortive, and too often the last state of the patient is worse than the first, for the general dyscrasia seems to assert a constant protest against healing. It is for the relief of pain, and for the better evacuation of sinuses and pockets, rather than to cure the fistula, that operations on phthisical subjects are justifiable.

Only 5 of the 49 anal cases were females. This sex is comparatively exempt from the disease. The explanation may be that the parts are much more protected by fat and connective tissue from external causes of irritation, such as bruising, or exposure by sitting on cold or damp seats, than those of the male.

The exemption would also seem to point against the internal origin of fistula, except from ulcerative disease, as phthisis, for there is no want of connective tissue in females in which abscess might spontaneously originate; and as for constipation, which is sometimes said to be the cause of fistula, there is every reason to believe that they are more prone to it than males.

There were 39 operations by the knife and 4 by ligature. The elastic ligature was used in two of the cases and the silk ligature in two. Both of the latter were cured, while one, in which the elastic was used, was a failure. There were 23 cures resulting from the knife operations, 14 were relieved, 1 stationary, and 1 died, the latter from phthisis. The knife is the favorite with all of the surgeons. Where there is much boggy or inert tissue, as is the case in old fistulæ, the parts, after division, are often swept over with a stick of caustic potash, and this is quickly followed by an oiled rag, which is placed so as to come in contact with the whole of the diseased and wounded surfaces. Where the fistula has a high inner opening the elastic ligature is safe and effectual. Sometimes it may be made to accomplish a double purpose, as in one case where, besides the fistula, there was a stricture of the rectum. The ligature was carried above the stricture, and, after cutting its way through, there was a complete recovery from both conditions.

Perineal Fistulæ.—These all occurred in males. The writer does not remember to have seen a perineal fistula in the female from any cause whatever, but is informed by Dr. Agnew that he has seen two cases. The rarity is of course readily explained from the anatomical relations of the parts.

9 of the 13 cases were operated on by external urethrotomy. Of these, 3 were cured, 4 improved, and 2 were stationary. One case of bad urethral fistula, owing to the patient having lacerated his penis and removed his testicles while under mania a potu, was ultimately cured by amputating what remained of the organ.

A very interesting case of recto-vesical fistula occurred. The patient was a male, aged 27. No cause could be given for the disease. The tract of the fistula was proved to be very high by throwing an injection of carmine into the bladder. The fluid was returned through the rectum, but entered it so far up as to be without the range of touch or vision. The patient left the house unrelieved.

William Hunt.

ASYMMETRY OF LIMBS.

HISTORY OF THE DISCOVERY OF ASYMMETRY IN THE LENGTHS OF THE LOWER LIMBS.

FROM an examination of the records of the hospital it will be observed that there have been treated nearly 5000 cases of fractures of the inferior extremities, and during the past five years 256 cases of fractures of the femur. Considerable interest has from time to time naturally been manifested by the surgical staff in the measurements of many of these cases, since formerly, at least, it was the rule to make the record of shortening when the patient was discharged.

In 1873 I had under my care a case of fracture of the thigh, which recovered without an unfavorable symptom, but on making a careful measurement, when the cure was completed, I found, to my surprise, that the limb which had been fractured was *an inch longer* than the other. This was very puzzling, and led at once to an investigation of the cause. Dr. Wm. C. Cox was at the time resident surgeon in my ward, and he then measured not only those who had had fractured limbs, but sound persons, those who never had any injury.

In the April, 1875, number of the *American Journal of the Medical Sciences*, Dr. Cox published a table, giving the results in fifty-four sound persons measured in the hospital. The variations were from one-eighth to seven-eighths of an inch, and only six persons in the list presented the lower limbs of equal lengths.

My colleague, Dr. Hunt, on discussing the subject of inequality in the length of limbs, based on Dr. Cox's measurements, in the *Philadelphia Medical Times* of January 16, 1875, under the heading "Clinical Notes and Reflections," stated, "It is well known that bilateral symmetry may be said not to exist as to breadth, but has it ever occurred to any one to state as a law that bilateral symmetry as to length is exceptional?"

In 1879, Dr. Hunt also, in an article on asymmetry,* says,—

"During the past two or three years frequent notices of this discovery and of its importance have appeared in the medical, scientific, and popular journals, and also in the newspapers." Dr. Wight, of Brooklyn, New York, has made contributions to this interesting question, his papers appearing subsequently to those above alluded to.†

"We wish, therefore, to make a claim of priority for the Pennsylvania Hos-

* *Amer. Journ. of the Med. Sciences*, January, 1879.

† *Archives of Clinical Surgery*, February, 1877; *Proceedings of the Medical Society of Kings County*, January 21, 1878.

pital, and especially for Dr. Cox, in *the discovery and full surgical appreciation of the fact that asymmetry as to length of the lower limbs of the same person is the rule and not the exception.*

"Dr. Frank H. Hamilton was at first unwilling to admit these important new facts as to measurements. He is now fully convinced of their truth. He writes, 'I think the subject of sufficient importance and so creditable to American surgery as to entitle it to a more conspicuous notice and a faithful historical record.'"

In order to settle the matter more positively, I recently (December, 1879) made an application to the Directors of the City Trusts of Philadelphia, through the Hon. Henry M. Phillips, the president of the board, for permission to measure the boys in the Girard College. The results of the measurements of 513 boys, ranging from 8 to 18 years of age, are not only interesting and instructive, but, I think, conclusive. This work was principally conducted by Drs. Cox and Charles H. McIlwaine, lately resident physician in the hospital, with the assistance of my son.

Instrument used in Measuring.—Many years ago I discarded the ordinary tape, and devised an instrument which has proved very satisfactory. An account of the apparatus was published by Dr. Stacy B. Collins in 1877, and it was with this instrument that all of the measurements of the lads at the Girard College were made.

"In attempting to accurately estimate the shortening which may take place after fracture, coxalgia, etc., it seems necessary to have, first, a simple instrument that can be used on a patient in bed without disturbing any applied apparatus. Second, to have it of material which cannot be stretched. Third, to have both limbs measured from one fixed point, thus avoiding inaccuracies resulting from obliquities of the pelvis, differences in height of the iliac bones, and similar causes. Fourth, to have an instrument that not only shows the inequality, but accurately measures it at the same time.

"To meet these requirements, Dr. Morton introduced (into the hospital) several years since an apparatus, which, with some late improvements, I shall now describe.

"It consists of two parts: first, a frame, with movable arms, intended to hold the body and legs in exactly the same right line, or, in other words, to make the patient lie perfectly straight; and, secondly, a measuring-rod extending from this, which passes between the legs, and by means of two short arms touches the internal malleoli at exactly the same relative point at the same time on both sides. When this is done the measurement is accomplished, as a scale on the extended rod at once shows and records any existing difference. The frame consists of three small steel bars, each three feet long, lying parallel to each other, and joined at the ends by means of two steel cross-bars at right angles to each other, and nine inches long. This makes a rectangular frame nine inches by thirty-six inches, and divided into two equal parts by the third long bar running longitudinally.

"From each side of this frame three arms extend, which, when it is in position on the person to be measured, drop perpendicularly, and by means of a right- and left-hand screw, held in position by the central bar, can be extended or contracted through brass boxes playing on the outside bars, so as to grasp the body firmly.

"The boxes through which the arms run can be moved along the frame so as to clasp the person at any desired point, or, in the case of children, allow all the arms to be moved to one end, thus practically shortening the frame. It is evident that if this apparatus be laid on the body, and the arms screwed up (see figure) so as to hold it tightly under the armpits, at the hips, and

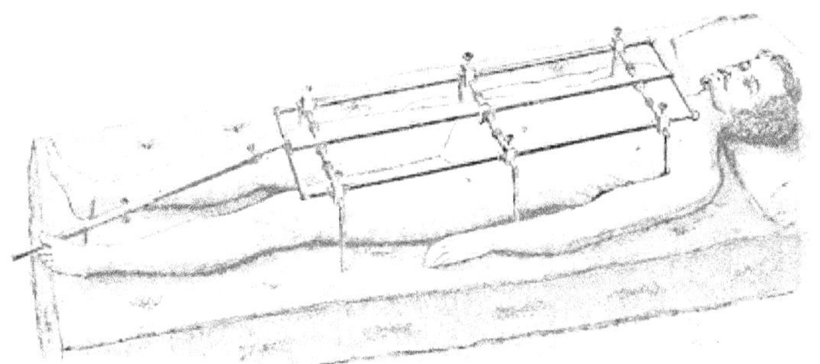

Apparatus for measuring any irregularity in the lengths of the lower extremities.

about the knees, any lateral motion is impossible, and the body is held immovably straight. This is necessary, because we have found from many experiments that the slightest deviation from the straight line at once produces an apparent shortening in the limb opposite to the side to which the body is bent. A hinged bar drops from the centre of the lower end of the frame and extends between the legs. Over this runs another bar, which can be pushed up or down as the length of the legs requires, and held in any desired position by means of a screw at the top. From either side of this outer bar extend short arms, which, by means of an elbow with a long fenestra fitting on a steel button, can be moved for a short distance up or down, a screw forming the top of the button, allowing it to be fixed firmly at any spot, while small pointers attached to the arms, and sliding up and down with them, run over a scale on the upper face of the outside bar, and show any existing shortening to the one-sixteenth of an inch.

"The manner of using this apparatus is sufficiently simple, and when it is carefully made, all sources of error (save possibly from an actual attempt to deceive) seem to be eliminated.

"Let us suppose the case of a patient lying in bed with extension applied. We do not wish to get him out of bed, or disturb the apparatus, at the same time we are desirous of knowing whether we have sufficient weight applied and the limb accurately adjusted in length. The clothes are turned back, and the frame is placed over him, so that one pair of arms clasps him under each axilla. The second pair firmly holds his pelvis a little below the iliac crests, while his legs are strongly pressed out on each side above the knees, against the third pair of bars, which are approximated so as to allow about two inches between the ankles.

"Now the sliding-bar is moved down a convenient distance, and one of the small arms is brought against the lower edge of one malleolus, and the other is moved upwards or downwards into a corresponding position on the other side.

"The body is now held immovable. Any shortening can be at once diagnosed, and its amount read from the scale without further trouble."*

Of the 513 boys examined, 272 showed inequality of the limbs, and in 241 there was no appreciable difference. It should be stated that each boy was interrogated if he had at any time had fracture or any bone or joint disease, and it is understood that all were normal limbs, unless otherwise stated. The ages of the boys ranged from 8 to 18 years. 91 showed a difference of one-eighth of an inch; 100, of one-quarter of an inch; 41, of three-eighths of an inch; 22, one-half of an inch; 12, five-eighths of an inch; 2, three-quarters of an inch; 2, one inch and one-eighth; 1, one inch and five-eighths. In 1 case there was a shortening in the right limb of three and one-quarter inches. This lad had, we found, suffered from a fracture of the thigh some years ago. It was also found that of the 272 cases of inequality, the right limb was the longer in 198 cases; in the remaining number, 74, the left limb was the longer. At the time the examinations were made, we found that the boys even with the greatest shortening were not aware of the fact, although in most instances an examination of the trousers showed that they were much more worn out on one side than the other, and this fact at least had been recognized by each individual.

From these measurements it will be seen that asymmetry as to the lengths of the lower limbs of the same person existed in more than half of the cases examined. The fact that asymmetry may be looked upon as the rule and not the exception is a most important fact, and the discovery has already had its influence in settling conclusively at least one medico-legal case, in which an action was brought against a medical gentleman in the interior of this State. The plaintiff claimed that his boy was irretrievably injured on account of a shortening of three-quarters of an inch, which resulted from a fracture of the thigh. The fact that inequality in the lengths of the lower limbs was not unfrequent was brought out at the trial, and this, with the measurement of a lad who had never had any injury, and who was in the court-room at the time, showing a difference of three-eighths of an inch, at once led to a nonsuit. In Dr. Hunt's words, "Thus ended the first trial for malpractice, in which the

* *Amer. Journ. Med. Sciences*, 1877, p. 141. Description of an Apparatus devised by Dr. Thomas G. Morton for Measuring any Irregularity in the Length of the Lower Extremities. By Stacy B. Collins, M.D., formerly Resident in Hospital.

new facts as to measurements were brought with great effect before a legal tribunal."

Since asymmetry is believed to be the rule, rather than the exception, we rarely deem it necessary to measure cases of fractured thighs at the hospital, for, as we have already remarked when referring to this class of fractures, "that as there is no possible method of insuring accuracy, when the injured leg may have been half an inch or more longer or shorter than the other before the receipt of the fracture, we simply trust to the appearances of the thigh, and avoid deformity by accurate inspection, measurement, or otherwise."

The following table gives the ages, the number of boys, and the inequality of both sides, and the number of normal lengths in each age:

Table showing the Results of Measurements of 513 Boys at the Girard College, Philadelphia, Pennsylvania.

Age	No. of Cases.	Right Side.									Left Side.					Normal Lengths.
		[illegible] inch.	[illegible] inch.	[illegible] inch.	[illegible] inch.	[illegible] inch.	[illegible] inch.	[illegible] inches.	[illegible] inches.	[illegible] inches.	[illegible] inch.	[illegible] inch.	[illegible] inch.	[illegible] inch.	[illegible] inch.	
8	7	...	1	...	1	...	...	...	...	...	...	1	...	...	...	4
9	23	2	2	3	2	...	...	...	...	...	1	1		...	...	12
10	67	8	10	6	...	1	...	...	...	1	2	4	1	...	...	24
11	97	15	13	5	5	...	1	...	...	...	6	3	1	...	...	48
12	86	10	16	4	3	6	1	...	...	...	3	5	...	1	...	37
13	65	7	7	6	4	...	...	...	...	...	7	4	...	1	...	29
14	67	4	12	4	1	1		1	...	...	8	5	...	...	...	31
15	40	6	5	1	2	3	...	1	...	...	2	5	2	...	1	12
16	45	5	1	5	...	...	...	...	...	...	3	2	3	...	...	26
17	15	1	1	...	2	...	...	...	1	...	...	2	...	...	...	8
18	1	1	...	...	...	...	...	...	...	...	...	...	...	...	...	
	513	59	68	34	20	11	2	2	1	1	32	32	7	2	1	241

Thomas G. Morton.

GANGRENE AND FROST-BITE.

SINCE 1873 we find that only 34 cases of gangrene and frost-bite have been recorded, but this will not include all who received attention during this time, since a very large number of these cases are constantly applying for relief during the severe weather of the winter months, but after the first dressing they are generally consigned to the out-patient department, the rule being that only the more severe cases are admitted into the wards.

To the degrees of frost-bite, we need not refer. The subjects are usually seamen, and others who, during intoxication, have suffered from exposure to extreme cold. The patient who is recorded as Case 8 of this table (and 26 of Amputations), was a remarkable instance of cruelty and severe exposure. Both legs were frozen, and subsequently were successfully amputated. In many cases intemperance, with cold, has been the cause of the accident, fingers and toes being generally the parts affected.

The treatment usually adopted consists in puncturing the bullæ, which are seen soon after the exposure; the surfaces involved are then covered with an emollient dressing. For this purpose there is nothing better than starch poultices, or lint saturated with carbolized oil. Occasionally we have used a dressing of chloral, turpentine, salicylic acid, or lead-water and laudanum. A mixture of equal parts of creasote and olive oil in cases of superficial blistering, or reddening of the skin, with intense burning, is often very soothing. Morphia is often required, and is given either by mouth or, more frequently, by the skin.

Most of the cases, however, require a considerable time. Poultices are often applied to hasten the suppurative action and separation of the gangrenous part, and during this period stimulants, tonics, and disinfectants are of great service.

Amputation is generally deferred until the patient has become thoroughly habituated to his invalid condition and his surroundings.

Of the 34 cases, all were cured except two. One of these was an old man 75 years of age, who was admitted for senile gangrene of the foot. Amputation of the great toe was performed; he was removed by his friends. The other case (the only death) had acute gangrene of the leg, following an injury of the leg. Gangrene was present when he was admitted. The case terminated fatally on the third day.

We never amputate during the acute symptoms, but make at once free, deep, longitudinal incisions into the parts, and evacuate all collections of gas, serum, or blood, and favor the process of separation by warm applications, the limb at the same time being immersed in a bran dressing, with some disinfecting powder. In one case of amputation of the leg after gangrene there were no vessels which required ligaturing, the inflammatory process having effectually obstructed all the vessels.

Gangrene and Frost-Bite.

No.	Age	Sex	Cause.	Part involved.	Treatment.	Result.	Days in Hospital.	Remarks.
1	54	M.	Exposure at sea; insufficient food.	Hand.	Partial amputation of the hand.	Cured.	28	85-day midwinter trip from Antwerp.
2	22	M.	Exposure at sea; poor food.	Index finger.	Amputation of finger.	Cured.	12	Same voyage as No. 1.
3	29	M.	Exposure to intense cold.	All of the toes of both feet, left heel.	Poultice, carbolized oil and creasote and olive oil. All the toes were amputated.	Cured.	128	
4	23	M.	Walked 30 miles during intensely cold weather.	Both feet; vesicated superficial sloughs.	Salicylic acid.	Cured.	80	
5	25	M.	Severe cold.	Foot and ankle.	Amputation of leg.	Cured.	128	(See Amputations, Case 5.)
6	20	M.	Severe cold.	Hand.	Amputation of hand.	Cured.	67	(See Amputations, Case 6.)
7	61	M.	Exposure.	Toes.	Amputation	Cured.	31	
8	21	M.	10 days exposed, during severe weather, in a swamp.	Legs.	Legs amputated.	Cured.	151	(See Amputations, Case 26.)
9	32	M.	Exposed during intoxication.	Feet.	Double amputation.	Cured.	174	(See Amputations, Case 45.)
10	15	M.	Severe injury.	Leg.	Amputation of leg.	Cured.	99	(See Amputations, Case 38.)
11	50	M.	Exposure.	Fingers.	Carbolized oil.	Improved.	4	
12	29	F.	Intense cold.	Fingers.	Lead-water and opium.	Cured.	11	
13	27	M.	Exposure.	Toes.	Oxide of zinc ointment.	Cured.	52	
14	21	M.	Exposure on an oyster-boat.	Toes.	Chloral and oxide of zinc ointment.	Cured.	49	
15	40	M.	Exposure.	Feet (partial); skin in places sloughed.	Carbolized oil, charcoal poultices.	Cured.	68	
16	28	M.	Sugar-house burn.	Thumb and index finger.	Amputation.	Cured.	49	
17	28	M.	Exposure.	Superficial, of both feet.	Lead-water and laudanum.	Cured.	5	
18	40	M.	Rupture of aneurism.	Knee and leg.	Amputation of thigh.	Cured.	102	(See Amputations, Case 30.)
19	17	M.	Cold and exposure.	Feet.	Tonics, zinc ointment.	Cured.	21	
20	32	M.	Cold, intemperance, exposure.	Slough of foot.	Tonics, carbolized oil.	Cured.	10	
21	75	M.	Senile gangrene.	Toe.	Amputated.	Unimproved.	11	Removed by friends.
22	68	M.	Exposure, intense cold.	Legs.	Double amputation of legs.	Cured.	267	(See Amputations, Case 54.)
23	25	M.	Exposure to cold.	Toes.	Amputation.	Cured.	30	
24	29	M	Exposure and cold, cutting ice.	Feet (skin only).	Lime-water, cosmoline, and chloral dressings.	Cured.	15	

Gangrene and Frost-Bite.—(Continued.)

No.	Age	Sex	Cause.	Part involved.	Treatment.	Result.	Days in Hospital.	Remarks.
25	45	M.	Exposed while driving.	Fingers.	Lime-water.	Cured.	8	
26	29	M.	Exposure at sea.	Fingers.	Turpentine and olive oil, then chloral.	Cured.	13	
27	39	M	Exposure at sea.	Fingers.	Turpentine and olive oil.	Cured.	26	
28	20	M.	Exposed to great cold unloading vessels.	Finger.	Cosmoline. Amputation of little finger.	Cured.	66	Little finger became greatly hypertrophied, and somewhat resembled keloid. It measured 3½ inches in circumference.
29	58	M.	Exposure to cold; skin sloughed on instep.	Foot.	Salicylic acid.	Cured.	39	
30	48	M.	Exposed 5 months before.	Deep ulcers on great toe, involving the bone.	Amputation of great toe and part of metatarsal bone.	Cured.	174	
31	27	M.	Exposed to cold.	All toes sloughed off.	Amputation. Zinc ointment.	Cured.	52	
32	9	M.	Railroad crush, involving main artery.	Leg and knee.	Amputation of thigh.	Cured.	98	(See Amputations, Case 96.)
33	21	M.	Exposure and cold.	Fingers and toes.	Chloral dressing.	Cured.	50	
34	27	M.	Followed injury; pipe weighing 700 pounds fell on leg.	Leg.	Stimulants and anodynes.	Died.	3	On day of admission there was gangrene of the leg.

THOMAS G. MORTON.

TUMORS.

Among the 39 cases coming under this title in the note-books, a few will be found in the list that do not properly belong there. Again, cancers and ovarian tumors have mostly been brought under separate headings. The cases have been put down precisely as they occur in the notes, which are sufficient for all clinical purposes. From the list it would not be difficult for the purist in nomenclature to arrange them under their more accurate titles. Operations were performed on 28 of the cases. Of these, 19 were cured, 5 relieved, and 4 died.

10 cases were subjected to constitutional and palliative treatment. 7 of these were cured, 1 died, 2 were stationary. One patient eloped.

In most operations for the removal of morbid growths the knife is used as little as possible after the first free external incisions are made. Separating and tearing with the fingers, or with some blunt instrument, are generally found to answer the purpose, and at the same time to greatly diminish the risk of troublesome and sometimes dangerous hemorrhage. Modifications of the Esmarch method, by elastic rings and cords, are sometimes resorted to. This method, particularly in certain pendulous growths, is capable of many useful adaptations, which will suggest themselves to the operator, and cannot be arbitrarily stated. It is sometimes surprising to see even the best surgeons harassed by an excess of venous bleeding and oozing, owing to the fact that their contrivances to stop hemorrhage are actually keeping it up, by interfering with the reflux of blood to the centres. Under these circumstances all pressure should be removed, for a short time at least, in order to find out whether it is the source of the trouble.

In the following pages the principal points of practical interest in the 39 cases are communicated in a tabular statement, in which the nature of the growth, the treatment adopted, and the result are succinctly stated, and occasional remarks made upon the course and complications of the disease.

Tumors.

No.	Age.	Sex.	Nature.	Cause.	Treatment.	Result	Remarks.
1	37	M.	Epithelioma of lip (?).	....	Removed by V incision.	Cured.	In hospital 12 days.
2	26	M.	Synovial bursa of knee, near insertion of sartorius.	No history of injury.	Rest and compress, with soap plaster and bandage.	Cured.	In hospital 11 days.
3	27	M.	Sebaceous, over left eyebrow. Had been operated on a year before admission.	....	Opened, emptied, and caustic potash applied, poultice.	Cured (?).	In hospital 13 days. Returned 6 weeks afterwards; was again opened, and cauterized with nitrate of silver.
4	19	F.	Cystic, of left eyelid.	....	Excised, silver suture, and water dressing.	Cured.	In hospital 7 days.
5	22	M.	Hæmatocele, of left buttock.	A blow on the part from a fall.	Rest and iodine paint.	Cured.	In hospital 14 days.
6	22	M.	Sebaceous, of scalp.	....	Had already opened; emptied by compresses.	Cured.	In hospital 45 days.
7	25	M.	Large adenoid, lying over parotid, on right side.	....	Removed by operation; four vessels ligatured.	Cured.	Had been growing for 8 years. Tumor extended from whole front of ear, downwards and behind it, and thence to angle of jaw.
8	32	F.	Exostosis, of lower jaw.	....	Removed by scalpel and chisel.	Relieved.	Left the house for out-treatment on same day.
9	40	M.	Suppurating adenoid, of neck.	....	Tinct. iodine paint; potass. iodid. internally; opened by knife and pus evacuated.	Improved.	In hospital 41 days. Was readmitted, and remained a short time.
10	28	M.	Large glandular tumor of groin.	....	Earth dressing, pressure; tumor removed by operation.	Died.	Died on the 17th day after operation, of fever and exhaustion.
11	53	F.	Large glandular tumor of neck, 14 inches in circumference, extending up behind angle of jaw.	....	Removed by elliptical incision. Easily enucleated, as adhesions were not firm; some small vessels ligated; washed out with gr. x solution of chloride of zinc, and dressed with carbolic acid and ung. zinci oxid.	Cured.	In hospital 38 days. There was some facial paralysis following the operation.
12	31	M.	Large tumor of groin, malignant.	....	Removed by operation; profunda artery tied.	Cured (?).	This patient was discharged apparently well, but the disease returned in the same spot; there was very rapid growth, and he died within a year from the operation.
13	28	M.	Large tumor of neck, bronchocele (?).	....	Wet clay dressing, under which appeared to diminish slightly.	Unimproved.	In hospital 9 days. Patient left hospital on a pass and did not return.
14	56	M.	Fibrous, of right side of nose.	....	Removed by operation; dressed wound with gauze and collodion.	Improved.	Tumor consisted of layers of fibrous tissue, with irregular cells. In hospital 6 days.
15	59	M.	Malignant, of groin.	....	Enucleated, chiefly by tearing after first incisions; very little hemorrhage.	Died.	Disease returned in cicatrix, and was rapidly fatal. In hospital 27 days.
16	50	M.	Large cystic, of neck.	....	Lugol's solution, emp. de Vigo cum mercurio, etc.	Died.	At first improved, but afterwards disease rapidly advanced. In hospital 62 days.
17	18 mos.	F.	Cystic, of neck.	....	Partially evacuated, and then excised by écraseur, and pedicle ligated.	Died.	Patient was taken home by parents, and died there.

18	63	M.	Fatty, of arm, circumference 12 inches.	....	Removed by operation.	Cured.	In hospital 22 days.
19	48	F.	Fatty, of side.	....	Removed by operation.	Cured.	In hospital 27 days.
20	55	F.	Fatty, of right shoulder.	....	Removed by operation.	Cured.	In hospital 26 days.
21	25	M.	Sarcoma, attached to lower part of testicle.	Patient gave a minute history of a gunshot wound received 10 years before, and that the ball descended into the scrotum. The tumor certainly felt like a bullet, but proved to be nothing of the kind.	Removed by operation.	Cured.	In hospital 25 days.
22	20	F.	Of left infra-maxilla (epulis?).	....	Removed, together with three of the teeth.	Cured.	In hospital 8 days.
23	17	F.	Of left cheek.	Inherited syphilis (?).	Constitutional. Locally, tinct. iodine, etc.	Improved.	In hospital 5 months, 14 days.
24	30	F.	Cystic, of face.	....	Two of the tumors removed by operation; carbolic dressing.	Improved.	In hospital 25 days; removed by friends. Had been operated on 13 years before, when six tumors were removed.
25	22	M.	Adenoma, of cervical lymphatics.	....	Potass. iodid. internally. Hypodermics of tinct. iodine into tumors every other day.	Improved.	In hospital 22 days; removed by request. Patient returned, and improved further under same line of treatment.
26	35	F.	Fatty, over right deltoid.	....	Enucleation.	Cured, to out-patient department.	In hospital 20 days.
27	39	F.	Uterine, fibroid.	....	Rest and palliatives.	Improved.	In hospital 47 days; discharged by request.
28	65	F.	Encephaloid (?), of breast.	....	Removed by operation.	Cured.	In hospital 85 days.
29	60	F.	Ovarian cyst. M. W.	....	Palliative, tapping.	Improved.	In hospital 18 days.
30	36	M.	Axillary.	....	....	Eloped.	In hospital 1 day.
31	14	M.	Of left testicle.	....	Rest, iodoform, zinc ointment.	Improved.	In hospital 17 days.
32	45	F.	Of groin (?), fecal.	....	Purgatives, etc., enema.	Improved.	In hospital 10 days. Patient admitted on own statement for tumor, but on examination was found to be due to fecal accumulation.
33	45	F.	Of left breast.	....	No operative treatment.	Discharged by own request.	In hospital 7 days.
34	28	M.	Enchondroma, of testicle.	Thought to be from injury.	Testicle removed by ecraseur.	Cured.	In hospital 1 month.
35	45	F.	Fibroid, of vagina, hanging by pedicle 2 inches long	....	Excised.	Cured.	Tumor was size of an orange. In hospital 2 weeks.
36	12	F.	Hydrocele, of neck.	....	Emptied by incision. Suppuration induced by seton of lint.	Cured.	In hospital 26 days.
37	...	M.	Fibrocystic, of testicle; 12 months' standing	....	Removed by incisions and écraseur.	Cured.	In hospital 18 days.
38	48	F.	Cystic, of cheek, size of walnut.	....	Removed by knife.	Cured.	In hospital 46 days.
39	3	F.	Hydrocele, of neck.	....	Excision.	Died.	

Two cases of large cystic tumors of the neck.

CASE 1. *Cystic tumor of unusual size; paracentesis of the cyst, and removal of the solid portion by the écraseur; death.*—The child, aged 3 years, was admitted with the statement that a few days after birth a slight enlargement of the left side of the neck was observed, which gradually increased, until at time of admission into the hospital it was larger than the child's head. (See figure.) The tumor extended from the ear to the clavicle, and at various places presented fluctuation upon palpation. It had been tapped, and treated unsuccessfully by a seton. The child was anæmic, and there was a general feeble condition of health. Dr. Levis punctured the various cysts, but there was a considerable solid mass, which, after long acupressure-needles were passed through the base in various directions, was encircled by the chain of the écraseur, and then removed. The wound was dressed with carbolized oil. Morphia and chloral were administered internally to produce sleep. Four days after the operation a serious diarrhœa supervened, which could not be controlled, and the child sank, and died from exhaustion.

CASE 2. *Large hydrocele of the neck; simple tapping; again tapping and suppuration induced; cure.*—Margaret I. N., aged 12, was admitted April 2, 1873. Two years ago a tumor was first noticed, about the size of an olive, above the sternum, which gradually increased in size, until on admission it measured three inches in length by two in breadth.

April 5.—Dr. Morton made an incision at the most dependent point. The fluid was evacuated, and the wound was united by silver wire.

April 9.—The tumor having again filled, an incision was made, and a plug of lint was inserted well into the cavity, and allowed to remain three days. Free suppuration ensued.

April 28.—Discharged, nearly well.

WILLIAM HUNT.

EXCISIONS, AND OPERATIONS FOR UNUNITED FRACTURES.

THE manner in which the accompanying table is made up from the note-books and other sources, will illustrate the rather vague way in which the terms excision and resection are used.

The word excision has a definite meaning, while resection seems to be indifferently applied to all sorts of operations upon bones and joints. We really see no need for the latter word whatever, and agree with those who would drop it altogether. Excision of a joint and excision in continuity, it seems to us, would express all that is required to designate the respective operations. The term resection is a puzzling one to the student, and no wonder, for besides conveying the idea of the repetition of an operation, he hears it, even to this day, indiscriminately applied to the cutting out of joints, to the removal of diseased or broken ends of bones, and to drilling and refractures. These two latter operations, unless excisions are actually part of the proceedings, should come under distinct heads.

We find, in culling excisions proper out of our list, that 8 have been of the hip-joint, 7 of the elbow,—*i.e.*, some portion or all of its joint surfaces,—7 of the knee, 7 of the ankle and calcis, 2 of metatarsal joints, 4 of inferior maxilla, 1 each of the humerus, carpus, head of fibula and patella. The malar bone and superior maxilla were removed, and there were two excisions of portions of the femur in continuity, and four operations of the same kind upon the tibia and fibula.

The record is a surprise to ourselves, as showing the unfrequency of excisions for recent injury. In the whole list of joint cases there are but two of this kind, both of them being of the elbow. All of the others, involving the great joints at least, were operations for chronic conditions. When we consider that this is a great accident hospital there is still more reason for surprise. The difference between times of peace and of war is well illustrated in the nature of the injuries, for it was gunshot wounds that made us so familiar with excisions of joints in recent cases, whereas the crushings and tearings of railroads and machinery are mostly so extensive and disintegrating, that there is rarely an opportunity to practise conservative surgery. While discussing the question of recent compound injuries to the larger joints, and the course to be pursued in their treatment, we will state what our experience, and observation of cases afterwards, has led us to do in practice.

Excisions of the Shoulder.—Having determined that an attempt to save the limb is justifiable, we should first excise the head of the humerus, and then remove the fragments from all sources at once. The rapidity of recovery and the use of the limb are often alike wonderful, and much to be preferred to the necrosis, anchylosis, and impaired movement that are almost sure to follow the ultra-conservative method of waiting for the dead bone to separate.

Excisions of the Elbow.—Now, strange as it may seem, we are less enthusiastic as to excision when the elbow-joint is the subject of compound fracture. Here amputation having been rejected, the only other choice may be excision; but suppose all of the parts are not irretrievably injured, and that there is a fair chance for recovery, with more or less anchylosis, with no other operative interference than that of taking away the necrosed portions of bone as they separate; by following this latter course we are confident that, when the parts are placed at a proper angle, a more useful and stronger member will often be secured than that which will result from immediate excision.

Among other cases strikingly illustrative of this point there was one requiring especial mention. The patient was a young woman, a daring and noted performer upon the "flying trapeze." One night she missed catching the cross-bar, and fell heavily upon the stage. The chief injury sustained was a very bad compound fracture of the left elbow. The bones were protruding, the condyles comminuted, and the olecranon broken. The radius escaped.

The broken parts were manipulated into position, and the limb was placed upon a pillow, and fixed at a rather acute angle. Free drainage was secured, and the wounds were dressed with carbolized oil.

Frequent consultations were held during the progress of the case, but non-intervention, except to remove loose fragments of bone as they separated by normal processes, was always decided upon.

The result was a recovery, with anchylosis of the humerus and ulna, but with rotation of the radius, and complete use of the wrist and hand. So complete indeed was the use of the limb that the patient actually resumed her former business for a while, but finding this rather dangerous, she devoted herself to the performance of tricks of sleight-of-hand, in which she was also an adept. We are confident that no such use of the limb would have followed excision. We have on several occasions compared cases that have been treated in both ways, and we have mostly found that the anchylosed limb, provided the proper angle had been maintained, was the superior of the other both as to strength and to control of movements. Therefore we would say as to the elbow, in acute cases, when the question is between excision and amputation, excise; when it is between immediate excision and waiting, wait.

Excisions of the Wrist-Joint.—An injury of the wrist-joint is one that will bear temporizing with to advantage. When we reflect that the whole of the complicated machinery of the upper extremity is really subservient to the thumb and fingers, and that without these the limb is little more than a lever or paddle, we appreciate the importance of efforts to save any connections and supports which may possibly contribute to the delicate and varied movements of the parts. As we approach the fingers these connections are concentrated, especially about the wrist-joint, so that a complete excision of it would give little promise of good as compared with the other method.

We regard the lower extremity with much less veneration than the upper in reference to questions of risks to be taken in efforts to save it. The remark of Maclise is as true now as it was when first written, that "no mechanical ingenuity can fashion an apparatus capable of supplying the loss of a finger, or even of one of its joints," whereas the lower extremity, being for progression and support, is, when lost, resupplied by that same ingenuity to a very fair extent. We have had a patient under care who had lost both hands by accident. He was a very pitiable object, and his disabilities may be much more readily imagined than described. We have, on the contrary, had several cases of the loss of both legs, where the patients were restored to health and usefulness, and in some the deformity was entirely hidden by machinery. The lesson learned, therefore, as to the lower extremity is, that when, in recent cases, the risk of an attempt to save will involve greater danger to life than any merely probable good that might accrue to the patient should he survive those dangers, that risk should not be taken, but immediate amputation should be advised.

Excisions of the Hip-Joint.—We should probably qualify the above remark as to the hip-joint, for the dangers to life of amputation are as great in that position, if not greater, than any other proceeding of operative surgery.

We do not remember to have seen a case of recent injury to the hip-joint requiring excision, except from gunshot, and in times of peace cases of the kind are rare. We have seen and known of manglings and dreadful crushings in this region by railroad and machinery, and squeezings between ponderous stationary and moving objects, such as a steamboat and the wharf, or an elevator and the floor or ceiling, but never has the injury been so defined in character as to call for immediate excision. Some remarkable recoveries have taken place, and possibly some have become subjects for excisions after having reached the chronic state. Should the opportunity occur, however, and the well-known conditions for the practice of all excisions, such as the integrity of the blood-vessels, etc., be present, we should advise the operation in preference to either amputation or a purely expectant treatment.

Excisions of the Knee-Joint.—The knee-joint is a treacherous part to deal with after acute injury, and the question of operative interference is one of the gravest character. We have known of most excellent results after the expectant plan of treatment, and at present cannot call to mind a single case of immediate excision during the time we have been connected with the hospital.

The injuries to the part are frequent, but cases appropriate for excision, that is, those intermediate between amputation and expectancy, have rarely, if ever, occurred.

Adults do badly as a rule, judging from the cases of excision of the knee-joint after gunshot wounds that we have seen and have read of, and the prospect seems little better for children after recent injury. The dangers to life certainly equal those following amputation, and the average time for recovery, or trying to attain it, is much greater. Not unfrequently amputation has to be the final resort. Nevertheless, with all these discouraging facts in view, we should not hesitate to advise and practise excision of the joint for recent injury in appropriate cases. Thoroughly antiseptic surgery may accomplish great results in this field.

Excisions of the Ankle-Joint.—We are very decided in opinion as to what to do here in acute cases when operative interference is necessary: amputate rather than cut out the joint. A bad compound fracture of the bones, with exposure of and breaks into the joint, is not to be temporized with. We have elsewhere stated in this book that regret rather than congratulation will mostly follow attempts to save cases of this kind from immediate amputation. Sometimes a compound luxation of the astragalus may be successfully treated by the removal of the whole of the bone.

Excisions of the Tarsus.—When we get below the ankle-joint, excision again rises in favor. The os calcis, the cuboid, in fact, any bone of the tarsus not directly entering into that joint, may be cut out wholly or partially with a very fair prospect of success, and often with the saving of a remarkably useful foot. When we consider also that the record of amputations through the tarsus shows quite a high rate of mortality, we are still further encouraged to adopt the conservative method.

In regard to excisions of the smaller joints after acute injury, we have to say that the usefulness of the special part involved must or should determine the course to be taken as to risk and time. All ingenuity possible should be expended in trying to save portions of the hand and fingers, and especially the thumb, while it would scarcely be worth while to devote the same energies to a smaller toe; not that the latter may not be worth saving, but that a clean amputation is much to be preferred over tedious and painful efforts to preserve what after all may be of doubtful use, or which may ultimately become even an impediment.

Excisions in Continuity are sometimes performed in the hospital for recent injuries, especially for compound fractures of the leg, with protrusion and comminution. They are operations that are oftener performed from necessity than from choice; though on the part mentioned with very good average results. For the femur they are not encouraging, and when done on this bone for comminution, as from gunshot, the chances of success are very small, and certainly no better than the ultra-conservative course of waiting, keeping up free drainage, and only interfering to remove necrosed pieces as they separate.

The humerus and the radius and ulna will bear such operations upon them with a very fair prospect of success.

The whole aspect of excisions changes when they are performed for the relief of *chronic conditions;* mostly arising from apparently trifling, neglected, or unsuspected injuries, though some undoubtedly have their origin in disease. The great joints bear them with a surprising degree of toleration, although it might well be said in many cases that it is really no joint we are removing, but a mass of diseased and dead material that bears no semblance to its original structure, and that has lost all of its original function. The broken-down sufferer that gets rid of the burden often recovers his health and liberty, with a great and sometimes almost perfect use of the restored member. The four great joints, the shoulder, elbow, hip, and knee, are operated upon most frequently and with the most success. The wrist and ankle, from the same causes that have been mentioned in the remarks about them as to recent injuries, often obstinately refuse to respond to the efforts in their behalf. Excision of the tarsal bones is often successfully done.

By our table we find that about the same number of operations were performed upon bones for vicious union and for ununited fractures as for excisions proper.

The table sufficiently sets forth the methods employed. When we consider the great severity of many of these operations the mortality after them is very small, for, of 42 cases, we find that 33 were cured, 7 were relieved or improved, and 2 died. One of the deaths followed drilling and refracturing of the bones of the ankle, and the other occurred to a case of ununited fracture of the femur.

The whole summary of the results of these severe operations upon the bones and joints is very satisfactory. Of the 99 cases, 75 were cured, 12 were relieved or improved, 10 died, 1 was unimproved, and 1 was removed by friends before the result was ascertained.

Of the excisions proper there were but 6 deaths, viz.: 2 of the hip-joint, 2 of the knee, 1 of the elbow, and 1 of the astragalus. A few additional cases have been added to the table since the first part of this article was written. Among them there are no deaths following excisions of joints to report. One death occurred after cutting out a portion of the radius and ulna, and one followed a like operation on the tibia.

Notes of Cases of Excisions, and Operations for Ununited Fractures.

Resection of the femur in continuity. Case 15.

CASE 15. *Necrosis of femur: ununited, spontaneous fracture of shaft; resection; recovery.*—Charles G., aged 8 years, was admitted August 19, 1869. Severe contusion of thigh in May was followed by abscess, and a few weeks after by spontaneous fracture of the femur. Attempts by extension and bandaging to secure reunion had failed. Sinuses were discovered leading to dead bone; soft parts infiltrated with pus.

On September 1, Dr. Morton, by a long longitudinal incision, turned out the ends of the necrosed bone, and removed half an inch from the upper fragment and about one inch from the lower. An excellent recovery followed, and the patient was discharged with firm union.

June 21, 1870, an artificial support was supplied (see figure), which enabled him to walk with entire comfort.

Case 16. *Necrosis of the os calcis and astragalus; excision; recovery.*—Dennis C., aged 13, was admitted September 27, 1869. In January, 1868, he received an injury of the left heel. An abscess formed; the entire foot became much swollen. Four fistulous tracts existed when admitted, and a probe readily detected extensive necrosis of the os calcis and astragalus.

Excision of the os calcis and astragalus. Case 16.

October 2, the usual operation was performed by Dr. Morton. No artery required ligation. A rapid recovery followed. The foot was shortened an inch. (See figure.) Several years afterwards this patient was examined, and the foot was perfectly well.

Case 17. *Complete osseous anchylosis of knee, with extreme flexion; excision of a wedge, including part of condyles, head of tibia, and patella; recovery.*—William H. M., aged 26 years, was admitted November 26, 1869. Nine years previously, accidentally inflicted upon himself an incised wound just below the knee, probably opening the joint. Using the joint too early suppurative action ensued in the knee, obliging him to keep his bed for fourteen months. At the end of this time there was firm bony anchylosis in complete flexion, so that he could only walk by almost squatting upon the ground. The leg was well developed, and inflammatory action had entirely ceased when he was admitted, although its results were seen in numerous cicatrices, showing the position of former sinuses around the joint.

On December 1, 1869, Dr. Morton, with the saw, excised a V-shaped portion of the bone, measuring four and one-half inches across its base and two and one-half inches in depth. A portion of the head of the tibia and of the condyles of the femur, with the patella, were removed; all traces of joint structure had disappeared. The hamstring tendons did not require division. At the end of three months fair union had taken place. Some fragments of bone were removed in June, 1870, and he was discharged, cured, in September.

A year afterwards the patient wrote that he was well, and could walk upright without crutch or cane. Anchylosis was firm in the improved position.

Case 23. *Necrosis of the head of the femur; excision; recovery.*—George W. H., aged 17, was admitted August 29, 1870. Patient in very bad condition from necrosis of head of femur and also of the ilium. The disease followed a gunshot wound, the gun being loaded with shot. The accident had occurred many months before admission. Dr. Hunt made a large and deep semilunar incision over the trochanter, and removed the head of the femur and a portion of the border of the ilium, with edges of the acetabulum. The limb was supported with sand-bags, and free drainage secured. The patient made a rapid recovery. His general health was excellent when he left the hospital, and he walked with a cane.

Case 24. *Necrosis of the os calcis; excision; recovery.*—Maggie G., aged 4 years, strumous, was admitted August 31, 1870. An injury five months before was followed by abscess, which left several sinuses around the heel. The os calcis was found diseased. A curved incision around the inner malleolus, made by Dr. Morton, permitted the removal of the entire os calcis, which was quite necrosed. She was discharged, cured, October 31, 1870.

Case 28. *Fractured femur united with great deformity; drilling; refracture; recovery.*—John H., sailor, aged 37, was admitted February 19, 1872. This patient came in to have his leg amputated, for he considered it useless. It had been fractured at sea, and had united very firmly in a position of great angular deformity. The inner border of a small portion of the front of the foot could be brought to the ground by great effort. The easiest position was that of equino-varus, and no effort could bring the heel to the ground. The patient had been in an English hospital for a long time, but no operative measures were deemed advisable. It was determined to make an effort to save the limb rather than to amputate at once. Dr. Hunt drilled the bones in various directions, and then, with the assistance of Drs. Hewson and Morton, broke them over the edge of the operating-table. The tendo Achillis was then divided subcutaneously, and the limb was put up with pasteboard splints and suspended in a fracture-box. The recovery, after passing through the various troubles incident to such a case, was complete. The patient is now employed on a farm, and has been often seen walking without a limp.

CASE 31. *Necrosis of inferior maxilla following scarlatina; removed in two operations; recovery; subsequent development of permanent teeth.*—Maria S., aged 8 years, was admitted several times in 1872. When first admitted, in 1870, the left side of the lower jaw being found sequestrated, was removed in several fragments by excision. She subsequently submitted to another operation, in which the necrosed right half of the bone was removed. The specimen, now in the museum, was lying loosely as a sequestrum, enclosed in the newly-formed bone. The teeth were not disturbed, and subsequently a permanent set have been developed, although showing some irregularities.

In 1877 she was examined, and found to be entirely well. The jaw was of good shape, and a few scars only remained to indicate the operation.

Pathologist's report.—The right half of the bone on its outer surface is smooth and of normal appearance, except near the angle, where it and also the inferior border are eroded. The inner surface at its middle portion is entirely destroyed.

The interior of the bone, from the inferior dental foramen, which is well preserved, forward, is hollowed out into an extended cavity, bounded by thin layers of bone, composed of the outer and inner surfaces and some spongy masses adherent to them. The neck, and the inner and lower part of the articular surface, have likewise suffered considerable necrosis. The angle is very obtuse, so that the line is nearly straight, the coronoid, which has also suffered some necrosis, making a sharp projection from the upper surface.

CASE 39. *Hip-joint disease; excision; recovery.*—Maria M., aged 14, was admitted March 30, 1873, with disease of the joint, which had existed for many years. Numerous fistulous tracts extended into the head of the bone. Dr. Morton made the usual incision, and removed the diseased bone; an excellent recovery followed. (See figure.)

Excision of the hip-joint. From a photograph taken six years after the operation. Case 39.

CASE 44. *Fracture of femur, fibula, forearm, and humerus; improved.*—George B., aged 13, was admitted April 26, 1873, with recent fracture of femur, also a fracture of humerus and forearm. Sand-bags applied to the thigh, internal angular splint and binder's board cap to upper extremity. Extension was applied May 1, when a fracture of the fibula was detected, and soon after an abscess followed. June 3, humerus and forearm united, and good union was taking place in femur, but he was restless and persisted in sitting up in bed. July 10, a silicate permanent dressing was applied. Pretty firm union on August 2, when the dressing was removed, but on the 4th he fell down and refractured the bone.

He was discharged, September 2, to return to his home, with ununited fracture of femur.

CASE 48. *Necrosis of ankle; resection of joint; relieved.*—Charles J. S., aged 17, was admitted September 30, 1873. At two years of age sustained a contusion of left foot, followed by abscess, which only gradually healed. Nine years later again injured the part, and abscess recurred. In January, 1873, the ankle commenced to swell again, and finally suppurated. Sinuses were found on admission. October 8 the astragalus was excised, with the ends of the tibia and fibula. No bleeding; dressed with carbolized oil. November 8, allowed to walk with crutches, and a posterior wire splint on the foot and leg. April 7, 1874, discharged, relieved.

CASE 52. *Deformity following fracture of humerus and femur: refracture.*—Stefano S., aged 15, was admitted February 27, 1874. Fell from the mast while out at sea, three weeks before, fracturing arm and thigh; bones pretty firmly united at an angle. Shortening of leg one and one-half inches. March 4, bones refractured by Dr. Hunt; lateral splints and extension; and on the 16th immovable plaster dressing was applied from the waist to the ankle. April 7, plaster removed; one inch shortening; extension reapplied.

He was discharged, cured, May 30, 1874.

CASE 53. *Deformity following fracture of leg: refracture: recovery.*—William L., aged 27, sailor, was admitted July 6, 1874. Fourteen weeks before, broke his leg while at sea, followed by overlapping and bad line of union. The bones were refractured by drilling and force; the overlapping portion was sawed off.

He was discharged with a useful limb, December 3, 1874.

CASE 58. *Necrosis of femur followed by incurable fracture.*—James J. H., aged 39, was admitted May 10, 1875. Necrosis of the right femur began at three years of age. At sixteen years he fractured this bone, and a piece of sequestrum was discharged. At twenty-four years he refractured the bone by a fall, but after recovery he fractured it again recently. All efforts to obtain union were useless, and he was discharged incurable.

Drills.

CASE 59. *Gelatinoid degeneration of knee-joint, with anchylosis: excision: recovery.*—Maggie S., aged 4, was admitted June 7, 1875. Received an injury of knee about two years before. This was followed by suppuration, the formation of sinuses, moderate atrophy

FIG. 1.

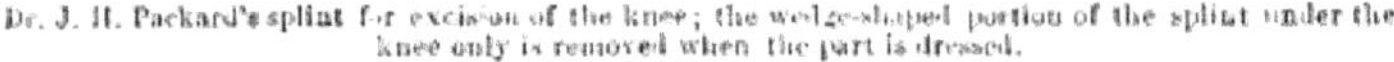

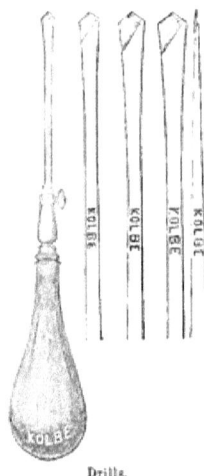

Dr. J. H. Packard's splint for excision of the knee; the wedge-shaped portion of the splint under the knee only is removed when the part is dressed.

FIG. 2.

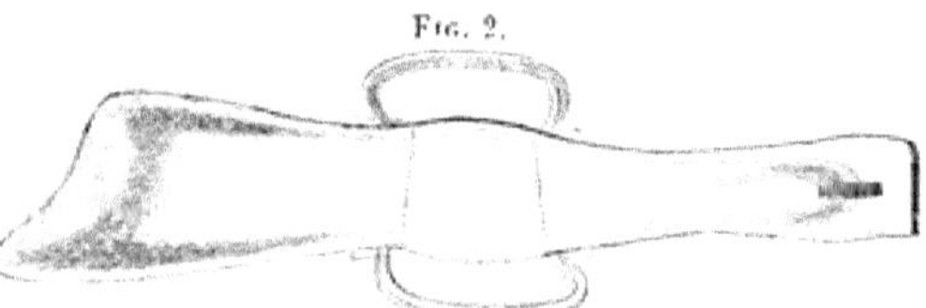

Side view of the splint. The dotted line shows the wedged portion.

of muscles, the knee assuming the position of complete flexion, in which it became anchylosed. On June 9, Dr. Levis made an anterior, oval flap, and removed half an inch from the articular surface of the femur, and the same from the tibia. The limb was then straightened and put in the fenestrated splint of Dr. Packard. (See Figures 1 and 2.) The case progressed without any bad symptoms. Anchylosis took place, and the wound entirely healed. The pathologist, Dr. Longstreth, gives the following description of the portion removed by operation:

Pathologist's report.—The specimen shows the articular surfaces of the femur and tibia. The femur has been divided about one-half inch above the prominences of the condyles. The surface of the section shows in its central portion a small area of cancellous tissue, which is much softened and has a dirty yellow color. The periphery of the section, perhaps about one-half inch in breadth, is composed of unossified cartilage. The capsular ligament and portions of the crucial ligaments remaining are very considerably thickened. The condyles on the articular surfaces are denuded of cartilage over about half their extent; the other portions are covered with tough fibro-cartilaginous material, apparently the result of an inflammatory process. The denuded portions of the condyles have an osteo-cartilaginous consistence, and show many small openings penetrating to some depth into the bone. Around these openings the deposit of bony matter is more abundant than at other portions of the articular surface. The head of the tibia shows very much the same condition as already described. There is no portion of it, however, denuded, but the amount of inflammatory matter covering its cartilages is much greater in amount.

CASE 63. *Scrofulous synovitis of the knee; excision; recovery.*—John H., aged 10, was admitted July 6, 1875. An excision was performed, but extensive suppuration ensued, and finally the necrosis became so extensive that an amputation was performed. (See Case 13, Amputations.) He was discharged, cured, October 25, 1875.

Pathologist's report.—The specimen shows the lower end of the right femur, the upper portion of the tibia and fibula, with the patella, attached by its ligament. The articular surface of the femur is partly denuded of its cartilage, especially on the posterior aspect of the condyles, and at the upper border anteriorly. The cartilage itself is swollen and doughy. The exposed end of the bone is a pinkish-yellow color, somewhat eroded and softened. The head of the tibia has lost its cartilage both on the inner and outer articulate surface. The crucial ligament from the inner side has entirely disappeared, while the one from the outer side exists only as a stump. The patella has a spot the size of a penny on its under surface denuded of cartilage.

CASE 64. *Hip-joint disease; excision of the head of the femur; recovery.*—David H. M., aged 23, was admitted August 25, 1875. Mother died of typhoid fever, father of phthisis. He was admitted into hospital in 1874 for synovitis, and was much improved when discharged. During this time was treated by extension. After this he considered himself quite well, and, until two weeks previously, had no further trouble; then extensive inflammation and abscesses appeared around the joint. Upon admission several fistulous tracts existed on the outer part of the thigh, with all the signs of hip-disease in the suppurative stage. The general health was much impaired, requiring tonics, stimulants, etc.

Excision of the hip-joint, Case 64.

On the 11th of September it became a question of life or death. The discharge was excessive, and the patient was very weak. A long excision was now made, and the joint was excised by Dr. Morton. The head of the bone was found surrounded by pus, the round ligament having been destroyed. The femur was divided one and one-half inches below the great trochanter. The digital fossa was involved in the necrosis. On September 14, extension was applied.

The temperature was elevated upon admission, and soon after reached 104°. It continued high until the day of operation, when it was 102½°. From that time it gradually fell, and fluctuated between 99° and 101½°. Discharged, cured. (See figure.)

Pathologist's report.—The specimen shows the femur divided just below the trochanters. The cartilage covering the head of the bone is softened and doughy, and easily indented by the finger or peeled off from the head of the bone. The round ligament was separated, and had nearly all disappeared, leaving a depression in the head of the bone the size of the end of the thumb, which is only partially covered by cartilage. The neck of the femur shows but slight alterations, there being at but one or two points a thin shell of bony tissue developed in the periosteum. The upper part of the shaft is considerably roughened in places, has numerous surface changes, existing either as thin shells of bone or as rough spongy growths, on its surface. The cut surface of the bone shows considerable increase in the thickness of its compact surface. The inner layer of the compact surface around the medullary cavity is becoming softened and spongy. The medullary canal is filled with marrow, whose consistence is less than normal, and having a pinkish-yellow color.

CASE 65. *Partial anchylosis of knee; excision; recovery.*—Michael D., aged 9, was admitted September 24, 1875. Consumptive antecedents. Had strumous synovitis of left knee, which came on without apparent cause three years before. The joint discharged for a year, and then healed up, leaving partial anchylosis. Poor health upon entering the hospital. Left knee flexed to an obtuse angle; slight motion in the joint; neighboring surface shows scars of old sinuses; patella adherent; hamstring tendons contracted.

September 25, the knee was excised by Dr. Levis. Patella firmly adherent to external condyles; strong ligamentous union between tibia and femur. The end of the femur, the patella, and the articulating surface of the tibia were removed as one piece, the bones being sawn partly through and then forced apart. The extremity was placed in a Packard splint, suitable dressings wet with carbolized solution being applied.

During the next few days there was considerable fever, with diarrhœa and delirium. The edges of the wound sloughed. On October 1 it was noted that the slough had separated; the patient's condition was improving. He was again etherized, and the end of the femur was removed; the wound was dressed with a poultice. From this time he did well until, on December 3, the wound was entirely healed. After remaining in the ward for some time as assistant, he was finally discharged entirely well.

CASE 66. *Compound fracture of right ulna, opening the elbow-joint; the left tibia and fibula dislocated forwards; resection of elbow; death.*—James P., aged 25, was admitted October 29, 1875. A wall fell upon him a few hours before admission. The olecranon process of right ulna was broken, and the humerus and ulna protruded through the wound. The left knee being dislocated (forwards), it was reduced. The patient being in shock, nothing in the way of operation to the arm was attempted, but antiseptic solutions were applied. Quinia and stimulants internally. Having reacted well, on October 31 the joint was resected, and carbolized oil dressing applied.

On November 4 he had stiffness and pain in the jaw. Ordered morphia hypodermically (gr. ¼), and chloral (gr. x) every two hours.

November 5, could only open mouth a short distance; arm much swollen; discharge free; muscles of abdomen and neck were rigid and tender. The next morning he was delirious. Temperature 101° in the evening, pulse 140, respiration 36. On November 7 he died of exhaustion, but not in a condition of spasm.

Autopsy was only partial. The vessels upon the posterior surface of the cord in the pia mater were distended with blood; very little fluid in the arachnoid cavity; dura mater white and glistening; slight sub-arachnoid effusion; no evidence of lymph; vessels of pia mater full, but more particularly in the lower part; cord deficient in firmness.

CASE 72. *Necrosis of the elbow; excision; recovery.*—Henry D., aged 35, was admitted April 29, 1876. Rheumatic pains in right arm during the winter. Three weeks before admission elbow began to swell, and compelled him to leave off work. Upon admission the elbow was greatly enlarged. An abscess had formed around the joint, in which a free incision had been made; partial anchylosis demonstrated the disease to be within joint; great pain in motion. A splint was applied, and general constitutional treatment ordered, including cod-liver oil.

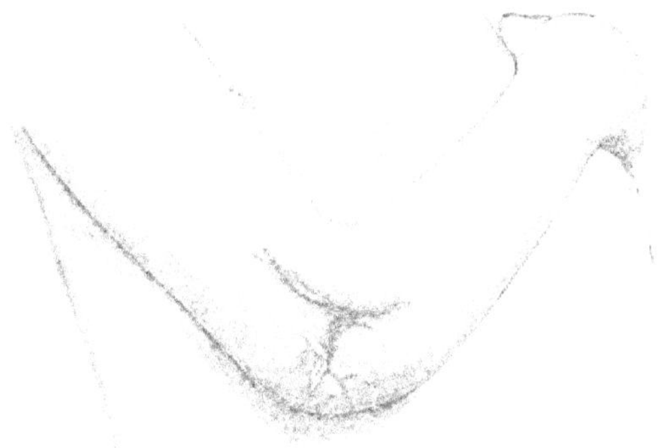

Showing excision of the elbow after recovery. Case 72.

On September 6, necrosis having occurred, resection was determined upon, and before Prof. Lister and other members of the International Congress, Dr. Morton made an incision five inches long on the outside of the arm. The internal condyle was found separated from the humerus. Upon bending the elbow, so as to expose the joint, the humerus gave way about the middle. Resection of the elbow was then completed. Carbolized charpie was applied, and an internal right-angle splint. No unfavorable symptoms followed the operation, and a rapid recovery, with a useful limb, resulted.

He was discharged November 28, 1876.

CASE 74. *Disease of the hip-joint; excision; recovery.*—Harry T., aged 17, was admitted August 12, 1876. No constitutional taint. Two years before he injured his leg by having it caught in the wheel of a wagon and badly wrenched. He walked home after the accident. Nine months afterwards the hip began to swell and an abscess was opened, which discharged freely. His health had steadily declined; emaciation was marked; appetite poor. Under tonics, oil, and good diet he greatly improved, although the discharge persisted, and there was some œdema of the limbs.

On November 11 the joint was laid open and head of bone excised. All the parts in the vicinity were softened. He was much depressed after the operation, but revived after the administration of champagne, which also relieved his distressing vomiting. He steadily improved with careful nursing, notwithstanding an attack of diarrhœa, and a cough that was evidently connected with structural pulmonary disease.

July 3, 1877, after having been about the institution for several months, he was discharged, cured.

CASE 77. *Necrosis of os calcis; removal; recovery.*—Frederick R., aged 20, was admitted September 12, 1876. Had been under treatment last March for abscess of the ankle, but remained in the hospital only four days. Although dead bone was detected, no operation was thought advisable on account of the patient's health. He had consumptive parents. Six months before his admission he sprained his ankle, and an abscess subsequently appeared followed by persistent sinuses. Pain, swelling, induration, and discoloration, and all the indications of necrosis of joint were present when he applied for readmission. He was placed on tonics and good food, and on the 20th of October the greater part of the os calcis was removed with a gouge. The disease continuing, on January 9, 1877, the remainder of the calcaneum was removed, having been found in a softened condition. The tarsal bones (cuboid, etc.) in immediate proximity were also chiselled, as their articular surfaces were involved. The astragalus was not diseased. The foot was afterwards kept at a right angle until union occurred. On February 5 lateral binder's board splints were applied, and he was allowed to go about.

April 2, 1877, he was discharged, cured.

CASE 82. *Deformity following fracture of femur: refracture: good result.*—Bridget T., aged 19, admitted April 9, 1877. Fractured femur in middle on September 26, 1876. Deformity from overlapping followed: two and three-quarters inches shortening upon admission. The patient can walk, however, without a cane or crutch. April 18 the femur was refractured, and Dr. Morton's apparatus applied for extension. Recovered with a useful limb.

Discharged July 2, 1877.

CASE 83. *Acute hip-disease from injury: rapid progress of the affection: great destruction of the joint and adjacent pelvic parts: spontaneous luxation of the femur on the dorsum ilii: excision: death.*—James McK., aged 17, was admitted May 12, 1877. Family history good. Always a healthy boy until four months ago, when he fell into a culvert. Soon after this he felt soreness in hip. In one week was confined to bed with constitutional symptoms; fever and rigors and emaciation. In less than three weeks a large abscess was opened, the pus burrowing down the thigh, and discharged at several points. On admission into hospital he is very much emaciated, pale and anaemic. There are bad sores over sacrum, elbow, and left scapula. The hip is flattened and the veins of abdomen prominent. There is a free discharge from opening over crest of ilium, and numerous sinuses on outer aspect of thigh. Bowels disposed to be loose. No fever. Pulse fair but frequent. Urine normal. Ordered syr. phosph. comp., ʒii t. d.; whiskey, ℥ii daily.

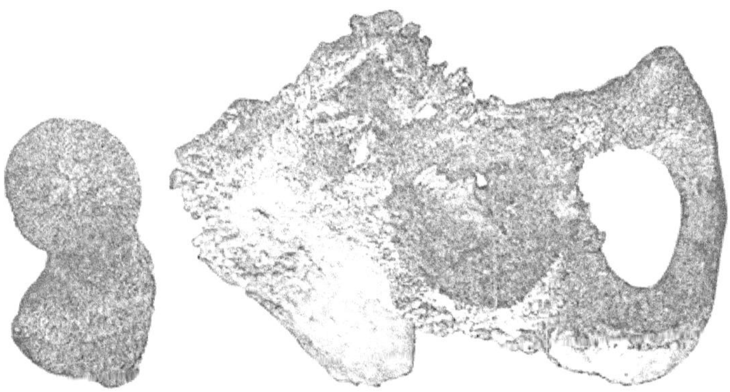

Excision of the head of the femur, from Case 83; also diseased pelvis, removed after death.

May 15.—Under ether, Dr. Morton excised the head of the right femur. A long incision was made to include several sinuses, and carried down to the trochanter. When the finger carried in revealed the head of the bone high up on the dorsum of the ilium it was turned out, and the shaft, about one and one-half inches below great trochanter, cut through by chain and straight saw. The acetabulum was much obliterated, and with the forceps a considerable portion of the crest of the ilium was removed, being found quite separated from the body of the bone. Wound dressed with solution of alum and compress.

May 16.—Dressed with carbolized charpie and extension.

May 17.—Dressing to be changed twice daily: doing well. Bowels not frequently opened, but stools watery. No vomiting or nausea. Opium suppository, gr. i, t. d.

May 18.—Commence again syr. phosp. comp., ℨii.

May 20.—Discharge excessive. Wound apparently healthy. Temperature 102°. Getting weak.

May 21.—Examination of urine negative. Patient is much feebler. Vomiting; ordered champagne.

May 25.—Opened small abscess at the right sterno clavicular juncture; escape of fetid pus, which is sucked in and bubbles out again with respiration. Temperature still about 103°.

May 27.—Died this P.M.

Autopsy.—Body much emaciated; the abscess of chest was found to extend down from sterno-clavicular articulation, which was much necrosed, involving the second right cartilage and rib, and to communicate with the anterior mediastinum, but not with pleural cavity. There were some slight adhesions in the left pleural cavity. Heart normal. Two ounces of clear serum in sac. Several infarctions in spleen, one the size of a walnut. Kidneys normal. Liver normal.

The pelvis, the ilium being boiled, showed on the diseased side very extensive ulceration, being riddled with holes, and great numbers of osteophytes were thrown out over its entire surface.

The acetabulum was not perforated, although involved in the ulcerative action.

The head of the femur is very much altered in shape, flattened from before backwards, and the anterior and inner part of the head has been considerably eroded. The articular cartilage has been entirely removed, and also the periosteum from the upper portion of the neck. These portions are in part covered by patches of inflammatory matter.

Clavicle shows at the sternal end an erosion, with a complete change in its form, presenting an irregular quadrilateral figure. The erosion extends deeply into the cancellous tissue, and there are thick layers of new bone, which have been deposited from the periosteum.

The whole of the right os innominatum is involved in the disease, except the ascending ramus of the ischium, the body and descending ramus of the pubes, the part of the ilium around the greater sciatic notch, and the posterior spinous processes. The diseased portions of bone are reduced to a rough, porous, spongy consistence; in and around the acetabulum the diseased process has had its greatest intensity, and from there has extended upwards to the spinous process and along the crest of the ilium. There is a thin shell of compact substance, about one inch long and one-half inch broad, in the hollow of the acetabulum.

The suppuration has extended deeply into the tissue of the bone, and the inflammatory process on its periosteal surface has led to the formation of numerous osteophytes, which in many places, and especially around the lower border of the acetabulum, have greatly increased the thickening of the bone.

CASE 84. *Suppurative synovitis of the knee joint; excision of the knee; subsequent necrosis, and amputation of the thigh; recovery.*—John C., aged 7, was admitted May 25, 1877. His family and personal history until the age of three were good, when without known cause there commenced trouble in the knee, which compelled the use of crutches. A week previous to admission the joint for the first time became red, painful, and swollen. On admission the leg was found completely flexed, so that the calf of the leg had made a depression in the muscles on the posterior part of the thigh. There were two sinuses in front over the head of the tibia, discharging a little creamy pus. On June 9, the head of the tibia and the lower part of the femur were removed. The bones were found riddled with holes and excavated by cavities. The hamstring tendons were divided, and the limb partially straightened.

After the operation the skin over the knee in front sloughed, but rapidly commenced to be covered with healthy granulations, to which skin-grafts were applied with success. The limb was gradually straightened more and more, and union of the divided ends was sufficiently firm by the end of September. During the following three months the pain, discharge, and swelling of the knee increased. For the subsequent seven months the patient was up walking about the wards and yard, and on the 1st of July, 1878, was removed to Children's Seaside Home, Atlantic City.

The patient was readmitted to the hospital September 17, 1878. There was a large ulcerative surface in front of the knee and in the popliteal space, extending to the calf of the leg. The discharge was less than previously, but the bones, both the femur and tibia, showed that the disease was extending farther in their length.

On November 16 it was concluded to amputate the thigh in the middle third.

After two secondary hemorrhages the patient recovered, and the stump was firmly healed by the first of the year. He was allowed to remain in the hospital until May 21, 1879, when he was discharged, cured.

Pathologist's report.—The specimen shows the articular ends of the femur, a portion of the head of the tibia, with the patella. The cartilage on the end of the femur is very considerably eroded, opening the cancellous tissue of the bone. The change is especially marked on the outer condyle, whilst the inner condyle and the inter-condyloid notch are covered with, or partly filled up with, firm fibrous material, having on its meshes thick inflammatory matter. The portion of the head of the tibia presents very much the same conditions. The ligament and fibrous tissue surrounding the patella are very much thickened. The under surface of the ligamentum patella is covered with a thick, velvety layer of synovial membrane. Nearly the whole under surface of the patella is denuded of its cartilage, and on the outer articular facet the bone is deeply eroded, extending nearly half-way through its substance: the inner facet has a thick layer of fibrous tissue covering it.

CASE 86. *Necrosis of head of humerus; resection; cured.*—Felix M., aged 9, was admitted September 11, 1877. Two weeks ago fell out of a wagon, striking on his shoulder. There is a good deal of swelling, pain, and loss of flesh. The tissues around the joint are glazed, tense, and infiltrated.

September 12.—Abscess opened at shoulder. Head of the bone was extensively diseased, but the patient was not in a condition for operation. Treatment: poultice, and good diet and stimulants.

September 16.—Chill, followed by fever. Ordered quinia, gr. xx, and brandy, ℥ij, every two hours.

September 19.—Appetite better; feels stronger.

October 8.—Very little discharge; gaining flesh.

October 10.—Etherized, and the head of humerus resected by an incision on outer side of shoulder five inches long. Wound dressed with carbolized lint and compress. Temperature 102°.

October 13.—Not much discharge; boy is gaining flesh.

November 7.—Wound healing slowly; arm hangs; there is some motion.

November 19.—Discharged, cured.

CASE 89. *Deformity following fracture of the tibia and fibula: refracture; recovery.*—Joseph G., aged 37, was admitted November 21, 1877. Three or four months ago he fell and fractured his leg at the lower third. He was in bed two weeks. Was then allowed to get up and walk about with crutches. On admission there is great deformity just above the external malleolus, foot being thrown on the side when standing. In walking he treads on the outer side of foot.

November 24.—Leg refractured, and placed in good position.

November 26.—Under ether leg was again straightened, and placed in fracture-box.

December 17.—Fracture-box removed, and lateral pasteboard splints.

January 3, 1878.—Splints removed: union firm. Leg bandaged, and ordered to use it.

January 10.—Is now walking about the ward; little or no deformity.

January 18.—Discharged, cured. Union firm; no deformity.

CASE 90. *Compound fracture of tibia and fibula; improved.*—Frank D., aged 27, was admitted December 18, 1877, with a recent compound fracture of the left tibia above the ankle. The bone projected over an inch from the wound. The fibula was also fractured about its middle. On the 21st, the limb appearing swollen and dark-colored, threatening gangrene, Dr. Morton made free incisions into the leg, and relieved the tension. Finding the ends of the tibia overlapping, he was obliged to resect the tibia, cutting off about two inches. Subsequent suppuration required a bran dressing, and extension was applied in the fracture-box. No union having occurred, on April 6, 1878, Dr. Hunt opened the fracture, and removed two fragments of dead bone. Moderately firm union having been secured, on August 27, a silicate bandage was applied to leg and thigh, which was removed September 9, owing to swelling of the foot, and the fracture-box was again resumed.

September 28, Dr. Morton drilled the lower fragment of tibia in four places with the dental engine. No benefit ensued.

On November 11, an apparatus for ununited fracture was applied, which enabled him to walk with ease. He was discharged with fibrous union of the tibia. The fibula had united more firmly.

CASE 91. *Necrosis of the head of the fibula; recovery.*—Mary W., aged 13, was admitted December 29, 1877. Family history was poor. She had had a discharging sinus at the head of the fibula for several years.

January 10, 1878.—The head of the bone was removed by Dr. Levis.

February 20.—Marked chill. Ankle and foot swollen, and erysipelas developing.

March 1.—Doing well.

April 4.—Discharged, cured.

CASE 93. *Ununited fracture of the leg; drilled; improved.*—Morgan J., aged 28, was admitted February 15, 1878. Six months ago was run over by a wagon, producing a fracture of both legs. The left united firmly. On admission there was slight union in the right, but great lateral motion, which prevents him walking. He was treated with lateral pasteboard splints and a fracture-box.

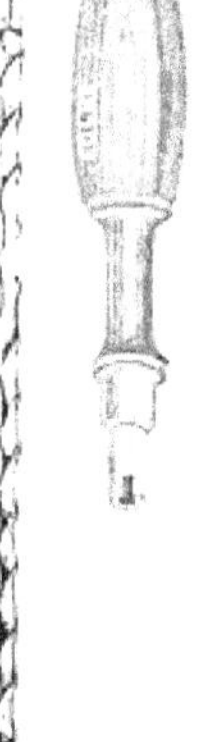

Gimlet.

February 23.—No improvement having occurred, a gimlet was inserted through the ununited ends of the bones. (See figure.)

February 24.—The tendo Achillis was divided to relax the tension.

February 27.—Gimlet removed.

May 18.—Bones drilled again, as there is some motion.

July 25.—Has been wearing a plaster dressing, and been about on crutches.

August 30.—Was discharged, improved.

CASE 94. *Acute hip-joint disease apparently beginning as rheumatism; resection; recovery.*—Patrick W., a sailor, aged 17, was admitted into the medical ward April 10, 1878, with the following history: Having been much exposed to bad weather, wet and cold, and poorly fed, he suffered from pains in the large joints, and was laid up for most of the voyage. Was anaemic and wasted when admitted, and unable to move on account of pain in right thigh, in which there was some effusion. Temperature 101°.

Under iron and quinine in decided doses, and good attendance, he had greatly improved by the 26th, except that he had still had pain on motion of the affected hip. The temperature gradually subsided to the normal on July 12. He had attacks of nausea and vomiting, with pains in the stomach, and occasional diarrhœa, which was relieved by acetate of lead and opium.

He was, on August 24, transferred to the surgical ward. The extensor and adductor muscles of the thigh were contracted, and the trochanter was very prominent. An abscess being detected around the thigh, it was opened, and discharged pus freely. Extension was applied to the limb.

December 10, the head of the femur was exposed, and, being extensively diseased, was removed by Dr. Hunter. The tuberosity of the ischium being rough, it was scraped with a burr and the dental engine. The parts continued to discharge until January 16, 1879, when a free incision liberated a large quantity of pus. From this time he improved, his progress being only slightly retarded by the appearance of erysipelas for a few days.

He was discharged, cured, July 15, 1879.

The histories of the following cases of operations for ununited fractures and anchylosis, taken from an old case-book lately found in the Medical Library, are considered of sufficient interest to have a more permanent record:

Ununited fracture of the os humeri united by a seton.—Isaac P., aged 28, was admitted in May, 1802, under the care of Dr. Physick. The year previously, while at sea, he sustained a fracture of the left arm above the elbow-joint. Three weeks after the injury he arrived at Alexandria, when it was found that the fracture had not been reduced, but that the bones were considerably overlapped and ununited. Various forms of treatment by extension and splints were tried without any success. It was not until December following his admission that an attempt was made to cure the case by operation. The original note states, "It still remained to decide by what means a bony union of the humerus might most probably be effected. In the year 1785 there was a case in our hospital similar to this one, in which an incision was made down to the extremities of the fractured bone, which were then sawed off. . . . No benefit was derived from this painful operation, and the arm was amputated. This case had made a strong impression on the mind of Dr. Physick, and he was unwilling to perform a similar operation. He therefore proposed that a seton-needle, armed with a skein of silk, should be passed through the arm and between the fractured extremities of the bone. . . . This operation was performed on the 18th day of December, 1802, twenty months after the accident had happened. The operation was succeeded by a moderate suppuration. The arm was now extended, and splints were applied. The dressings were removed daily for twelve weeks, during which time no amendment was perceived, but soon after, the bending of the arm at the fracture was observed to be not so easy as it had been. . . . From this time the formation of the new bony union went on rapidly, and on the 4th of May, 1803, was perfectly completed, and the patient could move his arm as well as before the accident. The seton was now removed. On the 28th of May, 1803, he was discharged, perfectly well, the arm as strong as it ever was."

Years after this successful result, Dr. Randolph, a son-in-law of Dr. Physick, was going along the street with his father-in-law, and asked the latter to go into a house with him to see a poor sick sailor, who in all probability would die. Dr. Physick at once recognized his old patient. The man died, and his restored arm-bone was obtained and was carefully preserved. It is now in the Wistar and Horner Museum, and both the case and the specimen are classical ones in the history of American surgery.

"*An account of a case of fracture in which union did not take place in the usual manner, and which was cured by the introduction of a seton between the ends of the fractured bones.*—Bryan M., aged 30, on the 20th of November, 1807, sustained a fracture of the right leg, about the middle, and the left thigh. The latter united readily, but the tibia, at the end of several months, remained ununited. The fibula, however, was firm. A seton was then passed through according to the plan recommended by Dr. Physick, but the surgeon wounded some large vessel, causing severe hemorrhage, and then desisted in the attempt, and the seton was removed. On the 24th of March, 1809, the patient came into the hospital, and Dr. Physick introduced a seton. In eight weeks a perceptible union had taken place. On the 20th of May union was complete. On the 28th of June the seton was removed, and the following day he left, with good use of the limb, having been without its use for two years and six months. The seton was in for three months and four days."

"*A case of artificial joint of the os femoris unsuccessfully treated by seton.*—Dennis J., aged 27, was admitted February 17, 1810, with an ununited fracture of the femur, just below the great trochanter. Eight months after the injury, Dr. Physick determined to try the effect of a seton. An incision was made down to the bone, and a ribbon was then passed through the ends of the bones; an extensive abscess formed in the thigh; fever and exhaustion from drain continued; diarrhœa supervened, and at the end of four months and four days the seton was removed, no union having occurred."

"*Artificial joint following a fracture of the tibia, and probably caused by excessive bleedings and insufficient food; cured by operation.*—Isaac H., aged 35, was admitted March 15, 1827, with an artificial joint of the tibia, about the middle of the bone. The history Mr. H. gave was as follows: Five hours after the injury upwards of a quart of blood was drawn from his arm. The next day a quart more was abstracted, and he was not allowed to taste animal food for six weeks. When admitted he was feeble, but after remaining in for some time and given tonics, Dr. Hewson exposed the bones at the place of non-union, and then removed a small portion of the ligamentous matter from between the ends of the bone; caustic potash was then applied to them, after which a slough formed. The limb was poulticed, and three months after the operation he was discharged quite well with good union."

"*A case of compound luxation of the ankle; reduction; amputation; hemorrhage; ligation of the femoral artery; subsequent necrosis of the knee; excision of the head of the tibia and fibula and lower part of the os femoris; recovery.*—Jacob Z., aged 31 years, who had been in the habit of drinking large quantities of ardent spirits, was admitted July 31, 1806. The tibia was dislocated from its connections with the astragalus and forced through the integuments, the internal malleolus was broken off, and could be distinctly felt under the tendo Achillis, the lower end of the fibula was fractured, and the astragalus was broken into four pieces. . . . Reduction was effected, and the limb was dressed in the common way. A consultation between Drs. Wistar, Physick, and Coxe, the next morning, was held, and amputation was performed.

"*August* 10.—This afternoon patient awoke with fright, having dreamt that he was falling, and in his sudden exertions displaced part of the dressing, and shortly afterwards hemorrhage from the stump was observed; upon exposure to the air the bleeding ceased, after losing from sixteen to twenty ounces of blood.

"*August* 11.—At 2 P.M. a bleeding was again observed, which ceased on exposing the stump.

"*August* 12, 6 A.M.—The stump was again observed to bleed; at 9 o'clock, Drs. Wistar, Physick, and Coxe met in consultation, and they determined on taking up the vessel; a large needle was accordingly passed deeply into the muscles and made to include a considerable portion of muscular flesh around the part from whence the blood seemed to flow (*acupressure?*); hemorrhage reappeared in half an hour; they now resolved to pass a ligature around the femoral artery.

"*August* 16.—Ordered to take a little animal food.

"*August* 25.—Hemorrhage was observed from the femoral artery, which was controlled by pressure.

"*August* 26.—Some hemorrhage continuing to-day, Dr. Wistar passed two other ligatures around the femoral artery, two inches above the former place, and the artery was divided between them.

"*September* 10.—The knee-joint shows inflammation; is much swelled; poultice applied.

"*September* 30.—On removing the poultice this morning the tendon of the rectus muscle and capsular ligament gave way and exposed the whole of the knee-joint.

"*October* 1.—Lower ligature of the femoral artery was taken away.

"*October* 4.—The remaining ligatures were taken away.

"*October* 11.—The head of the tibia and fibula were removed this morning by cutting at the back part; two inches of the femur is exposed in consequence of the destruction of the soft parts by mortification.

"*December* 31.—Dr. Physick removed to-day several inches of the os femoris. The patient made an excellent recovery."

Barton's Operation for Anchylosis of the Hip-joint in 1826.*

"John Coyle, a sailor, on the 27th of March, 1825, fell from a height of seven or eight feet on the outside of his right hip, which at the time gave him acute pain. He was kept in his hammock for eighteen days, and then removed to the Hospital of Porto-Cavello, where a consultation was held in his case by three medical gentlemen. They differed in opinion as to the nature of the accident, two supposing it to be a fracture, while the third considered it a luxation. About eight months after the accident he was sent to this city by the consul, and ere long admitted to the Pennsylvania Hospital, with the view of placing his limb in Desault's splint. They had no effect in straightening the limb, which confirmed the opinion that an anchylosis had taken place. Several consultations were held in the case, and it was finally decided that any further attempt to release the joint would be useless.

"The patient still felt that his case was not hopeless, and, after repeated solicitation on his part, Dr. Barton, with the consent of the attending surgeons of the house, performed, in the presence of a large concourse of medical gentlemen, the following operation on the 22d of November, 1826:

"An incision through the integuments, six or seven inches in length, one-half extending above and the other below the great trochanter. This was met by a transverse section of four or five inches in extent, the two forming a crucial incision, the four angles of which were to meet opposite to the most prominent point of the great trochanter.

". . . The fascia was then detached by turning the edge of the scalpel sideways, and at the same time separating anteriorly all muscular structure from the bone, without unnecessarily dividing their fibres. This was done behind and between the trochanters, and the bone divided transversely through the great trochanter and part of the neck of the bone by means of a strong and narrow saw made for the purpose. This being accomplished, the limb was extended.

"No blood-vessels were taken up; the edges of the wound were brought slightly together by adhesive plaster and light dressings. The operation lasted seven minutes. Desault's splints were applied after the patient was removed from the table."

Then follows a detailed account of the subsequent treatment; and the record states that on

"*March* 1.—The patient is in every respect well. His limb necessarily is weaker than the other. He dresses himself, and is in exercise the greater part of the day.

"By measurement from a straight line he can advance the foot 24 inches; in stepping backwards, 26 inches; in abduction, 20 inches; rotation inwards, 6 inches; outwards, 6 inches.

"Thus has a patient, who by a majority was condemned to deformity the remainder of his life, had the use of his limb and the symmetry of his body restored, by our skilful and respected surgeon, Dr. J. R. Barton, by an operation which stands alone in surgery.

"Vive le chirurgien!

"April, 1827." "CHS. MIFFLIN.

The following figures were "drawn by John Orght, an insane patient in the Pennsylvania Hospital, for Chs. Mifflin," who was resident physician in the hospital from 1826 to 1828.

* Copied from the original notes of Dr. Charles Mifflin, April, 1827; the illustrations were made at the time.

Illustrating Barton's Operation for Anchylosis of the Hip-joint.

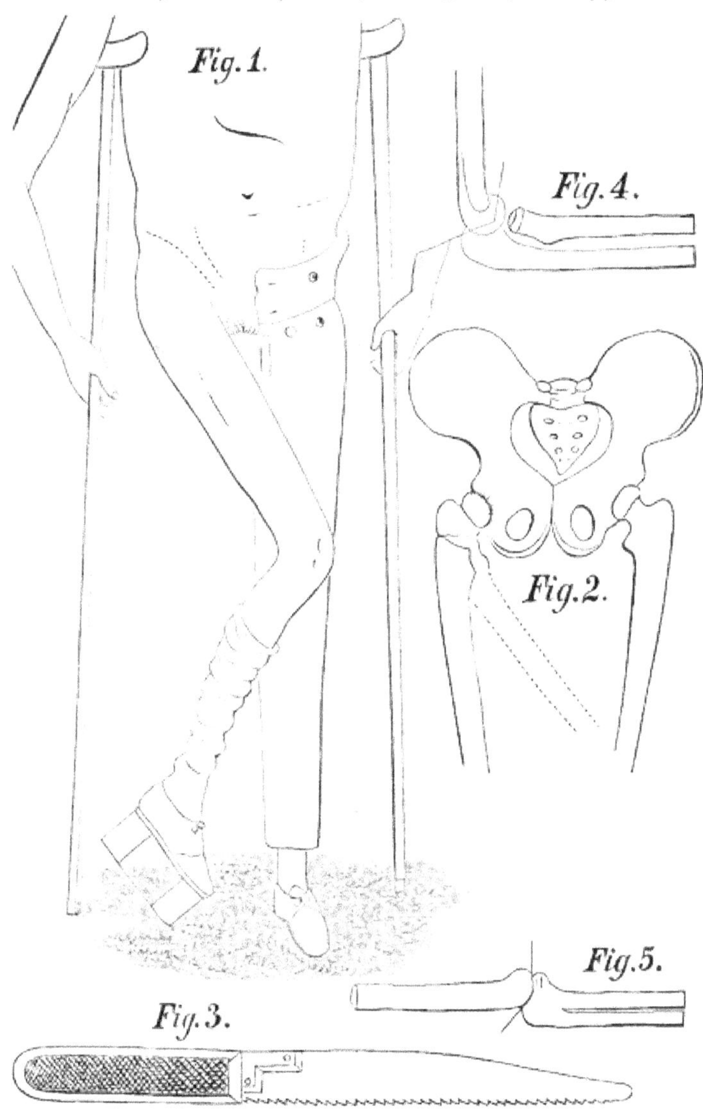

"Fig. 1. The patient with his thigh in an anchylosed state.

"Fig. 2 shows the direction of the femur previous to the operation, by the dotted line. The femur also shown after the operation in its proper direction.

"Fig. 3. The saw which was used in dividing the bone.

"Figs. 4 and 5. The first showing an elbow anchylosed at a right angle, with the direction of a division of the bones, for relief of the deformity.

"The same view of the elbow when anchylosed in a straight line."

Table of Excisions, and Operations for Ununited Fracture.

No.	Age.	Sex.	Nature of the Disease or Injury.	Date of Admission.	Duration.	Treatment.	Days in Hospital.	Result.	Remarks.
1	22	M.	Inferior maxilla, phosphorus disease of.	Jan. 11, 1865.	7 months.	Inferior maxilla excised.	76	Cured.	For history, see article on Phosphorus Necrosis, page 217
2	16	M.	Tibia, necrosis after compound fracture of.	May 20, 1866.	Recent.	2 inches of lower fragment resected.	110	Cured.	
3	52	M.	Humerus, necrosis of.	Nov. 24, 1866.	3 years.	2½ inches of humerus excised.	37	Cured.	
4	14	F.	Ulna, strumous disease of.	Nov. 28, 1866.	18 months.	Resection of olecranon.	62	Cured.	
5	14	M.	Femur, necrosis of.	Dec. 16, 1866.	Old.	2½ inches resected.	78	Relieved.	
6	17	F.	Ankle-joint, necrosis of.	Jan. 5, 1867.	2 years.	Resection of ankle.	107	Cured.	
7	52	F.	Forearm, comminuted fracture of.	March 12, 1867.	Recent.	Portions of radius and ulna excised.	7	Died.	
8	58	M.	Femur, ununited fracture of.	Aug. 25, 1867.	Several months.	Bones drilled and ends frictioned.	75	Cured.	
9	18	M.	Femur, ununited fracture of.	Oct. 7, 1867.	10 weeks.	Refractured and ends frictioned.	100	Cured.	3 inches short prior to operation.
10	15	M.	Forearm, compound fracture of.	Oct. 27, 1867.	Recent.	1½ inches of radius and ½ inch of ulna excised.	73	Cured.	
11	24	M.	Femur, deformity after fracture of.	March 25, 1868.	16 weeks.	Refractured.	106	Cured.	3½ inches short prior to operation.
12	15	M.	Fibula, ununited fracture of,	April 15, 1869.	8 weeks.	Bones drilled.	48	Cured.	Tibia, which was fractured at the time of accident, united well.
13	17	M.	Ankle, deformity after fracture of.	May 16, 1868.	12 months.	Bones drilled and refractured.	20	Died.	Pyæmia.
14	48	M.	Tibia, ununited fracture of.	June 10, 1869.	4 weeks.	Bones drilled and frictioned.	131	Cured.	Operation once repeated.
15	8	M.	Femur, spontaneous fracture of, after injury and abscess.	Aug. 19, 1869.	12 weeks.	Resected shaft in continuity.	306	Cured.	
16	13	M.	Os calcis and astragalus, necrosis of.	Sept. 27, 1869.	....	Excision of os calcis and astragalus.	252	Cured.	
17	26	M.	Knee, complete osseous anchylosis of, at a right angle.	Nov. 26, 1869.	9 years.	Excision of wedge of bone.	136	Cured.	
18	13	M.	Knee, strumous synovitis of.	Jan. 10, 1870.	6 months.	Excision of knee.	58	Died.	Pyæmia.
19	19	M.	Inferior maxilla, ununited fracture of.	Feb. 1, 1870.	5 months.	Drilled.	27	Cured.	
20	45	M.	Leg, ununited fracture of.	April 1, 1870.	4 months.	Gimlet introduced.	56	Cured.	
21	39	M.	Leg, compound fracture of.	April 21, 1870.	16 days.	3 inches of tibia removed.	110	Cured.	
22	29	M.	Femur, ununited fracture of.	April 14, 1870.	11 weeks.	Drilled, refractured, extension.	125	Died.	Slight attempt at repair; inflammation extended to joint of knee; abscesses; exhaustion.
23	17	M.	Femur, necrosis from old gunshot wound.	Aug. 29, 1870.	Old.	Head of femur and part of ilium excised.	141	Cured.	
24	1	F.	Os calcis, necrosis of.	Aug. 31, 1870.	5 months.	Excision of os calcis.	62	Cured.	
25	16	M. col'd.	Elbow-joint, synovitis and necrosis of.	March 24, 1871.	18 months.	Excision of olecranon.	241	Cured.	Afterwards disease reappeared; then amputated.
26	16	M.	Hip-joint, disease of.	June 7, 1871.	Many months.	Excision of head of femur.	122	Died.	Exhaustion.

27	25	M.	Patella, necrosis of.	Sept. 7, 1871	Many months.	Excision of patella.	51	Relieved	Slight sinus remaining.
28	37	M.	Leg, useless from vicious union after fracture.	Feb. 19, 1872.	Old.	Drilled and refractured.	241	Cured	
29	24	M.	Radius, necrosis of.	Aug. 12, 1872	Many months.	Head of radius excised	30	Cured.	
30	12	F.	Hip-joint, disease of.	June 17, 1872.	11 years.	Head of femur excised.	198	Cured.	
31	8	F.	Inferior maxilla, necrosis of	Several times in 1872.	1 year.	Lower jaw excised.	50	Cured.	
32	52	M.	Knee-joint, syphilitic necrosis of.	Oct. 17, 1872.	1 year.	Condyles of femur and head of tibia excised.	55	Died	
33	35	M.	Jaw, necrosis of.	Dec. 18, 1872.	1 year.	Wedge portion removed at the angle of jaw.	20	Cured.	
34	40	M.	Painful affection of foot.	Dec. 20, 1872.	6 years.	Excision of the fourth metatarso-phalangeal articulation.	37	Cured.	
35	50	M.	Carpus, caries of.	Nov. 27, 1872	2 years.	Carpus excised.	....	Cured.	
36	38	M.	Forearm, compound fracture of.	Feb. 27, 1873	Recent.	Radius resected.	43	Relieved	
37	23	M.	Humerus, ununited fracture of.	March 24, 1873.	Old.	Drilled, and screw left in.	56	Cured.	
38	36	M.	[illegible] compound fracture of.	March 29, 1873.	Recent.	Tibia resected.	31	Died.	
39	14	F.	Hip-joint disease.	March 30, 1873.	Several months.	Excision of the hip.	13 mos.	Cured.	
40	38	M.	Tibia, ununited fracture of.	April 2, 1873.	Recent.	Fracture-box.	76	Cured.	Delirium tremens for 2 weeks.
41	50	M.	Humerus and elbow-joint, compound fracture of.	April 6, 1873.	Recent.	Excision of condyles of humerus and the articulatory surface of ulna.	17	Removed by friends.	
42	46	M.	Tibia, ununited fracture of.	April 8, 1873.	20 weeks.	Bones refractured.	50	Improved.	Slight motion when discharged.
43	22	M.	Radius, ununited fracture of.	April 11, 1873.	Recent.	Drilled (gimlet).	97	Cured.	18 days after fracture there was no repair.
44	15	M.	Femur, ununited fracture of	April 26, 1873.	Recent.	Extension, fixed dressing afterwards.	149	Improved.	Carried around a shafting. Humerus, radius, ulna, and leg fractured at same time.
45	30	M.	Humerus and forearm, ununited fracture of.	June 25, 1873.	12 weeks.	Bones drilled and frictioned.	73	Cured.	Both fractures were compound and comminuted.
46	32	F.	Tibia, ununited fracture of.	July 23, 1873.	Old.	Drilled.	94	Cured	Had mania a potu afterwards.
47	62	M.	Leg, ununited fracture of.	Sept. 2, 1873.	Recent.	Drilled.	139	Cured.	
48	17	M.	Foot, necrosis of.	Sept. 30, 1873.	15 years.	Astragalus and the articulating surfaces of tibia and fibula removed.	185	Relieved.	
49	37	M.	Femur, ununited fracture of.	Oct. 30, 1873.	2 years.	Refractured, bones frictioned.	130	Relieved.	
50	24	M.	Great toe, metatarsal bone, necrosis of.	Jan. 10, 1874	4 months.	Joint excised.	48	Cured.	
51	20	M.	Tibia, ununited fracture of.	Feb. 13, 1874.	Recent.	Drilled.	106	Cured	
52	15	M.	Femur and humerus, ununited fracture of.	Feb. 27, 1874.	3 weeks.	Refractured and frictioned.	92	Cured.	Operation on both bones.
53	27	M.	Tibia and fibula, ununited fracture of.	July 6, 1874.	14 weeks.	Drilled and refractured, then resected.	150	Cured	
54	21	M.	Leg, ununited fracture of, injury compound.	Aug. 21, 1874	12 weeks.	Drilled, afterwards amputated.	302	Cured	
55	28	M.	Superior maxilla, phosphor-necrosis of.	Oct. 28, 1874.	....	Bone removed.	26	Cured	

Table of Excisions, and Operations for Ununited Fracture.—(Continued.)

No.	Age.	Sex.	Nature of the Disease or Injury.	Date of Admission.	Duration.	Treatment.	Days in Hospital.	Result.	Remarks.
56	24	M.	Femur, ununited fracture of.	Dec. 26, 1874.	Several weeks.	Drilled, refractured, and bones frictioned.	40	Cured.	
57	28	M.	Tibia, ununited fracture of.	Feb. 13, 1875.	Several weeks.	Frictioned and fracture-box.	81	Cured.	
58	39	M.	Femur, tuberculosis, ununited fracture of.	May 10, 1875.	9 years.	Necrosed bones excised.	131	Incurable.	
59	4	F.	Knee, gelatinoid degeneration of.	June 7, 1875.	2 years.	Knee excision.	92	Cured.	
60	23	M.	Humerus, ununited fracture of.	June 24, 1875.	12 weeks.	Drilled and bones frictioned.	116	Cured.	Compound comminuted fracture, 6 weeks before admission.
61	10	M.	Knee, scrofulous synovitis of.	July 6, 1875.	Old.	Excision of knee.	111	Cured.	Afterwards amputated. (See Amputations, Case 13.)
62	39	M.	Tibia, ununited fracture of.	July 24, 1875.	69 days.	Fracture-box.	44	Cured.	
63	20	M.	Tibia and fibula, ununited fracture of.	July 25, 1875.	75 days.	Refracture, and bones frictioned.	219	Cured.	
64	23	M.	Hip-joint disease.	Aug. 25, 1875.	Some weeks.	Excision of hip.	208	Cured.	
65	9	M.	Knee-joint, partial anchylosis and necrosis of.	Sept. 24, 1875.	Old.	Excision of knee.	144	Cured.	
66	25	M.	Elbow, railroad crush of.	Oct. 29, 1875.	Recent.	Excision of elbow.	9	Died.	Tetanus.
67	18	F.	Leg, deformity after fracture.	Dec. 30, 1875.	Old.	Refracture of the leg.	35	Cured.	
68	13	M.	Tibia, ununited fracture of.	Dec. 11, 1875.	Old.	Refracture, and rubbing ends of the bone.	74	Cured.	
69	30	M.	Leg, compound fracture of.	Dec. 19, 1875.	12 weeks.	Drilled, refracture, rubbing ends of bone.	231	Cured.	Kick of a horse.
70	60	M.	Fibula, ununited fracture of.	Jan. 25, 1876.	92 days.	Splint and fracture-box.	123	Cured.	
71	30	M.	Humerus, compound fracture of.	March 6, 1876.	Old.	Drilled, refracture, rubbing ends of bone.	63	Improved.	Very slight motion.
72	35	M.	Elbow, necrosis of.	April 29, 1876.	Many months.	Excision of elbow.	213	Cured.	During operation the humerus was broken; it promptly united.
73	19	M.	Forearm, compound fracture of (ulna and olecranon compound).	July 1, 1876.	Many months.	Elbow excised.	115	Cured.	
74	17	M.	Disease of hip-joint; necrosis of femur.	Aug. 12, 1876.	$2\frac{1}{4}$ years.	Excision of hip-joint.	325	Cured.	
75	52	M.	Tibia and fibula, ununited fracture of.	Aug. 14, 1876.	8 weeks.	Refractured on 34th day.	163	Improved.	Removed by friends.
76	33	M.	Tibia, compound fracture of.	Aug. 31, 1876.	Recent.	Drilled, then wired ends of bone.	133	Cured.	
77	20	M.	Os calcis, necrosis of.	Sept. 12, 1876.	6 months.	Excision of os calcis.	202	Cured.	
78	30	M.	Radius and ulna, ununited fracture of.	Nov. 3, 1876.	8 weeks.	Refracture after drilling.	42	Cured.	Afterwards amputation.
79	12	M.	Deformity after fracture of forearm.	Nov. 4, 1876.	10 weeks.	Refractured.	39	Cured.	

80	35	M.	Tibia, ununited fracture of.	Dec. 3, 1876.	5 weeks.	Drilled, bones rubbed.	215	Cured.	
81	40	M.	Leg, ununited fracture of.	Feb. 20, 1877.	....	Drilled, bones rubbed.	107	Improved.	
82	19	F.	Femur, deformity after fracture of.	April 9, 1877.	7 months.	Refractured.	84	Cured.	
83	17	M.	Hip disease.	May 12, 1877.	16 weeks.	Excision of hip.	15	Died.	
84	7	M.	Suppuration, knee-joint.	May 25, 1877.	4 years.	Excision of knee, afterwards amputation.	2 yrs.	Cured.	
85	30	M.	Leg, compound fracture of.	June 26, 1877.	Recent.	Tibia resected.	150	Cured.	
86	9	M.	Humerus, after abscess, necrosis of.	Sept. 11, 1877.	2 weeks.	Excision of the head of the humerus.	63	Cured.	
87	18	M.	Tibia and fibula, ununited fracture of.	Oct. 6, 1877.	Old.	Refractured and bones rubbed.	256	Cured.	
88	36	M.	Astragalus, necrosis of.	Oct. 23, 1877.	Old.	Astragalus excised.	39	Died.	Typhoid fever, in medical ward.
89	37	M.	Deformity following fracture of leg and luxation of ankle.	Nov. 21, 1877.	4 months.	Refractured and bones rubbed.	58	Cured.	
90	27	M.	Ununited fracture of leg after injury; compound fracture.	Dec. 18 1877.	1 year.	Resection.	365	Improved.	
91	13	F.	Necrosis, head of fibula.	Dec. 29, 1877.	Old.	Excision, head of fibula.	96	Cured.	
92	50	M.	Deformity following fracture of leg.	Jan. 19, 1878.	Old.	Wedge-shape portion of bone excised, bones wired.	228	Cured.	
93	28	M.	Ununited fracture of leg.	Feb. 15, 1878.	Some months.	Drilled, gimlet left in, drilled second time, 56 days after the first operation; tendo Achillis divided.	496	Improved.	Both legs were fractured; the left united firmly at time.
94	17	M.	Hip-joint disease.	April 10, 1878.	Several months.	Excision of the head of femur.	272	Cured.	Admitted in medical ward for rheumatoid arthritis.
95	45	M.	Tibia and fibula, ununited fracture after compound fracture.	March 20, 1878.	Recent.	Drilled.	210	Cured.	
96	25	M.	Tibia and fibula, ununited fracture after compound fracture.	June 25, 1878.	5 months.	Refractured, tibia resected.	150	Cured.	
97	29	M.	Superior maxilla, phosphor-necrosis of.	Oct. 28, 1874.	3 months.	Superior maxilla removed.	26	Cured.	
98	30	M.	Inferior maxilla, phosphor-necrosis of.	July 19, 1878.	4 months.	Inferior maxilla removed.	199	Cured.	For histories, see article on Phosphorus Necrosis, page 216.
99	25	M.	Malar bone, phosphor-necrosis of.	Sept. 5, 1879.	6 months.	Malar bone removed.	67	Cured.	

THOMAS G. MORTON.

WILLIAM HUNT.

OPERATIONS UPON VARICOSE VEINS.

THE milder forms of enlargements of the veins are treated by bandages or the elastic stocking. Of the 9 severe cases, 8 have been recorded where operations have been performed for the relief of this painful and often serious disease. Hemorrhage in 2 of the cases was the immediate cause for the patients being admitted for treatment. All of the cases operated on made excellent recoveries. 54 days was the longest, and 21 days the shortest period any of the patients remained in the hospital. In 3 cases the veins were exposed by incision and then portions were excised. In 4 cases, needles, armed with silver wire, were carried under the veins, which were then tied subcutaneously. In 1 case pins were thrust under the veins, and then were compressed by a silk cord. In only 1 case were any grave constitutional symptoms noted, and, as the case presented some points of interest, a brief history is deemed worthy of record.

CASE 3. *Long-standing enlargement of the veins of the leg: veins exposed: portions excised; cure.*—Jas. W., aged 47, was admitted June 19, 1875. Has had ulcers on both limbs for six years; veins have been varicose for twenty years. On admission there are four ulcers on the left leg, varying from three-quarters to one-half an inch in length. There is also an ulcer on the right leg one inch in diameter; the veins of the left leg are very varicose; those of the right less so.

June 30.—Excision of the veins performed on each leg; two portions of the vein, each about two inches in length, were dissected out of each leg, skin-flaps having been turned aside for the purpose; the open ends of the veins were treated by torsion; edges of the wounds brought together with gauze and collodion: both limbs were then wrapped in salicylic dry cotton and confined with a roller.

July 2.—At 2 P.M. patient had a chill. Evening temperature, hitherto not above 99°, is now 106°; pulse 104: tongue coated; anorexia and thirst; ordered quinia, gr. xii, in two doses.

July 3.—Temperature 103½°: the red line of angeioleucitis extends up to the groin on the left side and ends in a large bubo; when the cotton was removed pus was found in the wound: chloral solution (5 gr. to ℥i) applied: gr. xx quinine in two doses for three days. Evening temperature 103°; pulse 100.

July 4.—Better temperature, 101½°; pulse 96; swelling of groin less.

July 7.—Limb again enveloped in salicylic cotton.

July 8.—Dressing removed from right leg; wound well.

July 11.—Ankle of left limb found swollen, and an abscess was on each side, which was evacuated.

July 20.—Ulcers almost well.

August 2.—The ulcers which resulted from the incisions at the ankle are healed, and the patient discharged quite well.

Varicose Veins treated by Operation.

No.	Sex	Age	Duration or Complications.	Treatment.	Result.	Days in Hospital.	Remarks.
1	M.	65	Many years, ulcer.	Veins on leg exposed by incision, and two inches excised.	Cured.	37	Hemorrhage before admission
2	M.	39	Several years, with intense pain in the course of the veins.	Pins passed under the veins in each leg, and compressed by silk thread.	Cured.	25	
3	M.	47	For 20 years has had enlarged veins; ulcer for 6 years.	Veins exposed, and nearly two inches of vessel in each limb excised.	Cured.	45	
4	F.	27	Several years.	One large vein ligatured subcutaneously on tibial side.	Cured.	49	
5	M.	42	Years with ulcer.	Veins ligatured with silver wire in four places, subcutaneously.	Cured.	54	
6	M.	55	Several years; had phlebitis.	Veins ligatured subcutaneously in two places with silver wire.	Cured.	30	Hemorrhage from ruptured vein before admission.
7	M.	27	Injury, afterwards an ulcer for 5 years.	Two incisions, $2\frac{1}{2}$ inches long, exposed a great mass of veins; sections an inch long were excised at two places.	Cured.	21	
8	M.	43	From childhood.	Nine silver subcutaneous ligatures were used.	Cured.	21	

THOMAS G. MORTON.

THE CONTROL OF HEMORRHAGE DURING HIP-JOINT AMPUTATIONS.

Compression of the subclavian artery by the finger or a large door-key, in order to prevent hemorrhage during the operation of amputating the arm at the shoulder-joint, is a well-known expedient, and is generally referred to by systematic writers on surgery. In truth, it is commonly recognized as the proper procedure under these circumstances. The fact that the common iliac artery can be similarly controlled by digital compression is less widely known, and has only within a very recent period obtained any recognition among practical surgeons. In a short communication to the *American Journal of Medical Sciences* for January, 1874, the writer had the honor of suggesting the introduction of the hand into the rectum, during hip-joint amputations, for the purpose of reaching the iliac vessels as they pass over the inlet or brim of the pelvis, where they may be "compressed and controlled with as much certainty and almost as readily as the radial artery." The advantages of this procedure in the adult (*i.e.*, the class of cases where it would be most required) over the abdominal tourniquet, are too obvious to require discussion in this place. They have been briefly referred to in the article already mentioned, but we may state here that death has on more than one occasion resulted from this great pressure upon the sympathetic plexus and great vessels. Where such injurious compression can be avoided, and the object can be more quickly, certainly, and effectively attained by a simple manual procedure, it would seem as if, under ordinary circumstances, the use of the abdominal tourniquet during amputations of the thigh should be abandoned in favor of digital compression. The value of this, where the circulation is required to be checked for a brief period, in either one or both lower extremities, is sufficiently evident, but is enhanced by the fact that the compression is made by a sensitive fleshy pad directly upon the vessel, and the venous trunk is avoided if desired. As the patient is always anæsthetized in large operations, compression need not be instituted until the surgeon is ready to commence the external incision; it can be cautiously removed and instantly reapplied, if necessary, and is immediately removed when the vessel is secured, thus maintaining the pressure during the *minimum* of time. The manner of performing this was also indicated in the article referred to: "The bowel should be evacuated by a large warm-water injection previous to the operation. In controlling the right common iliac artery the right hand of the assistant will be the more convenient, and the left vessel will be more easily controlled by the left hand. In either case, the hand being anointed (with lard* in preference to oil), and the fingers folded into a cone, it is gradually introduced into the rectum, with its dorsum to the sacrum, until reaching the sigmoid flexure, where the hand may be pronated, and, as the vessels are right under the fingers, the main supply of blood to the limb may thus in a few moments be completely controlled." With the prior application of the elastic bandage from the foot to the thigh, the operation of hip-joint amputation may in this manner "be rendered almost as bloodless as some of the operations of minor surgery."

Since writing the above, I have been gratified to notice that direct compression of the iliac vessels is gradually growing into favor. Receiving the endorsement of Prof. Gross, who has included it as a legitimate procedure in his didactic course when lecturing upon hip-joint amputation, it was also favorably mentioned by Prof. Van Buren, in his address on the Treatment of Aneurism,† at the meeting of the International Medical Congress in Philadelphia in 1876. I have been informed that it was also referred to in terms of commendation by the late Mr. Callender.

Within the last year or two a modification of this method has been practised in the London hospitals with considerable satisfaction. Mr. Davy suggested the employment of a lever made of wood, which, being introduced into the rectum, could be made to bear upon the iliac vessels and temporarily occlude them, and he adopted this expedient in January, 1877, in the case of a young child, with perfect success. At a recent meeting of the Clinical Society of London,‡ Mr. A. Pearce Gould read notes of a case of amputation at the hip-joint, where the iliac artery was thus controlled successfully in a man 28 years of age. A number of cases in which the lever compression had been used with complete satisfaction were also reported during the discussion. The following advantages were claimed by Mr. Gould for the compressor over the abdominal tourniquet: (1) it disturbed the circulation less; (2) it did not interfere with the respiratory movements, nor was it interfered with by them; (3) its use was not prevented by obesity, rigidity of the abdominal walls, nor the existence of abdominal tumors; (4) the pressure required was less; (5) less liability to injury of the viscera and peritoneum; (6) greater ease and security in application; (7) greater cheapness and durability; (8) if the lever were not at hand its place could be more easily supplied.

The above advantages over the abdominal tourniquet are conceded to the lever, but in my opinion they belong, in a higher degree, to digital compression, which can be more readily and perfectly guided and controlled. It is

---

* Cosmoline, or vaseline (unguentum petrolei), or cerate would now be preferred.

† *Transactions Int. Med. Cong.*, 1876, p. 552.

‡ *Amer. Journ. of Med. Sciences*, July, 1879, p. 273, quoted from the *Lancet*, April 26, 1879.

evident that the lever is less discriminating and intelligent, and consequently more uncertain in its application, than the human fingers, and it is probable that in using the lever, at least in some cases, the pressure is brought to bear upon the junction of the common iliac vessels, thus causing venous thrombosis in the opposite limb, which was found to be the cause of death in Mr. Gould's case; although he was inclined to attribute the thrombosis to carbolic acid absorption rather than to the lever. This question must, however, remain for future experience to settle. Nevertheless, it seems reasonably certain, judging from reports that are furnished thus far, that immediate compression of the iliac vessels, either manually or by a lever, is destined to speedily supersede the use of the abdominal tourniquet in cases of ordinary hip-joint amputation, or operations high up on the thigh involving the first portion of the femoral artery, and in some cases of tumor or aneurism, where great danger of hemorrhage is thought to exist. A surgeon can always find an assistant with a small hand, to whom this part of the operation could be intrusted, and, if a little care be taken in the introduction, in order to dilate, instead of rupturing the sphincter ani, no trouble will result. Even a rupture of this muscle would prove only a temporary inconvenience, and would heal before the effects of the cutting operation were recovered from. Lacerations of the rectum are a more serious accident, but need not occur if ordinary care be exercised. Of course a stricture of the rectum, whether malignant or fibrous, would debar the patient from the benefits of digital compression, and the lever or tourniquet would have to be resorted to. But in the ordinary class of cases requiring amputation at the hip-joint, this expedient of direct digital compression of the iliac arteries certainly deserves a fair trial by surgeons generally, and the writer would be glad to receive for future publication any notes of cases in which this recommendation shall be carried into practical effect.

FRANK WOODBURY.

LACERATIONS, WOUNDS, CONTUSIONS, SPRAINS, ABSCESSES, AND ULCERS.

During five years, 1882 cases have been admitted into the hospital for the above causes. Of these, 332 are set down as being lacerations, 371 as wounds, 570 contusions, 221 sprains, 168 abscesses, and 220 ulcers.

Most of these cases were tabulated with the intention of publishing them under separate heads, but as our volume has already grown beyond our original intention, it was concluded to summarize them in this way.

There is certainly enough material for a volume upon any one of the subjects dealt with in this article, as the pile of papers before us, which we have been trying to condense, will amply testify.

As to **Lacerations and Wounds,** we find that scarcely any part of the body has escaped, and that the injuries inflicted include those of the most severe and even fatal character, as well as those of a trivial form. The causes have been cutting instruments, machinery, glass, china, bricks, bites, blows, falls, horse-kicks, elevators, wagons, engines, explosions, gunshot, etc. By far the greater number of the cases recovered, very many of them being ready in a few days to be sent to the out department or to be discharged cured.

We have nothing new or peculiar to state as to the general points of handling wounds when they come into the receiving ward. The arrest of hemorrhage is the first thing that demands attention. Ligatures, acupressure, torsion sometimes, pressure, styptics, ice, and cold or hot water are all resorted to, according to the demands of the case and the judgment of the dresser.

Hot water is something rather new, and to oozing wounds has been applied of late with great satisfaction.

A temperature of 130° or 140° is very readily borne. A good method is to wet a towel with the water and apply it to one's own cheek or forehead for a moment, rather taking the risk of scalding one's self than the patient, who is frequently under an anaesthetic. Then apply the saturated towel to the part with slight pressure, and renew frequently so as to maintain the temperature. A bleaching of tissue and an arrest of hemorrhage, although of a severe character, often takes place, which is remarkable.

The bleeding having been stopped, the cleaning of the wound, which may have been partially done during the first process, is completed, and all foreign bodies are carefully removed if possible. In cases of gunshot, a reasonable search is made for the ball, but if not accessible it is let alone, either to declare itself in the future or to become encysted, as the case may be. The same may be said of other foreign bodies that have passed out of reach or of ordinary search.

Officious probing of the great cavities is strictly forbidden and avoided. As to bullets, it seems impossible for the laity to comprehend that it is the mischief in most cases that the missile has done in its course that we have to deal with, and that the lately fatal object may be lying harmless beneath the skin. The latter, in such a case, must be removed, but the patient does not always respond to the hope raised by his friends when they say that the "doctors have found the ball." We remove the ball, of course, if found, and in a few cases the operation may be required at great risk, but generally there is no need to increase the hazard to the life of the patient by undue perseverance in what often may be an unsuccessful effort.

The dressing of the wound and its after-treatment is the next consideration. There is no subject of late years that has claimed so much attention as this, and it might be thought that the doctors are all at sea about it. The germ theory of the infection of wounds, whether accidentally inflicted or made by the surgeon, has deservedly gained a strong hold in professional estimation. Whether it will be fully established and recognized or not remains to be seen, for it is much easier to theorize about than to prove.

In the mean time, however, the practice of cleanliness and care as to both great and small matters, that has arisen out of it, is productive of much good. The articles that are used in dressing, having for their object the keeping out or destroying of the poisonous germs, are very numerous. In the Pennsylvania Hospital it would be safe to say that almost everything that is plausibly recommended has been tried.

There is one unfortunate influence that the theory in question has had, and that is in inculcating the belief that the sources of all evil to the proper healing of wounds come from without. Hence the possible constitutional origin of local trouble is often lost sight of. In a surgical sense we learn fully to appreciate by practice that as to the human body alone, "all flesh is not the same flesh." There is a flesh that heals and a flesh that will not heal, a flesh that ulcerates, a flesh that bleeds, and a flesh that sloughs. Weak central organs, diseased blood, vitiated secretions, and vicious habits may and do neutralize our efforts as well as bacteria and micrococci.

The *vis medicatrix* also asserts its place, and is now as ever among the foremost in the field of repair. In a general hospital, where surgeons having different views and methods are practising, it is not unusual to see an enthusiast triumphantly show the cure of a great laceration that has been treated with some favorite article and with more or less complicated dressing, while another will at once match the case with one equally severe and equally well, that has had nothing done to it save the necessary measures to keep its parts together.

The washing of our wounds is done with carbolized water through the hose from the reservoir on the ward carriage, so that even the same water does not go over the wound twice. After this whatever dressing may have been ordered for the special case is applied.

We may here say a few words as to some particular lacerations and wounds. First, of the scalp, we have a large experience. There is nothing whatever to justify the old and deeply-rooted prejudice against sutures in this part of the body. It is simply, as in any other position, the injudicious use of them that is objectionable. When, therefore, a suture is of service or necessary to support a part that otherwise would hang or gape, we use it. If the part can easily be kept in position by plaster and compresses, we reject the suture, and would be justified in so doing from humanity alone, as there is no excuse for inflicting unnecessary pain.

Incised wounds of the throat, involving the respiratory tract, are generally left open at first. When the patient gains sufficient control over the disabled muscles to get rid of the secretions, the parts are gradually approximated. By this method numbers of lives have been saved, some of them most ghastly and apparently hopeless cases. We have had them so bad as to require feeding through the gaping wound by a stomach-tube for the first two or three weeks.

A most capital way of getting a pint or quart of oatmeal gruel, milk, or soup into their stomachs is to put the food into a Thudichum bottle, and to attach this to the free end of the tube. When all is ready raise the bottle. Hydrostatic laws are fixed, and we often see a resisting fellow get a dinner with a rapidity that astonishes him.

Wounds of the face are carefully closed with reference to direction of approximation, so as to occasion the least possible deformity. The oblique-edged incision, when operating here, which has recently been advised by Dr. Packard, we have reason to believe will lessen the risk of deformity from cicatricial lines. An error is often made by putting in too many sutures, and especially by leaving them in too long. We have seen the suture marks sometimes left after almost all traces of the wound for which they were inserted have disappeared.

Our penetrating wounds of the chest are always promptly closed at once, and the injured side is made as immovable as possible by "shingling" with adhesive plaster. Sometimes the layer of this is doubled, the two crossing each other. Great comfort is nearly always experienced by the patient. By this method the chance is given for closure and healing by first intention. Rest may be more perfectly secured by the use of opium. Should the adhesions necessary to recovery not take place, nothing is lost, and the patient is subjected to other treatment, it may be to paracentesis or aspiration. Gunshot wounds are also often treated in this way.

In abdominal wounds with protrusion of contents, we are careful not to return bruised and dirty omentum. Sometimes the wound is brought together with a clean portion of the omentum stitched between its edges like the pedicle of an ovary. The favorite way, however, is to tie by a double ligature the offending part, and to cut this off on the distal side, or to let it slough away. We have no experience in tying such a mass with catgut ligatures, separating it, and returning the remnant completely. We see no objection to doing this, especially after the experience that has been had of a like procedure in ovariotomy, except that catgut will sometimes swell and untie the knot.

Of the 570 **Contusions** and 221 **Sprains,** many were of a severe character. The contusions are generally treated by rest and evaporating lotions. When the back is involved, there is nothing that expedites recovery and relieves pain so well as a good free cupping. Both wet and dry cups are used. The local hæmatoceles that frequently result after severe contusions are rarely interfered with by the knife, unless in some peculiar situations, or unless they are too large to be removed by absorption in a short time.

Sprains are, as a rule, much more boldly treated than formerly. Many of them require a good cupping or leeching, a measure that is nowadays too much neglected. After this, under the support of pressure, either by fixed or temporary dressings, the patient is allowed to use the joint at the earliest possible period, and, in case of the ankle, he is sometimes able to walk, and does so as soon as the dressing will permit. In this way we are certain that anchylosis is often prevented, and the confidence of the patient in his own powers is preserved. We have never had occasion to attribute any evil result to the treatment, whereas we all know about the harassing troubles that are apt to arise after treating a sprain of the ankle, for instance, by rest in bed for weeks, and maybe months. This is the case, particularly with females, who at times get into hysterical conditions under the affliction, which lasts for an indefinite period.

Abscesses.—The acute form is evacuated as early as possible, and often before pus is formed a free incision is made into the part. The bleeding and relief of tension often work great good, and sometimes determine resolution. The necessary cautions are observed when the opening has to be made in proximity to great vessels. Here, after dividing the skin, the matter is reached by breaking through the tissues with a director or forceps. When there is a doubt as to diagnosis, the exploring-needle is used.

The free opening of *chronic or cold* abscesses is by no means always followed with good results. In many a case we have observed the rapid decline of the patient through hectic and exhausting discharges to immediately follow an operation of the kind. Whether it is the admission of air or bacteria that causes the trouble, we do not know. In fact, some of us had seen so much evil follow active interference, that before the aspirator was introduced we were disposed to wait for spontaneous openings. Of the aspirator we must speak in the highest praise. Relief and cures are obtained by it very frequently, and its use rarely, if ever, interferes with more positive measures should they become necessary. When the abscess is open and its boundaries are defined, much good comes from daily distending the cavity with carbolized water for a few minutes, and then allowing it to drain away. When the discharge lessens and the parts have a healthy appearance this must be discontinued, and pressure may be applied with advantage.

Of **Chronic Ulcers** we have but little to say. They are classed as one of the "opprobriums of surgery." This is unjust, for surgery does a vast deal of good to them, even to those of the very worst character. It is the method of life, the necessity to labor, the poverty, uncleanness, and even filth of many of the subjects of them that are the insurmountable obstacles to permanent recovery. They are certainly rare and more tractable in the better walks of life. This is recognized in the fact that they are often named "almshouse sores."

Nevertheless, there is scarcely a case that does not improve, and often recover entirely, when put under appropriate treatment. It is the return to old habits and to old surroundings that brings about relapse, and eventually causes many of the cases to be classed among the incurables.

We have had experience in almost every method of treatment. Where the cause of the ulcer is specific, the appropriate internal remedies are used. The general health is carefully attended to. Rest, cleanliness, operative measures, pressure, support, position, astringents, caustics, and ointments are our aids in treatment. Varicose veins are sometimes tied, tension is relieved by appropriate incisions, callous edges are pared, and sloughs are removed. Pressure is applied by the ordinary roller, by adhesive plaster, or by the elastic bandage. When necessary, skin-grafting is resorted to. The effect of this is sometimes surprising, for we find that the forming islands of new tissue seem to stimulate the old and distant edges to reparative duty.

The new material, however, requires for a long time to be treated with great consideration. We have known it to melt away on slight provocation with the rapidity of an old cicatrix which has been subjected to injury.

WILLIAM HUNT.

MISCELLANEOUS CASES.

A *Comb forced into the pharynx behind the uvula and half arches: threatened suffocation: removal: recovery.*—Jane C., aged 46, was admitted August 15, 1872. In a fit of mania the patient pushed the comb, which she had used for her hair, into her mouth and then attempted to force it down the throat. The smooth arched top of the comb was pushed in first. Soon after the patient had accomplished its introduction swelling occurred, and intense dyspnœa came on. An unsuccessful attempt had been made to extract the comb. Dr. Morton was sent for, and, on looking into the mouth, was then only able to see the centre portions of four of the teeth of the comb, the rest was covered by the soft parts of the fauces and uvula. After some difficulty three of the teeth of the comb on one side were seized with a strong pair of long and narrow bone-forceps, and *version* was made, and the upper or smooth edge of the comb, which was downward, was turned upwards and outwards and delivered. With the exception of a slight inflammation of the fauces, no untoward symptom occurred, and the patient was soon after discharged, well. (See figure.)

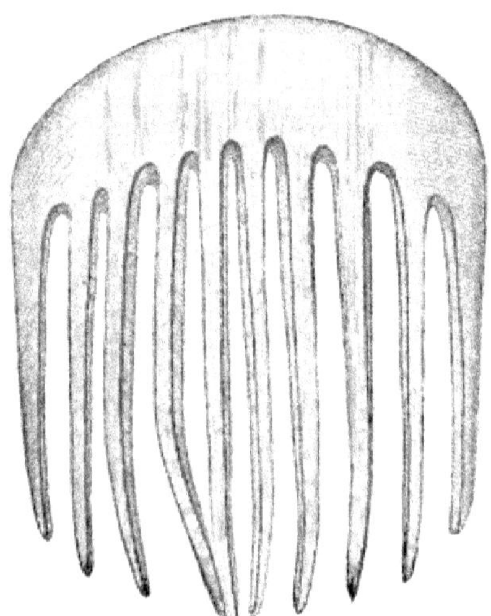

Actual size of the comb removed from the fauces.

Occlusion of the nares relieved by operation.—Margaret T., aged 18, was admitted May 21, 1874. In early childhood her face was badly scarred by small-pox, and complete occlusion of the anterior nares resulted. An operation was performed unsuccessfully, and repeated with no better result before she came to the hospital. On the 23d, Dr. Hunt opened the nares and trimmed the edges with curved scissors; the openings were distended with tents of oiled patent lint. During the treatment great tendency to contraction was noticed, but the surfaces had healed when she was discharged, July 2, 1874, with directions to continue the daily dilatation.

Congenital occlusion of the nares: cured.—John B., aged 8, was admitted for almost an entire occlusion of the left side of the nostril. The affection had existed from infancy. Had been treated on several occasions without any benefit. On being brought to the hospital, Dr. Morton found that a delicate probe could be pushed back between the sides of the nose at its lower part. A sponge tent was then introduced, which, after remaining in for four days, had stretched the parts sufficiently for the passage of a large catheter. The opening showed no disposition to close. The child was subsequently brought to the hospital for examination, and the result was eminently satisfactory.

Stricture of the urethra: false passages: inability to pass a catheter: retention of urine: aspiration: recovery.—Chas. H. B., aged 40, was admitted June 8, 1874. Had gonorrhœa several times. On two occasions had had retention, in 1858 and in 1864. Both attacks occurred in warm weather, after a hard day's work. Previous to admission the urethra had been badly manipulated, and false passages were made. After several ineffectual attempts to pass a catheter, an aspirator was introduced by Dr. Hewson and the urine evacuated. A poultice was applied, and a suppository of opium was given. The following day the patient was able to void the urine.

June 15.—Was discharged, well.

Necrosis of the patella, with ulceration of the tissues about the knee, in a case of locomotor ataxia; sudden bending of the knee, resulting in a compound fracture of the patella and opening the joint: death.—James A. F., aged 59, was admitted September 19, 1874. The patient, a subject of locomotor ataxia, had six months previously had an attack of erysipelas of the left leg, resulting in ulceration of the knee, which entirely denuded the patella, causing necrosis of its upper portion. The day before admission, while walking for exercise (first time since illness), in attempting to step from a curbstone, although supported by friends, he flexed his knee, causing a fracture of the patella, at the same time tearing open the tissues and exposing the joint. The fracture extended through the upper portion of the patella, and was transverse in character.

September 26.—Dr. Morton removed the patella and effected a free drainage.

The patient did comparatively well until the twelfth day, when a troublesome diarrhœa supervened, and gradually exhaustion: death supervened on the sixteenth day.

Congenital malposition of testicle.—Edward C., aged 19, was admitted under Dr. Morton's care in September, 1876. The patient stated that he had a rupture which had existed since infancy. On examination the right testicle was found in the scrotum, which was large and rather pendulous. The left testicle presented an anomaly of situation. It was found lying in the perineum, about one-half an inch in advance of the anus, directly in a median line. It was quite movable, could be pressed forward under the skin to a considerable distance, but on the relaxation of the pressure it at once resumed its former position. The organ was about half the natural size, and had never given any inconvenience in sitting or otherwise. The patient had been accustomed to ride on horseback and never had any difficulty, and he was quite unaware of the tumor being the testicle until he was so informed.

Cancer of rectum: colotomy (operation complicated by cyst of the kidney): death.—Mary S., aged 50, married, a housekeeper, born in Ireland, was admitted October 26, 1876, with constipation, nausea, vomiting, want of appetite, and general prostration. Upon examination a large cancerous mass was found compressing and apparently entirely surrounding the rectum a few inches above the anus. In the centre was an aperture barely large enough to admit the point of the little finger. She stated that about two years since she was seized with violent pain in the abdomen, followed by severe hemorrhage, and from that time her health failed and she was troubled with constipation. It being evident that any attempts at palliative measures would be useless, and as she was suffering great pain in the abdomen and had not had a passage for fourteen days, an operation was decided on, and, on November 1, Dr. Morton performed the usual operation for lumbar colotomy. Upon entering the abdominal cavity, a cyst, which, from its position, was supposed to be attached to the kidney, was encountered, a string passed through it, and it was pulled partly out of the wound while its contents, consisting of about ten ounces of a clear fluid without urinous smell, were evacuated. The colon, which had been pushed downward and forward by the cyst, was now reached, opened, and the edges of the wound stitched to the skin by interrupted silver sutures. Almost immediately a fecal discharge took place. The patient was removed to bed and stimulants and opiates were administered. She rested comfortably and seemed much relieved. At the end of forty-eight hours the bowels were fully evacuated, and this occurred once again before death, which happened on the fourth day from exhaustion, no signs of peritonitis having showed themselves. A post-mortem examination was refused.

Paracentesis abdominis; malignant growth in the abdomen, with ascites.—Mary D., aged 19, a native of Germany, unmarried. No history of the case was obtained beyond the fact that her friends had noticed a gradual increase in the size of her abdomen, coming on within a few months, and her body was becoming emaciated. She complained of difficulty in breathing, occasioned by the great distention of the abdomen, in which were distinguished solid and fluid contents. Since her health had begun to fail her menses had gradually disappeared. There was entire loss of appetite, but the bowels were regular. She was ordered Basham's mixture, which produced some absorption of the fluid, when a large, solid growth was detected, which filled up the right side of the abdomen. A sound passed into the cavity of the uterus only the normal distance. The diagnosis was made of tumor, probably malignant, springing from the right ovary. The abdomen having again become tense, Dr. Hunt, on the 24th of May, performed paracentesis, obtaining five pints of a clear serous fluid from the peritoneal cavity. The patient, however, grew progressively weaker, and died of exhaustion on the 5th of June, 1873. Before death the abdominal fluid again distended the cavity.

Autopsy made twenty-six hours after death, by Dr. M. Longstreth. The pleura of the right lung was tightly adherent to the parietes, and the lung was forced upward and compressed by the liver, which was displaced upward above the level of the nipple. In the apex of this lung were two calcareous masses. The other thoracic viscera were healthy. The abdomen contained about a bucketful of serum, and the peritoneum was adherent in many places. Springing from the side of the right ovary was a large solid tumor, which had attachments to the omentum. Its surface at other points was generally smooth, although somewhat irregular (lobulated), and some of the nodules were soft and doughy, though hardly sufficiently soft to be called fluctuating. At some points the surface was almost black (melanotic?), and the omentum in some places presented the same appearance. Numerous vessels were seen traversing its surface, giving it a pinkish aspect. On section the tumor in parts was firm, principally whitish in color, with ecchymotic patches of red; other parts were yellowish-white, more cystic in character, from central softening and escape of fluid contents. The other viscera were healthy and showed no traces of similar disease. The tumor was pronounced encephaloid by Dr. Longstreth.

Pseudo-membranous croup; tracheotomy; death.—Charles H., aged 4, was admitted February 6, 1869, with symptoms of suffocation from false-membranous croup; tracheotomy was performed the same day; the child had been ill for thirty-six hours. Two tracheal rings were divided and a canula tube was inserted; a good reaction followed. The child was greatly relieved for a time, but on the third day respirations became labored, and exhaustion supervened, and on the following day death occurred. At the autopsy, the false membrane and purulent matter were found in the smaller bronchial tubes.

Œdema of the larynx; tracheotomy; death.—Mary C., colored, aged 55, was admitted January 12, 1877, with great dyspnœa, cough, and sore throat. The winter before she had a similar attack, characterized by extreme difficulty in breathing, cough, mucous expectoration, and aphonia. Since then she had had good health until a fortnight before, when the old symptoms recurred with great severity. On admission the patient was very feeble, with fever, dyspnœa, and aphonia; respiration was noisy, harsh, and whistling in character. There was in addition to the œdema of the larynx a vegetation on the right side of the larynx connected with the vocal cord. The right and left arytenoid cartilages were scarified. Carbonate of ammonia in one grain doses was ordered every two hours, and quinine was freely given. During the succeeding eight hours the patient perceptibly lost ground.

January 13.—The symptoms being very urgent, Dr. Morton performed tracheotomy. The opening was made in the crico-thyroid space. A tube was introduced, and retained by two superficial silk stitches. Stimulants and beef-tea were freely given.

January 16.—Continues doing well. Tenacious mucus kept quite loose by the spray atomizer.

January 20.—A porous wet sponge is kept constantly over the tube, and serves a good purpose.

January 27.—Sat up, doing well.

February 1.—The case passed from Dr. Morton's care.

February 12.—The tube was taken out.

February 14.—The patient expressed herself as breathing better without the tube.

February 15.—Was found dead at five o'clock in the morning, and there was no evidence of the patient having had any struggle.

Self-amputation of penis.—Edward McB., a married man, was brought to the hospital with serious wounds involving the genitals. For several weeks the patient had been drinking, and had great depression of spirits. The day of his admission he had developed mania a potu, and was found in his room covered with blood. On examination the penis was found completely cut off close to the pubes. The skin of the scrotum had been so freely excised that the testicles were entirely denuded, but were not injured. Several arteries required ligation, and what little of the scrotum remained was stitched together, enclosing as far as possible the testicles. A catheter was introduced into the bladder. An hour after the patient's admission the parts which had been cut off were sent to the hospital under the impression that they could be sewed on. After a serious illness the patient recovered. A small silver cap was made, which covered the sensitive urethra, which was on a line with the abdominal walls. (The specimen is preserved in the hospital museum.)

Foreign body, which had been for years in the rectum, removed; recovery.—George K., aged 55, was admitted March 23, 1877. The patient stated that twenty years before he swallowed a peach-stone. Two years afterwards he had symptoms of rectal irritation, tenesmus, constipation, alternating with diarrhœa, and liquid stools, etc. These symptoms had continued ever since. His health had been markedly impaired. A digital examination revealed a hard, stony mass, two and one-half inches above the anus. Under ether, Dr. Morton divided the external sphincter, and, with a pair of bone-forceps, removed, with considerable difficulty, a good-sized peach-stone, which was lodged in the rectal tissues. The stone was very sharp at the ends, and had evidently lodged crosswise, and become imbedded. The patient was discharged quite well, and free from all symptoms.

Compound fracture of the patella; death.—Michael N., aged 46, a hard drinker, was admitted August 18, 1878. There was an incised wound three inches long over the knee, which extended into the joint, also a comminuted fracture of the patella. A part of the wound was open, in which a fragment of the patella was lying loose. This was removed. Water dressing was applied; quinine and morphia internally.

August 19.—Had a very restless night. Low delirium supervened. The knee is but little inflamed, but irrigation was substituted.

August 22.—Mania a potu developed, and has to be strapped in bed; slight erysipelatous blush extending up from the knee. Iron, whiskey, chloral, and bromide.

August 23.—Delirium has increased.

August 24.—Temperature, 102½°; pulse frequent, small; delirium active. Hypodermic of morphia.

August 25.—Gradually sank, and died with cerebral symptoms.

Heel of a shoe driven through the female perineum; recovery.—Catharine McD., aged 19, was admitted into Dr. Levis's ward December 21, 1878. The patient jumped out of a second-story window, and on reaching the ground her foot was turned under her as she fell. The heel of the right boot (large French heel) was driven through the perineum one inch from median line, midway between the anus and posterior commissure of labia majora. The wound extended into vagina one inch above external opening, in which the heel of the boot, which had been separated from the shoe, was imbedded, and projected into the vagina. With some difficulty it was removed. There was no other injury. Warm poultices were applied to the perineum, and the vagina was washed out frequently with carbolic acid solution, and then well anointed.

March 5, 1879.—Considerable leucorrhœa. Alum solution injected; no constitutional symptoms. The wound suppurated freely.

March 21.—Was discharged quite well.

Rupture of quadriceps tendon; partial recovery.—Thomas F., aged 50, was brought to the hospital with a complete rupture of the tendinous insertion of the common extensor; the accident was caused by muscular violence. The limb was placed on a straight splint, and an effort was made unsuccessfully to keep the ruptured end in contact with the patella. This patient has now only partial use of the limb, and requires a cane in walking.

Rupture of the quadriceps tendon; fair union; partial recovery.—William D., colored, aged 55, was admitted January 12, 1879. While intoxicated he fell and ruptured the tendon of quadriceps at its patellar insertion. A finger can be inserted under patella, and the latter lifted up. When the leg is extended there is a deep gap between patella and tendon. A slight attachment remains of the edge of the vastus externus muscle. The patient could extend leg but very slightly. The limb was placed on a Stromeyer's splint, and well covered with lead-water and laudanum.

March 20.—Space between ruptured ends filling up, and patient is allowed to walk with the aid of a crutch.

April 16.—The space between tendon and muscle getting smaller, and apparently filling up with fibrous tissue.

May 27.—Has almost perfect use of leg again, can walk up and down stairs well, and the union is apparently nearly complete, there being only a slight depression between end of ligament and muscle.

June 2, 1879.—Discharged.

Circular-saw wound involving the skull; recovery.—James B., aged 10, was admitted September 22, 1879, with a circular-saw wound of the head. He fell upon the rapidly revolving saw, and sustained a wound semicircular in character, which extended not only through the scalp, but through the external table of the cranium just above the frontal bone several inches in length. The brain was at first thought to be involved, but granulations soon filled up the wound, and no unfavorable symptoms followed; and one month afterwards he was discharged quite well.

The histories of the following cases, which were recorded in an old case-book, and only lately found in the library of the hospital, are thought worthy of a more permanent record.

"*Luxation of the femur upon the os pubis: reduction; recovery.*—In February, 1805, the patient was riding on a sled with his legs extended over its side; the left foot became entangled in the gears of a team of horses standing in the road; abduction of the leg and thigh was thus suddenly made, and the head of the bone was forced on the os pubis directly before the acetabulum. The head of the bone could be plainly seen and felt in the groin, under Poupart's ligament; the foot and knee were turned outward; the thigh was extended with the leg bent backwards." Dr. Physick reduced the luxation, and the patient left the hospital in three weeks quite well."

"*Dislocation of the femur upon the pubes.*—John C. was admitted October 27, 1831. On examination the patient was found lying upon his back with his knee flexed; the toes and knee were turned outward; a very great hollow was observable upon the outside of the hip, and the head of the bone could be distinctly felt on the pubes. Reduction was easily effected by Dr. Barton, by taking hold of the knee with the right hand, and drawing it upward and inward at the same time that the upper part of the bone was pushed forcibly outwards by the other hand. The accident happened from a bank of clay striking him upon his hip while in the erect position. Was discharged in good health."

"*Luxation of the femur into the thyroid foramen.*—David L. was admitted August 7, 1809, with a luxation of the femur into the thyroid foramen, which occurred three weeks and four days after the accident. On August 12, Dr. Wistar successfully reduced the dislocation."

"*Luxation of the femur into the thyroid foramen.*—John B., aged 15, was admitted January 31, 1821. The left thigh was fractured in two places. The right femur was dislocated into the thyroid foramen. Reduced by Dr. Hartshorne."

"*Dislocation of the humerus upon the dorsum scapulæ.*—John T., aged 55, was admitted October 14, 1834. Had been in the hospital three months before for a dislocation of the humerus of the opposite side in the axilla, which required the application of the pulleys. A deep depression existed under the acromion. The anterior edge of the scapula was thrown forward. The head of the humerus could be distinctly felt on the dorsum of the scapula, immediately below its spine. The elbow was directed forward with limited motion of the arm. The bone was reduced, and the patient discharged quite well."

"*Wound of the thorax; collection of blood in the pleural cavity; subsequent opening of the wound; death.*—Robert W., a carpenter, aged 21, was admitted April 27, 1803. With an intention of destroying himself he plunged his chisel into his left side, between the fourth and fifth ribs, about one-half an inch from the sternum. He was freely bled. The wound was stitched together. On the 29th suffered from difficulty in breathing; bleeding not relieving him, Dr. Wistar removed the sutures and opened the wound, and six ounces of blood ran out and subsequently more flowed. The symptoms were not relieved; death occurred three days after the injury."

"*Varicose veins treated in an original manner.*—Jacob W., aged 28, was admitted December, 1820. The disease was of nine years' duration. In 1819 the division of the large saphena had been unsuccessfully made. Dr. Hartshorne dissected out about three-fourths of the saphena vein, just below the knee. The ulcer healed rapidly, and the patient was discharged quite well."

"*Loose cartilage in the knee-joint successfully removed.*—This patient was admitted for an injury of the thigh, but suffered so much from a loose cartilage in the knee that, on August 19, 1812, an incision was made by Dr. Dorsey into the joint; the cartilage was removed without difficulty. On the fourth day the patient had a severe chill, after which he was well bled, and recovered without any other unfavorable symptom."

An extemporized bougie-pointed catheter in 1795.—A case is related of a man with fracture of both legs, in whom, upon the second day after admission, it was discovered that he had not voided any urine, nor had he done so since the accident. After an effort to introduce a silver catheter had failed, "Dr. Physick endeavored immediately to introduce a common-sized bougie, but was unable to pass it beyond the bulb of the urethra. A gum-elastic catheter was next tried, with no better success. The bougie was again had recourse to, and by bending the point of it to one side, he was enabled to pass it beyond the obstruction and into the bladder. He allowed it to remain some minutes in this situation, in the hope that when it was withdrawn the urine would follow it, but in this he was disappointed, and therefore attempted immediately to enter a gum-elastic catheter, but was not able to pass it beyond the obstruction at the bulb of the urethra. Under these circumstances Dr. Physick determined to cut off the end of the gum-elastic catheter and *tie on it two inches of the end of a bougie, because the flexibility of the bougie enabled him to bend it in the proper direction, which could not be done with the catheter.* The instrument thus formed was passed into the bladder with care and the urine entirely drawn off. The patient was afterwards relieved with this instrument twice every day for a fortnight, after which he became able to void his urine without assistance."

Immediate union after amputation. An antiseptic healing without antiseptics in 1804. Thus headed, "Curious circumstance attending the healing of a wound." —"Peter Drake, admitted for an injury of the index finger of the right hand near the joint of the first phalanx. The nature of the accident was such as to render amputation of the finger necessary, which was accordingly done by Dr. Wistar. No inflammation came on after the operation, and *nothing like pus was* observed during the healing of the wound. Granulations did not arise, neither was a cicatrix formed, but the stump was healed by an extension of the old skin. The patient was discharged six weeks after the operation."

Fracture of the cervical vertebra; extension.—A case of this kind is reported as having been admitted November 15, 1804. The man fell from the height of forty feet. "The patient was perfectly insensible except the head and neck. . . . The interesting point is that Dr. Physick ordered an apparatus to be applied by which the neck was kept extended. The feet of the patient were secured to the lower and his head to the upper part of the bedstead. About two hours after the dressing was applied he recovered in some measure the use of his arms." The report is accompanied with a drawing. An occipital and vertical strap are applied to the head, and from the latter there is a screw attached, which passes through the head of the bedstead and gets its fulcrum from it. The feet are firmly bound together, and attached to an upright at the foot of the bed. The patient died on the 17th. The fourth, fifth, and sixth vertebræ were fractured and pressed upon the spinal marrow.

A case of simple dislocation of the astragalus is reported as having been admitted September 26, 1831:

"The whole foot was found to be turned inwards. There was a hollow below the lower end of the tibia, and a considerable round projection on the outer part of the foot a little in front of the external malleolus. . . . From these appearances it was evident that there was a dislocation forwards and outwards of the rounded projection on the outer side of the foot, being the convex head of the astragalus, which articulates with the scaphoid. . . . Strong efforts were made by Dr. Barton to reduce the dislocation, but without any effect, after which it became a question whether the bone should be allowed to remain or should be removed. The latter course was determined upon after a consultation, in which all points were considered."

The operation was performed by Dr. Barton. The patient did not do well, and ultimately had to submit to amputation of the leg.

Had anæsthetics been in use at that time it is probable that such a severe measure as the excision would not have been necessary. A very analogous case was admitted in 1872, into Dr. Hunt's wards.

After the patient was thoroughly under the influence of ether the nature of the case was made out to the satisfaction of all present. With the aid of two assistants, one at the foot and the other at the leg, below the knee. Dr. Hunt had the good fortune to suddenly manipulate everything into place. The parts were then freely leeched, cold was applied, and the limb was put at rest in a fracture-box. The patient made a complete recovery, and was often seen afterwards at his old business, that of an express and car driver.

In the article on Excisions, and Operations for Ununited Fracture, a case is detailed where non-union apparently was the result of successive bleedings and a restricted diet, no animal food being allowed for six weeks; but it would not appear that excessive loss of blood alone interferes with the prompt union of fractures, for in the following case, which made an excellent recovery, no less than one hundred and two ounces of blood were abstracted before the patient came into the hospital, and ninety ounces were removed afterwards, during the treatment of the fracture of his thigh, for which he was admitted.

"*Fracture of the thigh.*—Isaac Y., aged 19, was admitted into the Pennsylvania Hospital, October 5, 1801, for a fracture of his right thigh. The accident happened a week previously to his admission, during which time he was almost constantly delirious, and had a pulse so active as to call for five bleedings, in all of which he lost ℥cii of blood.

"When he came into the hospital the parts adjacent to the fracture showed no marks of inflammation, notwithstanding the violence of the constitutional affection. The splint of Desault, as improved by Dr. Physick, was applied to the limb in the usual manner. The patient at this time had no appearance of delirium and complained of but little pain, though his pulse was strong and full; he continued in this situation until the evening of the 7th, when he became delirious; he was now bled to ℥xx. The good effects of the bleeding were very evident, for during the emission of the blood from the vein he fell asleep.

"He rested well the greater part of the night, and was easy the whole of the succeeding day (8th) until evening, when the delirium again came on. The bleeding was repeated to ℥x; towards the latter part of the night he fell asleep, and continued free from delirium until next evening (9th), when he was again bled to ℥x, afterwards passed the night as the last. He was after this delirious every evening until the 15th, and at each time the delirium came on lost ℥x of blood. From the 15th he recovered rapidly, and was afterwards discharged from the hospital."

THOMAS G. MORTON.

CONSTRUCTION OF THE HOSPITAL, VENTILATION, AND HOSPITALISM.

In 1750 the first step was taken towards the establishment of a hospital in Philadelphia; the following year a private house was taken and temporarily occupied, and in February, 1752, patients were first admitted. The corner-stone of the present hospital was laid on the 28th day of May, 1755, bearing the following inscription, which was prepared by Benjamin Franklin:

"In the year of CHRIST
MDCCLV.,
GEORGE the Second happily Reigning
(for he sought the happiness of his people),
Philadelphia Flourishing
(for its inhabitants were public spirited),
This Building,
By the Bounty of the Government,
And of many private persons,
Was piously founded
For the Relief of the Sick and Miserable.
May the God of Mercies
Bless the undertaking."

In December, 1756, patients were admitted, the eastern wing having been completed. The western wing was first used in 1796, and the centre in 1805. In 1851–52 the eastern wing was rebuilt, and at this time many important improvements were made throughout the building generally.

"The hospital, which is most substantially built of brick, faces the south, is two stories high, with commodious attics, and is placed in the centre of a plot of four acres. The grounds are, for sanitary purposes, beautifully cultivated, and adorned with flowers, trees, and shrubbery. Around the margin of the enclosure, at intervals of forty feet, stand lofty buttonwood-trees, which were planted more than one hundred years ago. The centre building is sixty-five feet in front by sixty-two feet and a half in depth; a balustrade surrounds its cupola, seventy-two feet from the ground. From the centre, east, and west extend wards, each eighty-one feet long, thirty-three wide, and twelve high. At the ends of these wards are wings, each one hundred and eleven feet long, otherwise corresponding in size with the wards they join.

Heating.—"Until 1821 the wards were warmed by large open wood-fires, which not only furnished sufficient heat, but also kept the air in fair condition. During a portion of the year, when fires were not required in the wards, the air was supposed to be sufficiently changed by simply opening windows and doors. In 1823 grates burning anthracite coal were introduced; but wood-fires in some of the wards were used until 1826. In 1846 the west wing was heated by air passing over steam coils; and this plan proved so satisfactory, that in 1851–52 the same arrangement for heating was introduced throughout the hospital.

Ventilation.—"In 1851–52 openings were made near the ceilings, some of which communicated with ducts leading directly to hot flues. In 1865 ceiling ventilation was abandoned, and openings near the floor were made.

"In 1876 it was determined by the Board of Managers that the best form of ventilation should be introduced, and accordingly a fan was placed in the basement of the west wing. The system of forced currents of air from a fan is more likely to give uninterruptedly, by day and by night, and at all times of the year, a regular supply of fresh air, and no system of ventilation can be regarded as perfect unless it is insured by some forcing power, either by means of heated chimney-stacks or a fan, and with the latter it makes but little difference where the openings for the foul air are placed.

"In the successful use of a fan the cardinal rule is that it should be always running from year's end to year's end, and that at the same rate of speed, for the heat of the wards should be regulated entirely by the steam-chamber, and any attempt to diminish the supply of pure air in cold weather strikes at the root of the system, and shows that its fundamental requirements have not been met.

"The Managers require the fan to run continuously, and it may be said that during the past year there has been but twelve hours that the fan was not running. The fan makes on the average ninety revolutions per minute, and distributes 4000 cubic feet of air per hour to every patient. In cold weather the air is driven into chambers, where the iron coils are heated by steam.

"Of old hospitals it has often been observed that they cannot by any means be kept healthy, and that, on this account, they should give place to cheaply-built pavilions, which, from their light construction and inexpensive character, may, so soon as they become injurious, be replaced at a comparatively trifling cost. Pavilion wards, or hospitals, may apparently meet all the requirements for the proper treatment of surgical cases, but it is doubtful if any better or as good results can be adduced from their statistics than are to be found in good permanent buildings.

"From 1842 to 1876, 37,272 surgical cases were treated. Of this number, 2459 died, showing a mortality of a little over six and one-half per cent. If, however, we take from this number 627 deaths which occurred from compression and concussion of the brain and spinal cord, rupture of the abdominal viscera, and railroad injuries, etc., deaths which immediately followed the accident as a result of shock, we have the mortality during the past thirty-five years reduced to barely five per cent.

"During the past twenty-five years there were 16,297 cases of fractures of all kinds, accidental injuries and gunshot wounds. Of this number, 1598 died, giving a total mortality, in all of these serious cases, of only a little over nine and eight-tenths per cent."*

Hospitalism.—The subject of "Hospitalism" has from time to time been much discussed, and occasionally with an especial reference to the Pennsylvania Hospital. The following (abridged) letter written to the editor of the *Philadelphia Medical Times* by Dr. William Hunt, and printed in the number of that journal for November 21, 1874, attracted much attention at the time, and made quite an incident in the history of the hospital. It is considered worthy of preservation in a more permanent form. There has been no reason whatever to change the views then expressed:

"After hearing Dr. ——'s address before the Public Health Association the other night, I asked myself, 'Are general hospitals, as heretofore and as at present constituted, pest-houses and abominations, the direful spring of woes unnumbered, or are they blessings?' The next day I searched for, and found in my office, the last two annual reports of the Board of Managers of the Pennsylvania Hospital to its Contributors, and, taking that of 1873, I made some approximative statistics. The year was not selected. I had intended to go over the last two years, but time prevented, and the work was great. It is often, however, a subject of remark how much one year is like another; and when we have the fact that the yearly mortality of the hospital from its very beginning ranges from eight and a half to ten per cent., I think nothing is wanting to establish my statements.

"See, then, the record of the oldest 'pest-house' in America (now well advanced in its second century) for the year 1873:

Summary of an Analysis of the Surgical Mortality of the Pennsylvania Hospital for the years 1872–73.

Whole number in house, 2155, of whom there died 172, or 8.59 per cent.
Of these cases, 1387 were surgical, and 768 medical.

The whole number of surgical deaths was		90
Of these there died within twenty-four hours, that is, *were killed* by the accident for which they were brought to the hospital	35	
There died in 2 days	7	
" " 3 "	4	
" " 4 "	1	
" " 5 "	2	
Of troubles essentially fatal, as spine- and brain-injuries, burns, wounds, etc., there died in periods from 6 to 93 days	9	
There died of cancer	3	
" " tetanus	2	
Cases fatal in themselves, in or out of hospital		63
Leaving a mortality of		27

to account for, of whom 4 are put down to pyæmia. The others were made up of very severe cases, such as compound fractures, exhaustive suppurative troubles, necrosis, etc., etc., and it is certainly fair to presume from the social status of most of them that their lives were prolonged and their sufferings were ameliorated as well within as without a hospital.

"Query: How many died of hospitalism?

"All will certainly agree that the thirty-five one-day cases should be elimin-

* *Amer. Journ. of Med. Sciences*, January, 1878.

ated altogether from the consideration of the mortality. The hospital, as a hospital, had nothing to do with the death of the remainder of the sixty-three, and as for the twenty-seven, there was enough to kill them in themselves, without going to hospitalism to account for it, except, maybe, the four pyæmia cases. I question whether in thirteen hundred and eighty-seven surgical cases, mostly very serious, in private practice, four cases of pyæmia would not occur; and I am very certain of it, if the thirteen hundred and eighty-seven had been treated in the places where they lived. Think of the material that this hospital gets within its walls,—that, ranging from the essentially fatal ones, we have to treat those of all grades of severity,—and you have a full justification of your statement made a short time since, that, for its size, it is one of the most active hospitals in the world. No one is more surprised at the record than I am myself.

"I wish to write a readable article, not burdened with statistics, but remember, I have the records of all, and I enclose you the list of the forty-nine who died during the first five days, if you choose to use it, to give some idea of the nature of injuries we have to deal with outside of this fatal line. As I have sometimes expressed it, the work we have to do on some black days is truly appalling. The list is as follows:

"1, compound fracture of skull, 5 days in hospital; 2, burn, 1 day; 3, burn, 2 days; 4, burn, 1 day; 5, compound fracture of both femurs, 5 hours; 6, fracture of spine, 1 day; 7, contusions of trunk, etc., 1 day; 8, burn, 1 day; 9, cut throat, 2 days; 10, compound fracture of humerus, 1 day; 11, laceration of legs, 1 day; 12, contusions of body, 1 day; 13, fracture of spine, 1 day; 14, lacerations of legs, 1 day; 15, fracture of scapula and contusions, 1 day; 16, fracture of ribs and legs, 1 day; 17, compound fracture of both legs, 1 day; 18, contusions of trunk, 1 day; 19, gunshot wound of chest, 4 days; 20, tetanus, 1 day; 21, fracture of skull, 1 day; 22, fracture of clavicle and ulna, and contusion, 2 days; 23, concussion of brain, 2 days; 24, lacerations and contusions, 3 days; 25, rupture of urethra, 3 days; 26, fracture of spine, 1 day; 27, compound fracture of both legs, 1 day; 28, compound fracture of thigh, foot, and arm, 2 days; 29, burn, 1 day; 30, contusion of head, 1 day; 31, compound fracture of leg, 1 day; 32, laceration of head, 2 days; 33, compression of brain, 1 day; 34, compound fracture of humerus, 1 day; 35, fracture of skull, 5 days; 36, compound comminuted fracture of femur, 1 day; 37, gangrene of leg from ulcer, 3 days; 38, burn, 2 days; 39, compound comminuted fracture of femur, 1 day; 40, compound comminuted fracture of femur, 1 day; 41, burn, 1 day; 42, fracture of pelvis, 1 day; 43, compound fracture of leg, 1 day; 44, fracture of skull, 1 day; 45, burn, 1 day; 46, burn, 1 day; 47, strangulated hernia, 1 day; 48, fracture of leg, Bright's disease, serous apoplexy, 3 days; 49, compound fracture of both arms, 1 day.

"I cannot help thinking that this matter of hospitalism has been very much over-stated of late. That there are preventable causes of mortality in almost all general hospitals, I have no doubt. That a *well-managed hospital*, of whatever construction, necessarily contains elements of fatality in the material of that construction or in its form, I very much question.

"I am in favor of the study of construction and management in all directions, in order to obtain the best results; but I protest against high authorities giving the weight of their influence in such a way as to favor popular prejudice against what, so far at least, are among the best institutions of the land. If what some say in regard to these institutions is true, every willing hand should withhold his contribution, and every bequest or legacy now standing in a last testament should be revoked.

"The ideas that a layman would get of hospitals from hearing and reading about hospitalism are, it appears to me, very analogous to the notions he would get of insane-hospitals through the authority of Reade's *Very Hard Cash*.

"Imagine a dictator, say, of the United States. Heretofore not having paid any attention to the subject, he by accident reads the famous novel. In righteous wrath he rises and issues a decree to free the inmates of all the insane-hospitals, to burn the buildings to the ground, and to hang all the Kirkbrides and Rays in his empire. And he would be right, if the novel were true; but it is not. Who that thinks about it does not know that if there is one thing that proves there is an advance in culture—one thing that proves there is, after all, humanity in man—it is to be found in the present treatment of the insane as contrasted with that of bygone years?

"The layman's opinion of a hospital is very much influenced by what he may happen to see or experience on a given visit.

"Let him be one of a committee, for example, going through the wards when there happen to be no very offensive cases. He will be delighted, and will report everything lovely in that institution. But in a short time another committee reports that a certain ward is very foul, and that the institution is badly managed. Our friend is surprised; but if he inquires he will find that a bad burnt case, or one of traumatic gangrene, has come in since he last went through, and, in spite of everything, it will make some smell. It can't be helped. Some of my friends will smile at my bringing in here my favorite illustration. We all know that a dead horse in an open field, with all the winds of heaven blowing about it, will contaminate the air for a certain distance in spite of the *perfect ventilation*, and make it unbearable for any ordinary human being.

"Now, what is a case of gangrene, or one of bad burn, in so far as the sloughing dead parts are concerned, but a mass of carrion, the same as portions, as it were, of the dead horse?

"If, then, by our arts we can make the emanations from such cases reasonably bearable, so as to be able to administer to the needs of the poor sufferers, should we not be pleased at our efforts, while we study to make them still more perfect? Don't have such cases, you say, in the wards. That is just one of the questions. They must be somewhere. If we crowd them together most or all of them will die, and maybe their attendants also. We must remember that, with the exception of the dead parts they are trying to get rid of, they are living bodies with immortal souls. Therefore some of us approve of scattering them, and I have no doubt this plan gives the best results. I do not like the idea of a special ward for such cases in the main building, for in time it will become a pest-ward indeed. For this class of cases, on account of the general comfort, I should advocate an out-ward for temporary purposes *until the sloughing stage is over*, then I would remove them and distribute them in the main wards.

"I do not intend to enter upon the endless subject of construction; of its great importance there is no doubt. Much is to be said on all sides, and, with thorough attention to a few radical principles, the purposes for which any given institution is wanted, and the space and funds that it has to carry out its object, will regulate the structure.

"I hope I have been fully understood. I by no means claim perfection for the Pennsylvania Hospital, for in it I know we have much to attain. But when we consider the nature of the cases that come into it, and the fact, I believe, that an annual mortality of two per cent. is considered as indicative of a healthy condition of a large community like our city, what has so-called *hospitalism* to claim in the institution under consideration, when, excluding the thirty-five killed outright on the first day of their casualty, and yet honestly put into the table, we can sum up a mortality of but 4.8 per cent. for the remainder?

"The hospitalism that inculcates strict attention to every detail of administration I believe in. The hospitalism that incites popular prejudice against magnificent charities, and that sees pestilence in the structure, from the chimney-pot to the foundation-stone, I utterly condemn, believing that what is wanting is less hospital on the brain and more brains in hospital."

Pyæmia.—During the past five years, from May, 1874, to 1879, *there has not been a single case of pyæmia:* the most carefully conducted post-mortem examinations by Dr. Longstreth having failed to demonstrate even an approach to this malady. This immunity we believe to be due not only to the great care taken in dressing our cases, but to the very perfect system of fan ventilation, which, by day and by night from one year's end to the other, is constantly in operation. The method of dressing wounds which has been universally employed in the hospital since 1866 claims some attention, for, from the period the "dressing-carriage" was introduced, this terrible disease has been much less frequently observed.

Fig. 1.

Dr. Morton's hospital dressing-carriage, front view.

This apparatus (see Figs. 1 and 2) was devised and introduced into the surgical wards of the Pennsylvania Hospital by the writer in 1866. The apparatus supplies flowing water for cleansing wounds, and thereon are carried all the materials which are used for ward purposes, and all the refuse dressings are emptied into an appropriate receptacle and carried from the wards. The apparatus consists of a table or stand, three feet high, three feet two inches in length, and eighteen inches wide, supported on three wheels; two of these are made similar to carriage-wheels, twenty-two inches in diameter, with a guiding-wheel in front nine and a quarter inches in diameter, the measurements in each including the rubber bands, which cover the rims in order to deaden noise when the carriage is pushed through the wards.

Upon the top of the table and in the centre is a copper reservoir, which holds a large supply of water. This can is twenty inches high and forty inches in circumference, with an outlet at its base to which a gum tube nine to twelve feet long is attached; at the end of this is fastened a metallic pipe four inches long, provided with a stop-cock to regulate the flow of water. On either side of the reservoir there are appropriate places for bandages, charpie, oiled silk, or waxed paper; jars with various ointments; salt-mouth jars are arranged back of the receiver for carbolized and olive oils, and such other preparations commonly used in the wards.

A copper can for hot water, with a spirit-lamp, is carried in the firm ring bracket in front of the carriage, for heating strips of adhesive plaster.

There is a large copper bucket, with a projecting lip, which occupies one-half the space under the table; into this all refuse dressings, discarded poultices, and soiled water after dressing the wounds, are emptied.

FIG. 2.

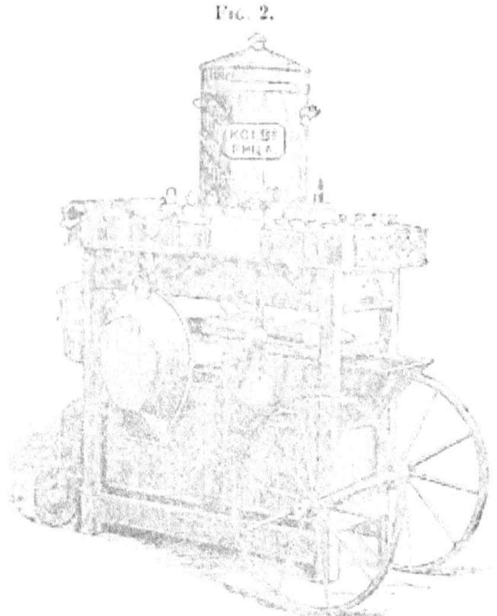

Back view of the hospital dressing-carriage.

Another copper can with two compartments adjoins this bucket; in one of these the soiled bandages from fractures, etc., are thrown; in the other are placed the oakum, cotton batting, muslin, and old linen which are used for the ordinary dressings. One copper basin serves for the entire ward or series of communicating wards, since this basin is merely the receptacle for the water and discharges flowing from the part during the process of dressing. The tin receptacle under the heating apparatus contains the towels. There is a rack upon the back part of the apparatus, made of wire and oak, for holding wax paper used by the resident surgeon, and the ward book. The lids to the various cups are fastened to the apparatus with brass chains.

"The contrast at dressing hours, before and since its introduction, is very marked. *Then* the head-nurse, with his regular assistant and one or two limping and clumsy convalescents, followed the doctor with hand-trays, water-cans, and slop-buckets, taking up and setting down at each bed, and, unless well drilled, awkwardly moving about and stumbling over the impediments, and not unfrequently upsetting them. *Now* the hands of all are free, and not so many of them are necessary; if anything is wanted the nurse hands it from the carriage, or the surgeon takes it at the required moment; one case is finished and the little vehicle moves noiselessly on to the next, leaving the new and clean behind it, and carrying away the old and soiled.

"One basin serves for all, but it is a mere slop basin; the water, either pure or medicated, that falls into it has already served its purpose, and is not intended to be used again. It flows from the hose over the part; a piece of clean oakum answers for a sponge to wash about the wound, and all the débris is thrown away.

"There is no doubt that erysipelas is a rare thing in the wards now. Pyæmia, too, is not nearly so common. That the ward carriage has diminished the liability of the patients to these diseases there is no doubt."*

This apparatus has been introduced into several hospitals abroad, and in this city into the hospital of the University of Pennsylvania, the Episcopal, and several others.

The Pathological Museum of the hospital, under the care of the curator, Dr. Morris Longstreth, is constantly increasing both as to the collection and as to usefulness. There is much work done in pathological microscopy.

New quarters have lately been provided for the Out-Patient Department, the demands upon which are constantly increasing. There are ample accommodations for the general work, and also rooms and appliances for especial examinations.

The ambulance, which is sent for through the district telegraph, is in almost daily use, and has contributed much to diminish the sufferings of the sick and wounded while being brought to the hospital.

Recently all of the attending surgeons of the hospital have been put in communication with one another and with the institution by the telephone. The usefulness of this arrangement is very great, and in speaking of it we have again to express our acknowledgment of the readiness of the managers to do all that, in the words of the original charter, may contribute to "the saving and restoring useful and laborious members to the community."

THOMAS G. MORTON.

* An account of the Ward Carriage, by Dr. Wm. Hunt, *Pennsylvania Hospital Reports*, 1869.

Medical Officers of the Pennsylvania Hospital in Lineal Succession, prepared by Morris Longstreth, M.D.

(Read downwards.)

1751	Lloyd Zachary	Thomas Bond	Phineas Bond	Thomas Cadwalader	Samuel Preston Moore	John Redman.		
1754	William Shippen.							
1759					Cadwalader Evans.			
1773			John Morgan		Charles Moore.			
1774					Adam Kuhn (resigned in 1781 and re-elected in 1782).			
1777			Thomas Parke	James Hutchinson.				
1778	William Shippen, Jr.			John Morgan.				
1779	James Hutchinson.							
1780						John Jones.		
1783				Benjamin Rush.				
1784		John Foulke.				William Shippen, Jr.		
1791								
1793	Caspar Wistar.						LYING-IN DEPARTMENT.	
1794		Philip Syng Physick			Benjamin Smith Barton.		Closed in 1854.	
1798						John Redman Coxe.		
1802						Thomas C. James, transf'd.		
1807						Joseph Hartshorne	Thomas C. James.	
1810	John Syng Dorsey							
1813				John C. Otto.				
1816		Joseph Parrish			Samuel Colhoun			
1818	Thomas T. Hewson.							John Moore.
1820					Wm. Price, appt. surgn.	William Price.		
1821					John Wilson Moore.			
1824						John Rhea Barton.		
1825			Samuel Emlen					
1827					John K. Mitchell.			
1828			Benjamin H. Coates.					Charles Lukens.
1829		Thomas Harris					Hugh L. Hodge.	
1832								
1834					William Rush.			
1835	Jacob Randolph			George B. Wood.				
1836						George W. Norris.		
1838					Thomas Stewardson, Jr.			Charles D. Meigs.
1839								
1840		Edward Peace.						
1842			William Pepper.					
1845					William W. Gerhard.			
1848	George Fox.							Joseph Carson.
1849								
							ADD'L SURGEONS.	ADD'L PHYSICIANS.
1852							John Neill.	
1854	Joseph Pancoast.							James J. Levick.
1856							Edward Hartshorne.	
1859			John Forsyth Meigs.	Francis Gurney Smith.				
1861		Addinell Hewson.				William Hunt.		
1864								
1864	Thomas G. Morton.						D. Hayes Agnew.	
1865				Jacob M. Da Costa				J. Aitken Meigs.
1868					James H. Hutchinson		Richard J. Levis.	
1871								
1877		D. Hayes Agnew.						Morris Longstreth
1879								

Abstract of the Surgical Cases treated from May, 1873, to May, 1878, prepared by Jonathan Richards.

	Cured.	Improved.	By Friends or Request.	Misconduct.	Eloped.	Died.	Remaining.	Total.
Fractures, compound, cranium and face	17	3	3	...	...	12	1	36
" " trunk and pelvis	1	1	...	...	...	6	...	8
" " superior extremities	67	18	13	..	3	12	8	121
" " inferior "	89	18	3	...	1	72	30	213
" simple, cranium and face	33	16	4	...	3	21	8	85
" " trunk and pelvis	82	19	11	...	2	25	2	141
" " superior extremities	272	57	34	1	8	5	23	400
" " inferior "	500	40	23	3	..	35	110	711
" vicious or delayed union	6	5	2	...	1	...	3	17
Luxations, simple	46	13	2	...	1	1	2	65
" compound	5	...	...	...		1	2	8
Anchylosis	12	9	7	...	1	...	3	32
Synovitis	37	19	5	3	1	1	12	78
Necrosis and caries	50	34	18	3	2	6	24	137
" " morbus coxarius	2	8	1	...	2	...	1	14
Osteitis and periostitis	8	7	1	1	...	...	2	19
Sprains	151	43	16	2	3	...	6	221
Coxalgia	5	7	2	...	...	3	6	23
Rupture of ligament	1		...	...	...	...	...	1
Fracture, thyroid cartilage	...		1	...	...	...	...	1
Club-foot	1	...	...	...	...	...	...	1
Crushed from railroad accidents	12	2	2	...	...	15	...	31
Wounds, incised	76	29	18	...	6	6	4	139
" punctured	44	9	7	...	3	2	1	66
" lacerated	182	80	31	2	6	7	24	332
" gunshot	70	38	16	...	1	20	7	152
" penetrating	26	3	2	...	1	2	2	36
" contused	415	77	40	2	7	9	32	582
Concussion of brain and spinal cord	20	4	6	...	...	17	3	50
Compression of brain	...	...	...	...	...	4	...	4
Burns and scalds	80	32	10	1	1	37	16	177
Frost-bite	18	2	...	...	...	...	9	29
Diseases of the eye and ear	56	33	17	3	2	1	17	129
Aneurism and varix	16	12	2	...	...	5	1	36
Hemorrhage	13	1	...	...	...	...	2	16
Tumors, not classified	23	11	17	...	2	5	7	65
" ovarian	1	1	1	...	...	...	...	3
Cancer, epithelial	16	18	12	2	...	4	2	54
" scirrhous	6	2	6	...	1	5	1	21
Stricture of urethra, etc.	280	123	25	4	20	6	40	498
Rupture of urethra	1	...	...	...	...	...	...	1
Hydrocele, varicocele, etc.	29	10	2	2	2	1	4	50
Cystitis and orchitis	20	15	3	...	1	1	3	43
Renal, vesical, and urinary calculi	6	5	2	...	1	1	...	15
Vesico-vaginal fistula	1	2	...	...	...	...	1	4
Exstrophy of bladder	...	...	1	...	...	1	...	2
Procidentia uteri	1	...	...	...	...	...	1	2
Retention of urine	2	1	...	...	...	1	1	5
Incontinence of urine	2	1	1	...	...	...	...	4
Testicle, enlarged, strumous, and cancerous	1	2	...	...	...	1	...	4
Irritable bladder	...	1	1	...	1	...	...	3
Enlarged prostate	1	...	...	...	...	...	...	1
Atresia vaginæ	...	1	...	...	...	...	...	1
Gravel	...	2	...	...	...	...	...	2
Hernia	26	12	3	...	...	13	2	56
Diseases of the rectum	52	30	20	2	4	3	14	125
Abscess	89	35	10	2	6	10	12	164
Carbuncle	9	...	...	...	...	1	1	11
Deformity	12	12	5	...	1	1	5	36
Inflammation	5	6	...	...	...	...	...	11
Tetanus	2	...	...	...	...	2	...	4
Gangrene	2	1	1	...	...	2	...	6
Ulcer	145	56	20	4	3	3	26	257
Adenitis	13	4	4	...	...	...	2	23
Elephantiasis	...	2	...	...	...	1	2	5
Cellulitis	4	...	...	...	...	1	...	5
Foreign body	3	2	...	...	...	...	...	5
Ingrowing toe-nail	2	...	...	...	...	...	2	4
Irritable stump	2	1	1	...	1	...	...	5
Onychia	...	1	...	...	...	...	...	1
Traumatic erysipelas	...	...	...	...	...	...	1	1
" neuralgia	1	...	...	...	...	...	...	1
" paresis	...	...	1	...	1	...	...	2
Paronychia	4	...	1	...	...	...	1	6
Podelcoma	...	...	...	...	...	...	1	1
Peritonitis	...	...	...	...	...	2	...	2
	3134	996	434	37	99	390	490	5580

www.ingramcontent.com/pod-product-compliance
Lightning Source LLC
Chambersburg PA
CBHW021403210326
41599CB00011B/992

* 9 7 8 3 3 3 7 0 2 1 5 2 8 *